RECENT DEVELOPMENTS IN TURBULENCE MANAGEMENT

FLUID MECHANICS AND ITS APPLICATIONS
Volume 6

Series Editor: R. MOREAU

MADYLAM
Ecole Nationale Supérieure d'Hydraulique de Grenoble
Boîte Postale 95
38402 Saint Martin d'Hères Cedex, France

Aims and Scope of the Series

The purpose of this series is to focus on subjects in which fluid mechanics plays a fundamental role.

As well as the more traditional applications of aeronautics, hydraulics, heat and mass transfer etc., books will be published dealing with topics which are currently in a state of rapid development, such as turbulence, suspensions and multiphase fluids, super and hypersonic flows and numerical modelling techniques.

It is a widely held view that it is the interdisciplinary subjects that will receive intense scientific attention, bringing them to the forefront of technological advancement. Fluids have the ability to transport matter and its properties as well as transmit force, therefore fluid mechanics is a subject that is particulary open to cross fertilisation with other sciences and disciplines of engineering. The subject of fluid mechanics will be highly relevant in domains such as chemical, metallurgical, biological and ecological engineering. This series is particularly open to such new multidisciplinary domains.

The median level of presentation is the first year graduate student. Some texts are monographs defining the current state of a field; others are accessible to final year undergraduates; but essentially the emphasis is on readability and clarity.

For a list of related mechanics titles, see final pages.

Recent Developments in Turbulence Management

Edited by

K.-S. CHOI
Department of Mechanical Engineering,
University of Nottingham, U.K.

SPRINGER SCIENCE+BUSINESS MEDIA, B.V.

Library of Congress Cataloging-in-Publication Data

Drag Reduction in Engineering Flows Meeting (5th : 1990 : London, England)
Recent developments in turbulence management : proceedings of the 5th Drag Reduction in Engineering Flows Meeting / edited by K.-S Choi.
p. cm. -- (Fluid mechanics and its applications ; v. 6)
Includes index.
ISBN 978-94-010-5560-4 ISBN 978-94-011-3526-9 (eBook)
DOI 10.1007/978-94-011-3526-9
1. Turbulence--Congresses. I. Title. II. Series.
TA357.5.T87D73 1991
620.1'064--dc20 91-35262
CIP

ISBN 978-94-010-5560-4

Printed on acid-free paper

In memory of

Professor Itiro Tani

Contents

III. Surface Roughness

IV. Compliant Surfaces

V. Polymer Additives

Preface

The European Drag Reduction Meeting has been held on 15th and 16th November 1990 in London. This was the fifth of the annual European meetings on drag reduction in engineering flows. The main objective of this meeting was to discuss *up-to-date* results of drag reduction research carried out in Europe. The organiser has adopted the philosophy of discussing the yesterday's results rather than the last year's results. No written material has therefore been requested for the meeting. It was only after the meeting the submission of papers was requested to the participants, from which 16 papers were selected for this proceedings volume. The meeting has attracted a record number of participants with a total of 52 researchers from seven European countries, U.K., France, Germany, the Netherlands, Italy, Switzerland and U.S.S.R. as well as from Japan, Canada and Australia.

The subjects covered in this proceedings volume include riblets, LEBUs (Large Eddy Break-Up device), surface roughness, compliant surfaces and polymer additives. Riblets seem to be one of the most extensively studied devices in the past years. Reflecting this situation in the European community, there are six papers on riblets covering their practical applications to aircraft and to a model ship, near-wall coherent structure of the boundary layer and effects of flow three-dimensionality. Possibility of heat-transfer enhancement with riblets and potential use in the laminar flow are also investigated. An analytical model is developed for the boundary-layer with a LEBU device. Physical mechanisms of turbulent skin-friction reduction with LEBUs are reviewed in the light of some recent studies. The d-type roughness is investigated in conjunction with riblets for drag reduction. A correlation method of roughness parameters with the drag penalties is also presented. This approach may have a potential for cross-fertilisation between the drag-reduction community and the surface-roughness community to exploit new techniques and methodology. Two papers are devoted for further theoretical developments of compliant surfaces in transition delay. Probably one of the most exciting recent developments in turbulence management is the use of compliant surfaces in the turbulent boundary layer. There are two papers describing some theoretical and experimental work carried out on this subject in the U.S.S.R. Some further studies on the effects of polymer additives are also presented.

It was very fortunate for the organiser of the meeting to have welcomed a large contingent from industries, in particular from aerospace and heavy industry. Their presence among the academics has made the transfer of these important technical developments and dissemination of scientific as well as industrial knowledge possible.

The pannel discussions at the end of the meeting were used as a forum to exchange views and plans on future research, Europe-wide collaborations and industrial applications of drag reduction techniques. An edited record of the panel discussions is provided in this proceedings volume.

The Fifth European Drag Reduction Meeting was jointly sponsored by BMT Fluid Mechanics who hosted the meeting with necessary skills and resources, and by European Research Community on Flow Turbulence and Combustion (ERCOFTAC) who provided with scholarships for young researchers enabling them to attend.

Kwing-So Choi

Nottingham, August 1991

I. Riblets

Experiments with a 1:4.2 model of a commuter aircraft with riblets in a large wind tunnel

J. G. TH. VAN DER HOEVEN and D. W. BECHERT*

Hermann-Föttinger Institut für Thermo- und Fluiddynamik
T.U. Berlin
Berlin, Germany

* *Deutsche Forschungsanstalt für Luft- und Raumfahrt*
Berlin, Germany

K.-S. Choi (ed.), Recent Developments in Turbulence Management, 3–24.

Abstract

Experiments with drag reducing riblet film on a 1:4.2 model of a DORNIER Do 328 commuter aircraft have been carried out in the large German-Dutch Wind Tunnel (DNW). 82 % of the aircraft model including large parts of the wings and the nacelles were covered with the riblet film. The leading edge of the wing, however, was covered with smooth plastic film. Data with and without a tripping strip on wings, fuselage and nacelles were collected.

With a tripping strip, drag reduction with riblets was found in virtually all test cases. Without this tripping device, however, significant drag reduction was found only if the fuselage and the nacelles alone were covered with riblets. We assume in the absence of detailed knowledge of the flow conditions on the wing that in that particular case the riblet film was not of the optimum size for the wing. The measurements suffered also from lack of repeatability, with deviations in the drag measurements of up to 3 % of nominally identical conditions. This discrepancy was independent of the riblet film and was probably caused by lacking reliability of the wind tunnel balance. Nevertheless, the data show drag reductions by 1 - 6 % with riblet film. The upper value of 6 % is clearly not realistic, but the data do confirm previous estimations of 2 - 3 % total drag reduction.

Besides this confirmation of previous data, there are a few other relevant observations: The riblet film has no negative consequences for the wing performance at high angles of attack near and beyond flow separation. In addition, there is no hysteresis at these high angles in the drag polars. As a matter of fact, both lift gradient and maximum lift coefficient are marginally increased by the riblet film, but only by 1 and 1.5 % , respectively. Because of the effect on the wing, we recommend to normalize drag reduction data not at constant angle of attack, but at constant lift coefficient of the aircraft.

1. Introduction

It is now a well-established fact that the turbulent shear stress on a plane surface can be reduced by about 6 % by surfaces with tiny ribs ("riblets") aligned in the streamwise direction [1 - 4]. The lateral spacing of these ribs has to be in the order of the thickness of the viscous sublayer. In most technological applications, and in particular on aircraft, this viscous sublayer is very thin, typically between one hundredth and one tenth of a millimeter. Thus, riblets are usually very small and not visible with the naked eye. The mechanism of how these riblets work has been explained in a previous paper [5].

Considering that about 30 - 50 % of the drag of an aircraft is produced by turbulent skin friction, the total drag could be reduced by about 2 - 3 % if the plane were be covered with plastic riblet film. As a matter of fact, a drag reduction of 1 - 2 % has been achieved previously with an A 320 Airbus commercial aircraft in wind tunnel and full-scale flight experiments. In order to demonstrate that these benefits may also be achieved for commuter aircraft, a realistic experiment with a 1:4.2 scale model of the Dornier Do 328 aircraft in the large German-Dutch Wind Tunnel (DNW) was carried out. Preliminary trials by the DNW staff with small drag-producing bodies attached to another model had suggested that the accuracy of the equipment would be sufficient for this purpose. For testing time in the DNW is rather expensive, the measurements were limited to two days (> 100 000 DM). The test sequence at the DNW was such that on the first day the drag of the model without riblets was measured. Then, during the weekend, the plastic riblet sheets were applied by 3M France. Finally, during the second test, the drag of the model with riblets was measured. All experiments were carried out with and without a tripping device. To study the properties of the tripping device, some specific experiments on this had been carried out before the DNW tests.

2. Experiments

2.1. Experimental set-up

The main part of the experiments presented here were conducted in the large subsonic atmospheric wind tunnel (DNW) in Marknesse, the Netherlands. It has a test section with 8x6 m^2 cross section. The configuration of the commuter aircraft, a DORNIER Do 328, is as shown in figure 1. The fuselage length is 4.95 m, the wing span is 4.76 m, and the wing chord length is 0.508 m in the centre of the wing. The wing area is 2.06 m^2 and the wing aspect ratio is $\Lambda = 11.0$. During the experiments some parts of the model were removed. So were the bulges of the wheel gear fairings and the propeller blades. The supporting rod was attached to the upper side of the fuselage, as can be seen in figure 2. This was because previous experiments by DORNIER had shown that a supporting rod on the lower side would interfere with the flow around the fuselage, particularly when the wheel gear fairings were mounted. On the other hand, with the supporting rod on the upper side, the empennage would be immersed in the wake of the rod. Consequently, also the empennage was removed altogether during our tests.

The aerodynamic forces on the model were determined with a 6-component strain gauge balance which was located inside the model. Forces normal and parallel to the model were measured. With the measured angle of attack, lift and drag can be easily calculated. However, not only the aerodynamic forces on the model were sensed by the balance, but also the weight of the model which is about 1 ton. In order to take the weight forces into account properly, a calibration at various angles of attack with zero wind speed was also carried out. Obviously, the precision of the angle measurement is of paramount importance for the accuracy of the drag measurement. We have estimated that a deviation of 0.01 degrees in the angle measurement (which is the nominal angle resolution) results in a deviation of 3.2 % of the drag measurement at 40 m/s wind speed and in a deviation of 1.1 % at 70 m/s. The actual measurements show an even poorer reproducibility between data taken at nominally identical conditions. This was probably caused by some unspecified mechanical hysteresis in the system.

This is supported by the observation, that the data shifts occurred, particularly, in those situations when the drag polars were recorded with slowly decreasing angle of attack, instead of the usual procedure with slowly increasing angle of attack. However, there was no conclusive evidence of this either. In addition, there was no hint of any aerodynamic hysteresis. Thus, balance and/or sting mechanism are believed to have been the cause of the problem.

As a general criticism, we would like to mention, that wind tunnel balances of the previous generation did not suffer from a lack of precision. There, large forces (lift, weight of model) and small forces, such as the drag, could be deduced from directly measured forces and did not interact with each other (apart from cross-talk corrections). This basic principle of measuring physics has been abandoned, encouraged by confidence in the improved accuracy of strain gauge balances and the high resolution of digital electronics and, perhaps, by a more convenient operation of these new devices. Hence, one can easily end up with an accuracy so low that it would not meet the requirements of an ordinary household balance.

2.2 Experiments with the tripping device

To prescribe the laminar-turbulent transition at a certain location we used a tripping strip. Due to limited testing time (time is money) at the DNW we chose an adhesive letter tape "DYMO" with V's raised printed on it, see figure 3. This particular tape is relatively easy to apply, to remove and to re-apply in a reproducible manner. In addition, the tripping effect of this tape was known to be reliable as it had been frequently used before in our institutes. Finally, this tape does stick even on plastic riblet film. The V's on the tape were pointing with their lower corner into the streamwise direction. The total thickness of this trip strip including the protruding V's was 0.55 mm. Incidentally, the same effect is produced if the V's are pointing in the opposite direction. The drag measurements showed that the tripping device increased the total drag of the aircraft by about 10 %.(1) This increase was almost entirely generated by the tripping device on the wing. However, this increase is not only due to the drag of this device itself. By its definition it eliminates also part of the laminar flow on the wing, a fact, which obviously causes an increased drag. For the tripping strip, we were very careful to produce an identical shape of the printed V's, because we were aware of possible deviations produced by different printing machines and by varying pressure of the printing process.

In order to determine an appropriate location for the tripping device on the wing, we manufactured a 1:1 replica of a central section of the model wing. It had a chord length of 0.50 m and a span of 0.2 m. This wing section was tested in a small jet flow facility at the DLR institute in Berlin. The jet diameter was 10 cm and flow speeds of 90 m/s could be generated, as in the DNW. For flow visualization on the airfoil surface we used a mixture of black copying pigment powder and vegetable oil. To compare the situation with and without a tripping device we applied the strip only on one half span of the wing model. During the flow-induced development of the black oil film on the airfoil surface, areas with relatively high skin friction (downstream of the tripping device) and with relatively low skin friction (at the laminar separation bubble) could be distinguished. As a measure for the skin friction we considered the speed at which the black oil film was driven in the downstream direction.

---

1 It should be mentioned here, that 10 % drag increase is typical for tripping devices on large aircraft models with low drag. One reason for the application of a tripping device is the generation of flow conditions on the suction side of the wing being comparable to the conditions at the higher Reynolds numbers of real flight. This is particularly important for high angles of attack near flow separation. The issue of tripping devices, however, has its pros and cons and therefore we also have collected data without a tripping device.

An example of this is shown in figure 4, which was obtained on the upper side of the airfoil at a flow speed of 80 m/s and at zero angle of attack. On the upper side of the airfoil we looked for the location where, at the highest angle of attack before separation (which occurs at about 15°), the tripping device would be just in front of the laminar separation bubble. We found that this was at about 8 % of the chord length from the leading edge. On the lower side laminar flow seems to prevail, in particular at higher angles of attack. Therefore, the position for the tripping device seemed to be less critical and it was attached at 11 % chord length. Finally, we checked whether or not the plastic riblet film would be blown off from the surface. Though we teared off on purpose the leading edge of the film, it was not blown away, not even at an air speed of 90 m/s.

On the fuselage, the tripping device was tested at the outset of the experiments in the DNW. In this case, we used for flow visualization a mixture of mineral oil, petroleum and titanium dioxide powder as a white pigment. To check the effectiveness, we removed some small parts of the strip, as can be seen in figure 5. Thus, the status of the flow with and without trip strip could be compared directly. In figure 5, the resulting pattern can be seen, obtained at the lowest wind tunnel speed of 40 m/s. In addition, the direction of the flow can be recognized in this photograph.

2.3. Plastic riblet film

Unfortunately, the DORNIER company was unable to provide data on the wall shear stress and the local flow direction on the model surface. As the testing time was limited, no Preston-tube measurements to experimentally determine the shear stress and no detailed flow visualization studies to determine the flow direction could be carried out. Based on the expected wind tunnel speed range, 40 - 90 m/s, some rough estimations were made to determine the "optimal" riblet dimensions. Assuming that the turbulent boundary layer on the aircraft grows approximately similar to the one on a flat plate, we can use for the local skin fiction coefficient [6]

$$c_f' = 0.0592\, Re_x^{-0.2} \tag{1}$$

with the Reynolds number Re_x being based on the distance x from the (virtual) origin of the turbulent boundary layer. This origin is assumed to be located at the nose of the fuselage or on the leading edges of the wing, respectively. Using its definition, based on the skin friction, we can derive for the local skin friction velocity

$$U_\tau = U \cdot \sqrt{c_f'/2} \tag{2}$$

In the case of the flat plate with zero pressure gradient, the optimal dimensionless lateral groove spacing is at $s^+ \approx 12$ for triangular riblets with equal height and spacing [2,7]. The quantity s^+ is defined as

$$s^+ = \frac{s \cdot U_\tau}{\nu}, \tag{3}$$

where s is the real lateral groove spacing and ν is the kinematic viscosity of the air. If we assume for the calculation of an average value of Re_x based on half the length of the fuselage (2.5 m) and for the wing half the center chord length (0.25 m), we obtain the required lateral groove spacing s

	s on fuselage (mm)	s on wing (mm)
U = 40 m/s	0.125	0.099
U = 90 m/s	0.060	0.048

Thus, we chose from the few available rib spacings a lateral spacing of s = 0.076 mm. Obviously, this is a compromise and not optimal everywhere.

Another essential issue is how the riblet film has to be aligned on the model. According to Walsh and Lindemann [2] and Bechert et al. [7], the drag reduction is not changed for cross flow angles up to ± 15 degrees. However, beyond that, the deteriorating effect of cross flow becomes important. For instance, for a cross flow angle of ± 25°, the drag reduction is completely lost for otherwise optimal conditions of the riblets. Therefore, and due to lack of more detailed information on the flow direction, the longitudinal axis of the commuter aircraft was used to align the riblets. The 3M company which manufactured the plastic riblet film, applied it also on the aircraft model. First, the model surface was de-greased. Then, a suitable piece of the plastic film was cut from sheets 0.28 x 0.86 m^2 in size and then the cover layer on the adhesive side of the film was removed. Subsequently, the film and the model surface were wetted with a water-detergent solution. This made it possible to place the film and position it properly. Finally, a "squeegee" (a soft plastic wedge, developed for this purpose) was used to remove the water being trapped between riblet sheet and model surface and, thus, to establish contact between adhesive layer and model surface. The different pieces of riblet sheet were overlapped, cut and butted together. The riblet film was between 0.16 and 0.20 mm thick. In order to avoid a sudden step for the flow at the leading edge of the riblet sheets on the fuselage and on the nacelles, a very thin adhesive tape, 0.05 mm thick and 19 mm wide (Tesa film plus) was used to cover and taper this step, see figure 6. Some butted joints have also been covered with this tape because of gaps in between. The plastic riblet film was applied to fuselage, wings and nacelles, see figure 2. As can be seen in the figure, the more spherical shaped parts of the fuselage and of the nacelles were left uncovered. This was due to the limited amount of stretching which can be tolerated by this film. Also, spherical shapes would imply patching of many little pieces with different directions of riblet orientation. The rear part of the fuselage was also left uncovered because of application problems. The leading edge of the wings was covered with a particularly compliant smooth film in two layers, each 0.08 mm thick. The two layers together had the same thickness as the riblet film, see figure 7.Also in figure 7 the location of the tripping device can be seen.

3. Results

In figures 8 - 10 drag polars <u>with</u> tripping device are shown. Here, the drag coefficient C_{DP} is plotted (2) as a function of the lift coefficient C_L. The drag coefficient C_{DP} is the conventional drag coefficient minus the induced drag from the wings

---

2 We have been requested by the DORNIER company not to give absolute values of the drag coefficient.

$$C_{DP} = C_D - \frac{C_L^{\ 2}}{\pi \cdot \Lambda} \tag{4}$$

In equation (4) Λ is the aspect ratio of the wing, which is $\Lambda = 11.0$ in our case. This particular choice of the drag coefficient C_{DP} was made for plotting reasons only. It is easier to plot small horizontal differences between curves if these curves are not too extended in the horizontal direction, i.e., the drag polars are compressed by this selection of the drag coefficient.

The drag polars show a clear shift towards lower drag for the condition with riblets, in particular, as expected, at lower velocities. There is little data scatter within one data set of a drag polar. However, it is worrying that the reproduction of nominally identical data is only possible within 0-3 % deviation. Thus, no clear statements can be made here on the actual drag reduction by riblets; it can be 1-6 %, whereas 6 % is clearly not realistic. Consequently, the previous rough estimations of 2-3 % drag reduction by riblets are considered more reliable. On the other hand, the present measurements are not at variance with these previous estimations.

However, some more instructive positive observations have been made. It was found that the application of the film on the wings influenced the lift force, as can be seen in figure 11. In this figure the lift coefficient C_L has been plotted versus the angle of attack α. Comparing the different curves with and without film applied, a small increase of the lift gradient of about 1 % was found for the riblet case. Also, a small increase in the maximum value of the lift coefficient was found (+ 1.5 %). Considering these data, it seemed more logical to us to use a constant lift coefficient to compare equivalent experiments, than a comparison based on a constant angle of attack. This is because if the lift is increased by the riblet film, that would also produce an enhanced induced drag on the wings. At constant angle of attack, this could look like an increased total drag. However, the aircraft would fly at a lower angle to produce the same lift. Hence, a normalization of data with equal lift is more meaningful.

In the following we are going to discuss the influence of the tripping device. Comparing the data with and without trip strip there is, of course, a shift in the drag of about 10 %. In addition, the shape of the drag polars is changed, see figures 12 - 14. However, without trip strip, the riblets seem to cause a drag increase. In order to investigate this effect, all plastic sheets (riblet and smooth) were removed from the wings. The data with this configuration with riblets only on fuselage and nacelles are also plotted in figures 12 - 14. Fortunately, drag reduction had returned, in particular at lower angles of attack at and below cruise conditions.

An explanation for this latter behaviour is easy to give. Drag reduction from fuselage and nacelles can be expected only if the riblets are well aligned with the flow. At higher angles of attack the cross flow on the riblets deteriorates more and more the riblet performance. But there seems to be a negative influence on the wing aerodynamics. One possible explanation may be this: Maybe that there are extended regions of laminar flow on the wing, according to the shape of the airfoil and according to our previous visualization studies. Hence, the little gaps between the smooth plastic film and the riblet film may have caused earlier transition and thus higher drag, for the case without trip strip. Another additional reason may be that the lateral riblet spacing might have been chosen incorrectly for the wings. This is not surprising because we had no reliable information on the actual flow situation and the shear stress on the wing.

Another point of interest is the behaviour at higher angles when the flow starts to separate from the upper surface of the airfoil. This has been tested for wind tunnel speeds of 50 to 90 m/s. The angle of attack had been gradually increased up to 17 degrees and slowly decreased again. Some representative results obtained at 90 m/s are given in figure 15. No hysteresis is found between the two curves. In addition, the riblets do not cause premature flow separation.

4. Conclusions

As mentioned earlier, the actual testing time of this investigation was limited to two days, for financial reasons. This constraint seems to be incompatible with the basic requirement of research to obtain well-grounded results. In case of data uncertainty, that would mean a search for systematic errors and repetition of measurements. In particular, the drag measurement error was significantly greater than one can possibly tolerate in such an investigation.

Nevertheless, the measured drag reduction of 1-6 % is encouraging and not at variance with previous estimations and measurements which were, however, more accurate. Clearly, a measured drag reduction of 6 % for the whole aircraft is not realistic and is undoubtedly a result of inaccurate drag measurements. The previous estimations rather suggested a 2-3 % total drag reduction, and they are believed to be more reliable.

Important, however, is the finding that riblets do not cause premature separation. In addition the marginal increase of lift gradient (+ 1 %) and maximum lift (+ 1.5 %) is interesting. As a consequence of this shift in the lift coefficient, a comparison between the drag with and without riblets, based on data referring to constant lift seems more meaningful than a comparison at constant angle of attack.

5. Acknowledgement

The film used, both smooth and grooved, were supplied by 3M at cost-price. The application was carried out by Mr. Kus and Mr. Delachanal (3M France) with a high professional quality, and free. The aircraft model was provided by DORNIER GmbH; the good cooperation with Mr. Lück and Mr. Böhme from this company was appreciated. The cooperation with the DNW staff was also very good. In particular, we would like to mention that Dr. Eckert suggested the data normalization based on the lift coefficient. The funding for the DNW measurements was provided by the DLR.

References

1. Walsh, M. J.,Turbulent boundary layer drag reduction using riblets. AIAA-paper 82 - 0169 (1982).

2. Walsh, M. J. and Lindemann, A. M., Optimization and application of riblets for turbulent drag reduction, AIAA-paper 84 - 0347 (1984).

3. Nitschke, P., Experimentelle Untersuchung der turbulenten Strömung in glatten und längsgerillten Rohren. Max-Planck-Institut für Strömungsforschung. Bericht 3/1983, April 1983.

4. Bechert, D. W., Hoppe, G. and Reif, W.-E., On the drag reduction of the shark skin. AIAA-paper 85 - 0546 (1985).

5. Bechert, D. W., Bartenwerfer, M. and Hoppe, G., Turbulent drag reduction by nonplanar surfaces - a survey on the Research at TU/DLR Berlin. Proc. IUTAM Symposium "Structure of turbulence and drag reduction" Zürich 1989, A. Gyr (Editor), Springer-Verlag Berlin, Heidelberg 1990.

6. Schlichting, H., Grenzschicht-Theorie. Verlag G. Braun, Karlsruhe, 1965.

7. Bechert, D. W., Gerich, D. A. and Hoppe, G., Short (internal) report on measurements with sawtooth riblets (3M plastic riblet film). DFVLR/HFI Berlin, 1987.

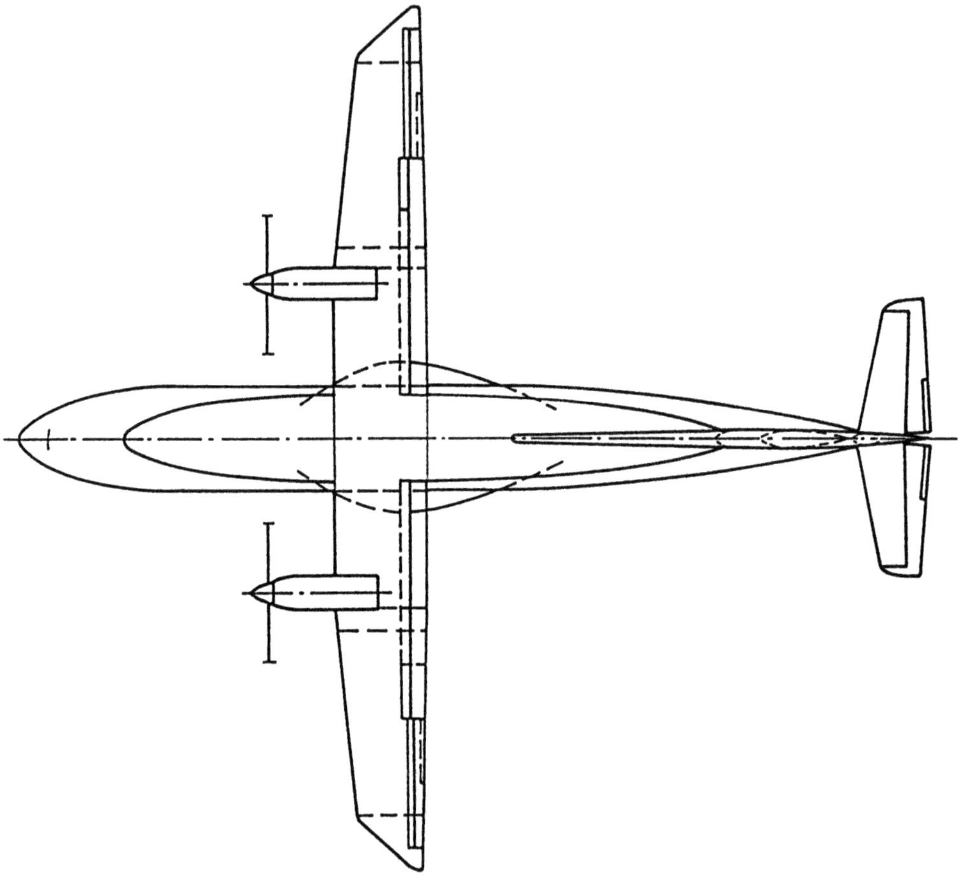

Figure 1. DORNIER Do 328 Commuter aircraft.

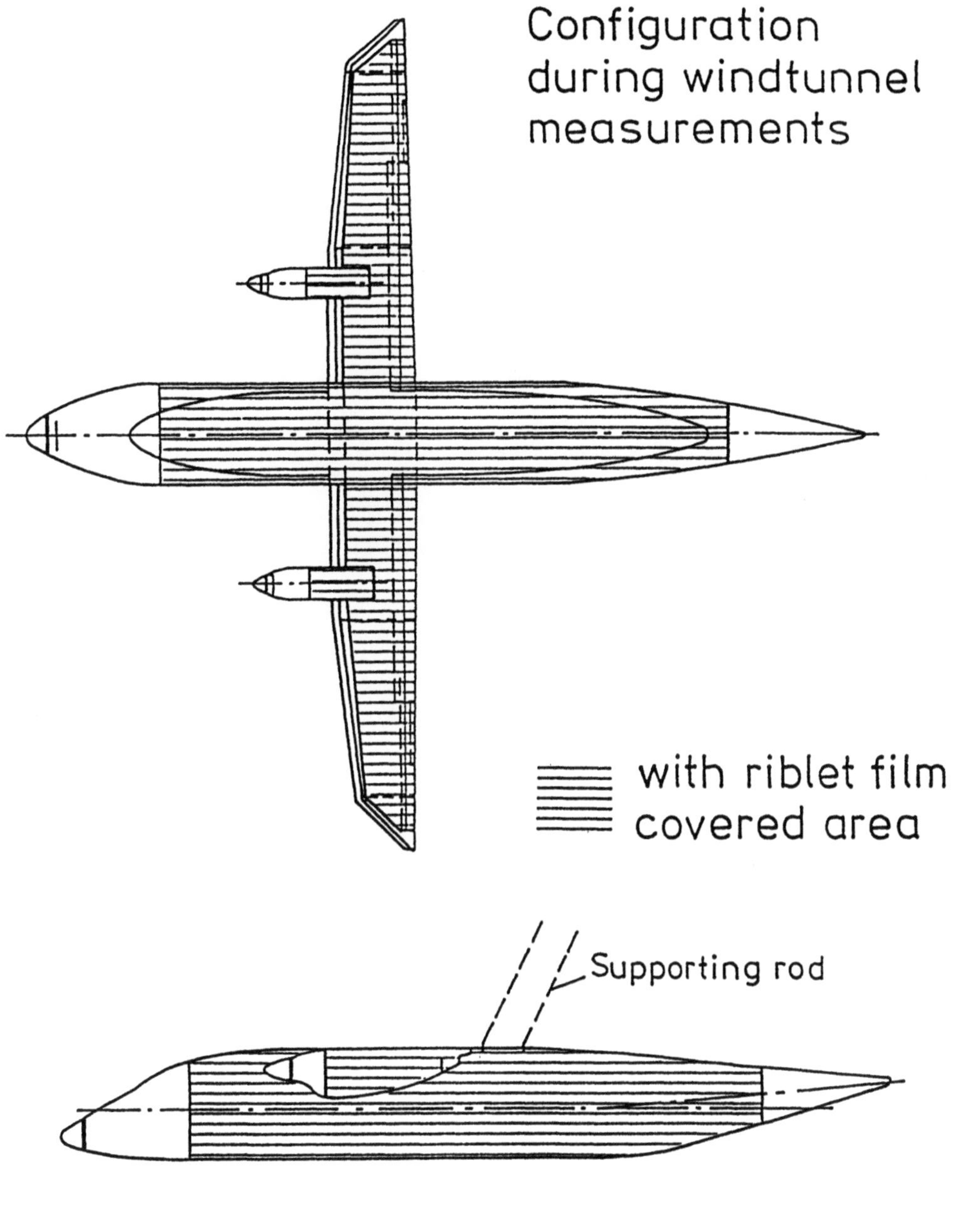

Figure 2. Distribution of riblet film and location of tripping devices on the model. The empennage had been removed during the tests.

Figure 3. Tripping strip to enforce transition to turbulent flow.

Figure 4. Test of tripping strip on the wing section in a small jet facility in Berlin. On the right hand side, the high turbulent shear stress downstream of the strip has swept away the black oil film. On the left hand side, the low shear stress under the laminar separation bubble kept the oil film there.

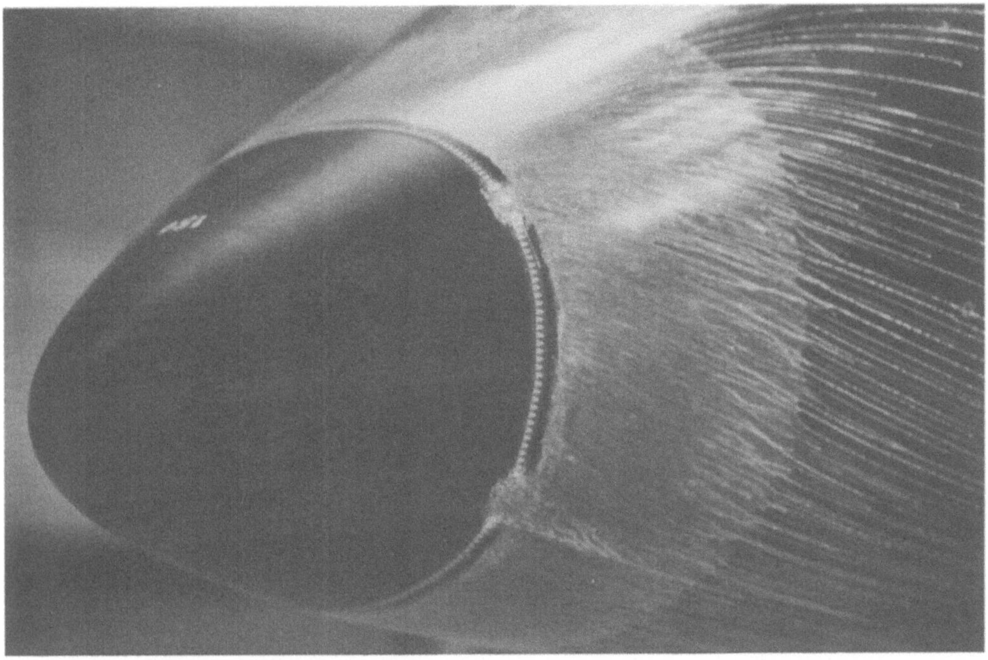

Test of tripping strip on the nose of the fuselage, prior to the measurements in the DNW.

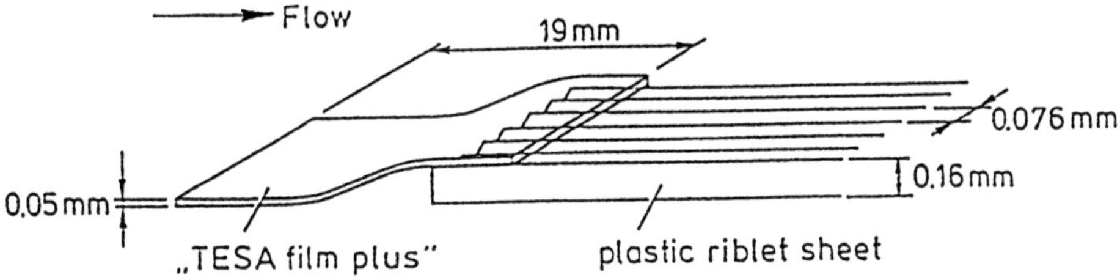

Figure 6. Leading edge of the riblet film on fuselage and nacelles. The step is smoothened by a thin covering adhesive film.

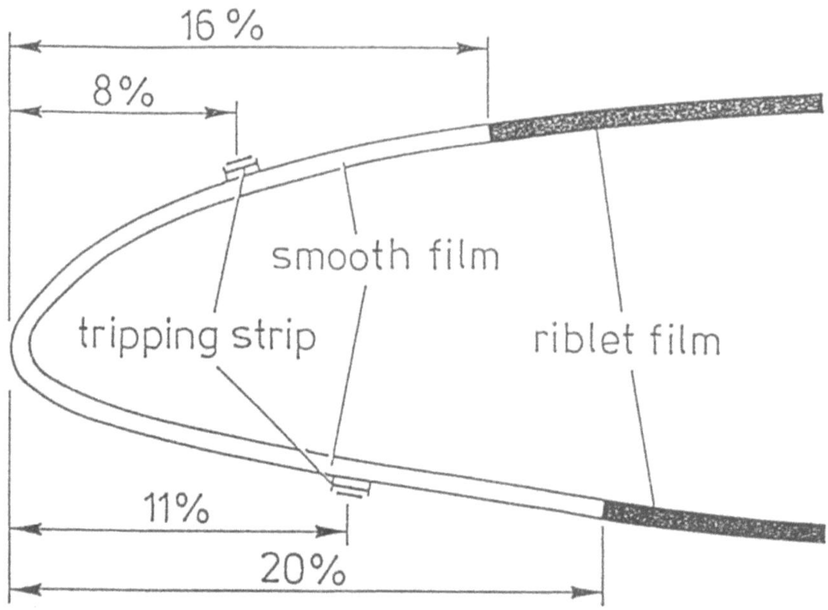

Figure 7. Leading edge of the central part of the wing. Location of the tripping devices and of the smooth cover on the surface. The distances are given in percent of the airfoil chord length.

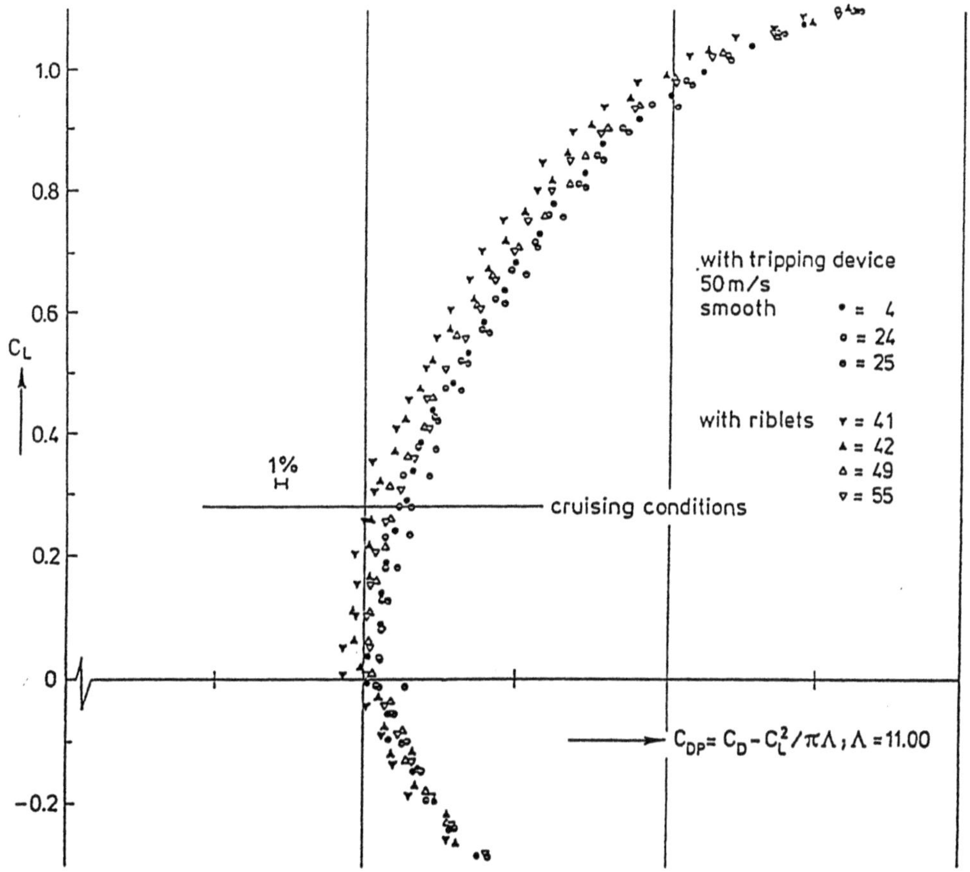

Figure 8. Drag polar at 50 m/s with tripping device. The dimensionless riblet widths, estimated with equation (3) are $s^+ = 8.9$ for the fuselage and $s^+ = 11.2$ for the wing, respectively. The numbers at the symbols refer to the individual drag polar numbers as measured consecutively in the wind tunnel. The zero position for the drag coefficient is not given, at the request of the DORNIER company.

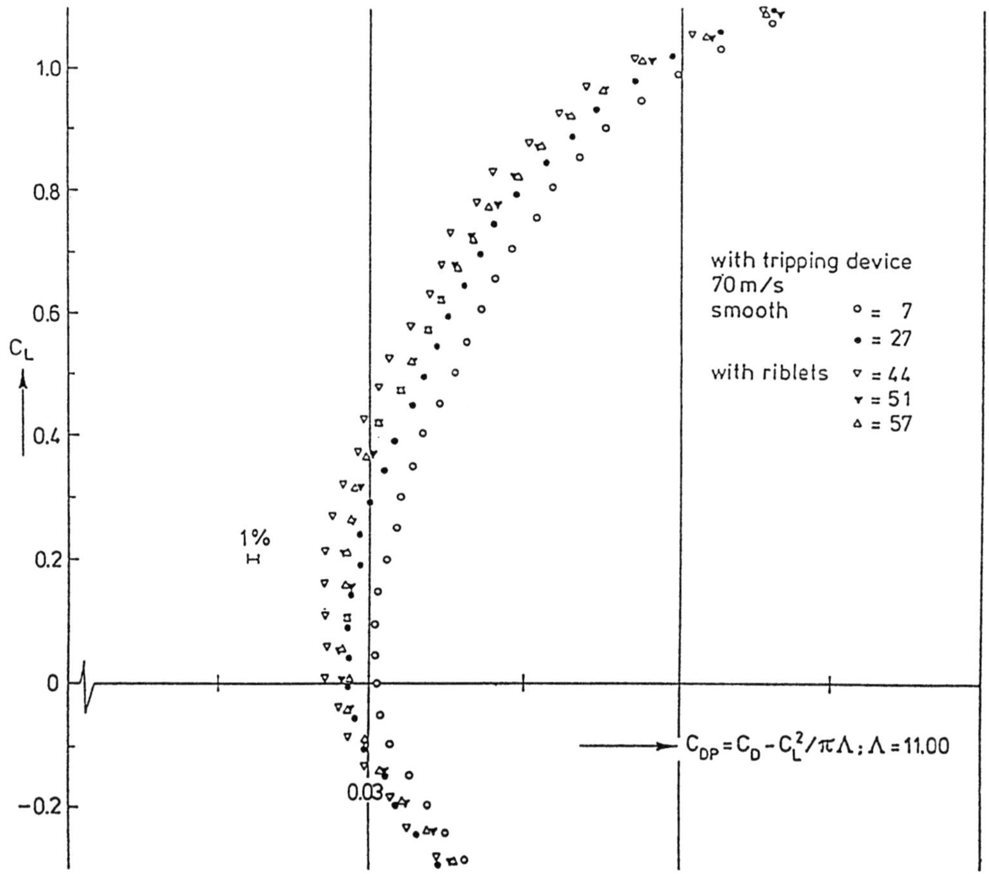

Figure 9. Drag polar at 70 m/s with tripping device. $s^+ = 12.1$ for the fuselage and $s^+ = 15.2$ for the wing.

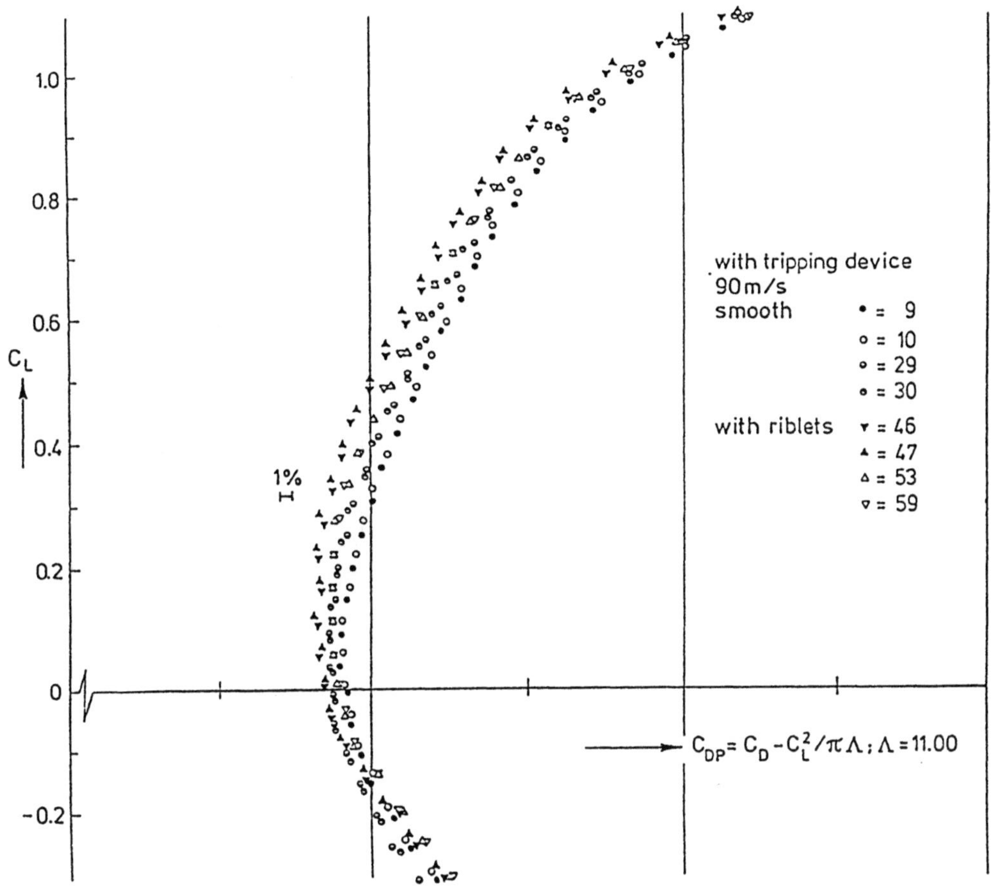

Figure 10. Drag polar at 90 m/s with tripping device. $s^+ = 15.1$ for the fuselage and $s^+ = 19.0$ for the wing.

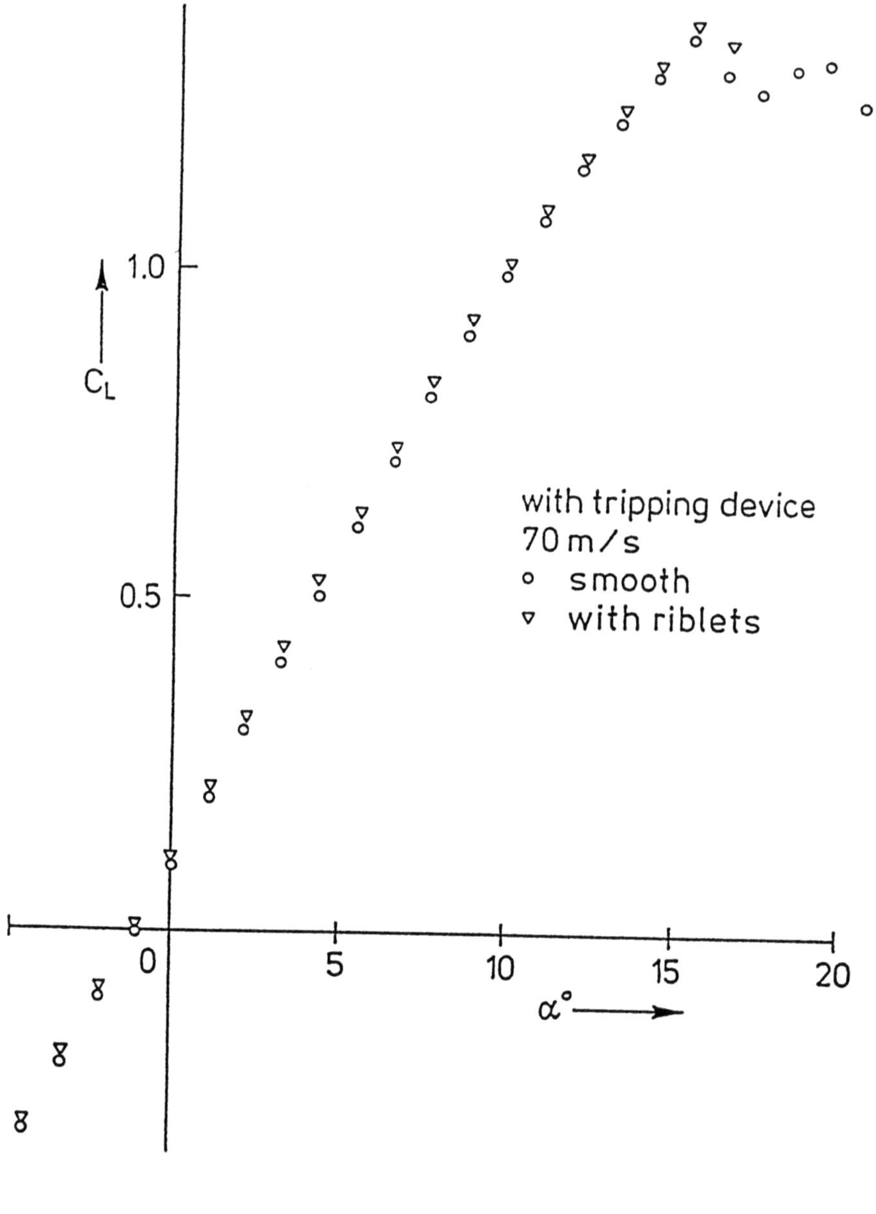

Figure 11. Lift coefficient as a function of the angle of attack, with tripping device.

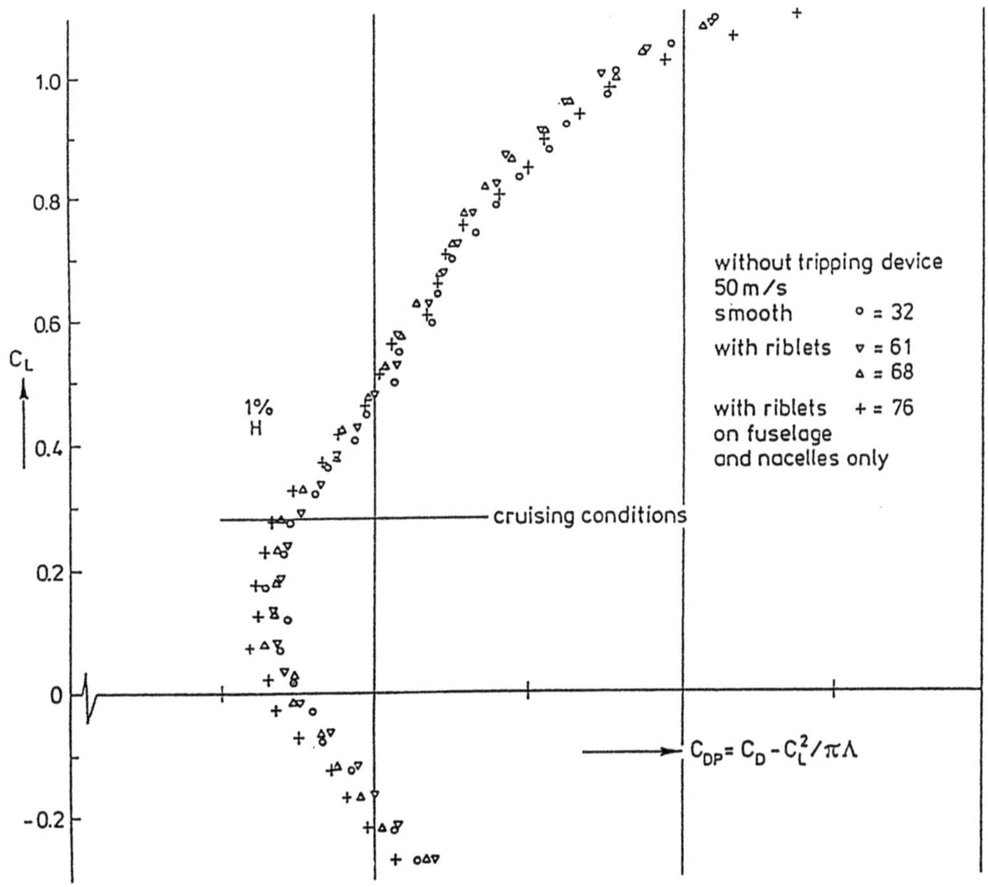

Figure 12. Drag polar at 50 m/s without tripping device. s^+ as in figure 8. Data with riblet film on fuselage and nacelles alone are also given.

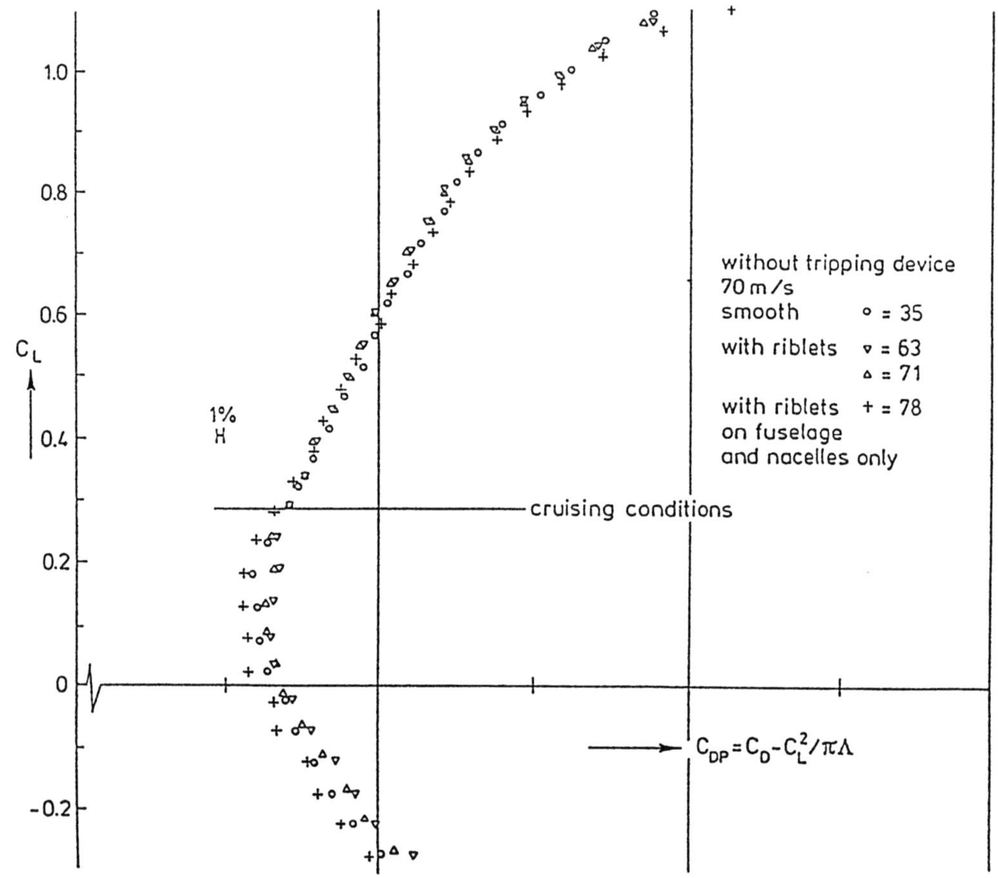

Figure 13. Drag polar at 70 m/s without tripping device. s^+ as in figure 9. Data with riblet film on fuselage and nacelles alone are also given.

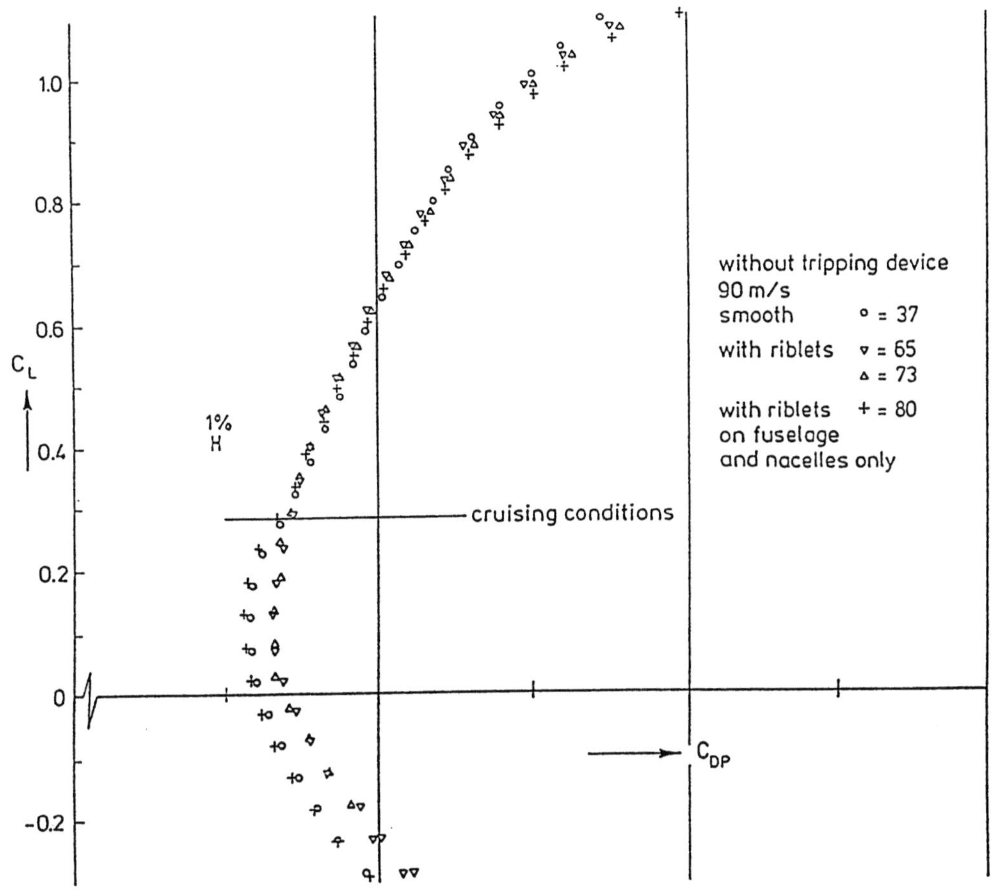

Figure 14. Drag polar at 90 m/s without tripping device. s^+ as in figure 10. Data with riblet film on fuselage and nacelles alone are also given.

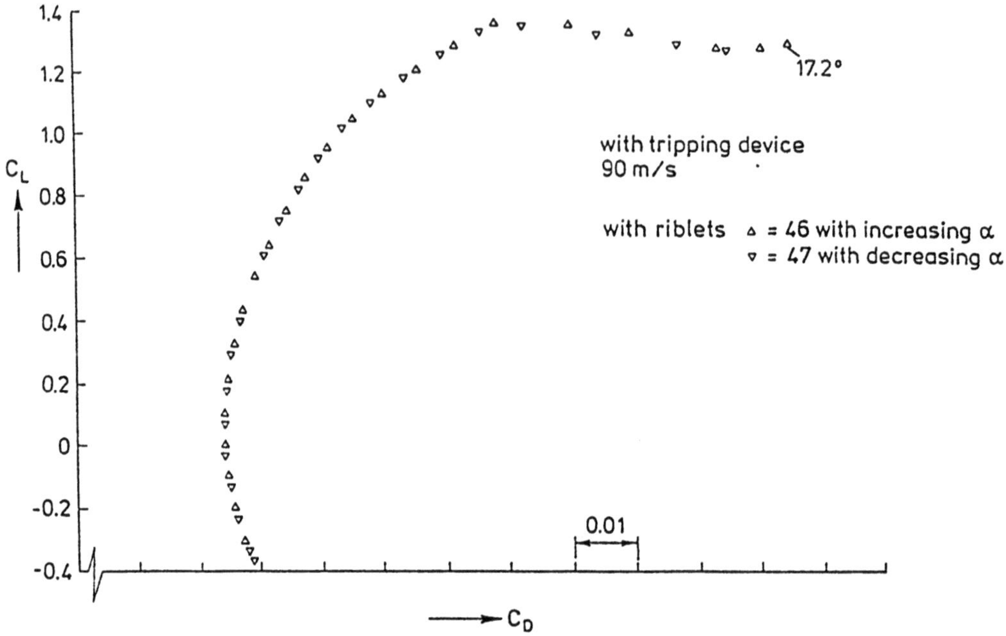

Figure 15. Drag polar at high angles of attack with tripping device and riblets. Velocity: 90 m/s. The data were taken at increasing and decreasing angle in order to detect a hysteresis which, as can be seen, does not exist. The zero position of the drag coefficient is not given, at the request of the DORNIER company.

Heat transfer study of riblets

K.-S. CHOI and S. HAMID*

University of Nottingham
Nottingham, U.K.

* *Defence Research Agency*
Farnborough, U.K.

K.-S. Choi (ed.), Recent Developments in Turbulence Management, 25–41.

Abstract

A wind tunnel study of the heat-transfer characteristics of surfaces with triangular profile riblets has been carried out in a two-dimensional turbulent boundary layer and the results compared with those of a smooth surface. The heat-transfer coefficient has been measured for the Reynolds number of 2.8 to 11.3 $\times 10^5$ based on the streamwise length, covering a range of dimensionless riblet height and spacing of 6 to 60 wall units. The experimental results indicate that triangular riblets can enhance the convective heat-transfer by as much as 30% compared with the smooth surface without a penalty of increasing the drag. During the experiment, the heat loss due to thermal radiation was shown to be significant at lower Reynolds numbers. An attempt has been made to nullify this loss by heating the surrounding walls to the same temperature as the test surface.

Nomenclature

h	heat transfer coefficient
I	current supplied to heater pad
k	thermal conductivity of fluid
Nu	Nusselt number ($= h.x/k$)
Pr	Prandtl number ($= \nu/\alpha$)
$\dot{q}$	heat flux from test surface ($= I.V$)
Re	Reynolds number ($= U_\infty.x/\nu$)
s	riblet spacing
s^+	non-dimensional riblet spacing ($= s.u^*/\nu$)
T_a	ambient temperature
T_w	test surface temperature at reference point
ΔT	temperature difference ($= T_w - T_a$)
U	mean velocity
U_∞	freestream mean velocity
u	velocity fluctuation
u'	turbulence intensity ($= \sqrt{u^2}$)
u^*	friction velocity
V	voltage across heater pad
X	streamwise distance from beginning of working section
x	streamwise distance from leading edge of test plate
Y,y	co-ordinate normal to test surface
Z,z	spanwise co-ordinate
α	thermal diffusivity of fluid
β	flap deflection angle
δ	boundary layer thickness
δ^*	displacement thickness
Θ	momentum thickness

ν kinematic viscosity of fluid
ξ streamwise distance between origins of geometric and thermal boundary layers
σ Stefan-Boltzmann constant

1. Introduction

Walsh and Weinstein [1] in 1979 briefly examined the combined drag and heat-transfer characteristics of small longitudinally ribbed surfaces in a two-dimensional turbulent boundary layer. Their results suggested that with triangular ribbed surfaces it may be possible to augment the rate of heat-transfer from the surfaces without a penalty of increasing the turbulent drag; the use of, say, cooling fins (large longitudinal ribs) in heat exchangers always increases the drag. Although Walsh and Weinstein's preliminary tests seem to have suffered from lateral heat conduction and heat radiation problems, their results were encouraging for potential application to, say, compact heat exchangers but no attempt appears to have been made so far to substantiate their findings.

The present study focuses on the heat-transfer aspects of triangular riblets, whose turbulent drag reduction characteristics are well-documented [2, 3, 4]. As in Walsh and Weinstein's work at NASA Langley the present study has been conducted in a wind tunnel using air as the working fluid but here the implications for the use of riblets in water are also discussed. The main objective is to establish if there is an enhancement of heat-transfer within or close to the drag reduction regime of riblets, taking precautions to minimize the experimental uncertainties due to lateral heat conduction and radiation from test surfaces. A non-dimensional riblet spacing s^+ (see Nomenclature for definition) ranging from 6 to 60 has been investigated using triangular riblets of unit aspect ratio (0.73 & 1.83 mm) at four freestream velocities (2.25 to 9.0 m/s). The Reynolds number based on momentum thickness was in the range of 990 to 1830.

2. Experimental Arrangement

2.1 The Wind Tunnel

The experiments were conducted in a suction type, low-speed open-return wind tunnel at the Osney Laboratory, Oxford University. The tunnel has a 3.0 m long rectangular test-section of constant height 310 mm with the vertical side walls diverging gradually to compensate for the boundary layer growth; the width at the inlet is 460 mm and at the end of the test section is 500 mm. Figure 1 shows a plan view of the tunnel working section. The tests were carried out at four different freestream velocities in the range U_∞ = 2.25 to 9.0 m/s. A vane anemometer at the centre of the test section at X = 150 mm (i.e. 150 mm downstream of the beginning of the working section) was used for setting and checking the freestream velocity but was removed from the tunnel during the hot-wire and heat-transfer measurements.

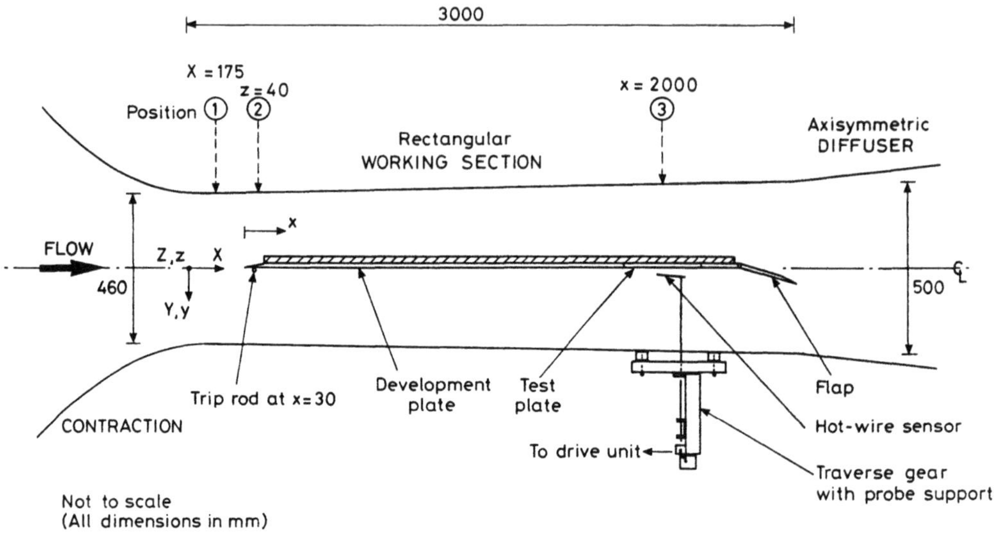

Figure 1 Plan view of wind tunnel working section.

2.2 The Test Plate

The heat-transfer tests involved the measurement of surface temperatures of a heated plate in a two-dimensional turbulent boundary layer. To ensure a fully developed turbulent boundary layer at the measurement station and also to attain reasonably high Reynolds numbers for the experiments the test plate extended almost the entire length of the test section (see Figure 1). The test plate consists of a 1.8 m long smooth flow development plate (see Figure 2), with a boundary layer trip rod 30 mm downstream of a sharp leading edge, followed by a 400 mm long slot which houses the interchangeable smooth and riblet heated test plates. The peaks of the riblets are level with the upstream plate surface. Downstream of this test plate there is a 200 mm long smooth (unheated) section followed by a 300 mm long hinged section of flap with a sharp trailing edge. The plate assembly has a constant span of 310 mm which is mounted vertically on the centreline of the test section with its leading edge 300 mm downstream of the beginning of the test section (X = 300 mm). Two aluminium L-sections are attached to the underside of the plate to ensure the flatness of the test surface along its length.

Three heated test plates, one smooth and two with triangular profile riblets (spacing, s = 0.73 mm and 1.83 mm, see Figure 3a) of unit aspect ratio (s/h = 1), have been used in this comparative study of heat-transfer characteristics. Each riblet plate is comprised of an array of small rectangular panels machined from solid aluminium, see Figure 3b. The panels, which are not in contact with each other, are bonded to a 3 mm thick gold plated heater pad sandwiched between the aluminium panels and a perspex base. The perspex base plate was chosen to minimize heat conduction. The smooth test plate is made of a perspex plate with a thin aluminium foil heater pad bonded to the surface.

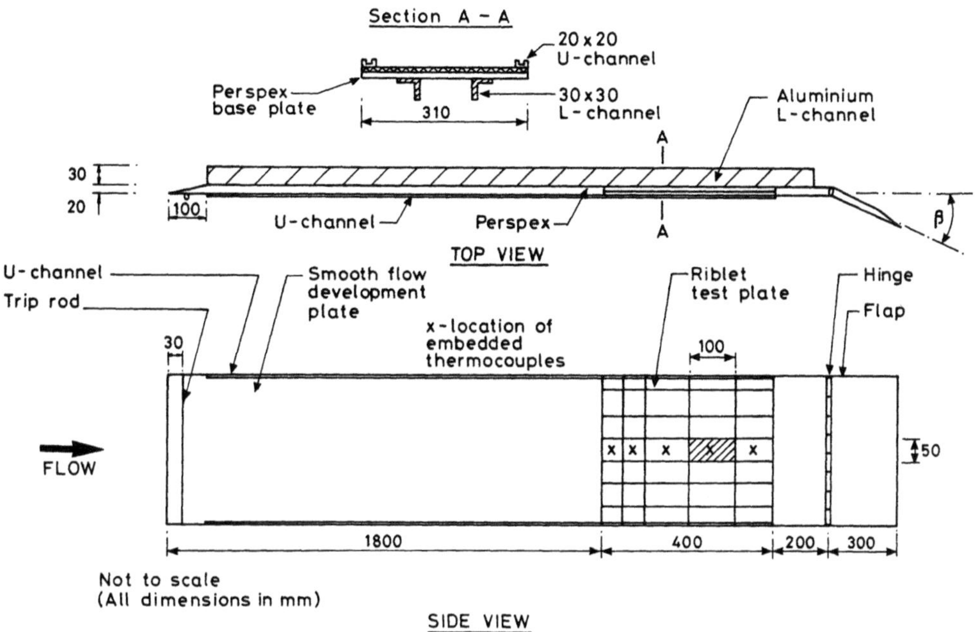

Figure 2 Schematic diagram of plate assembly.

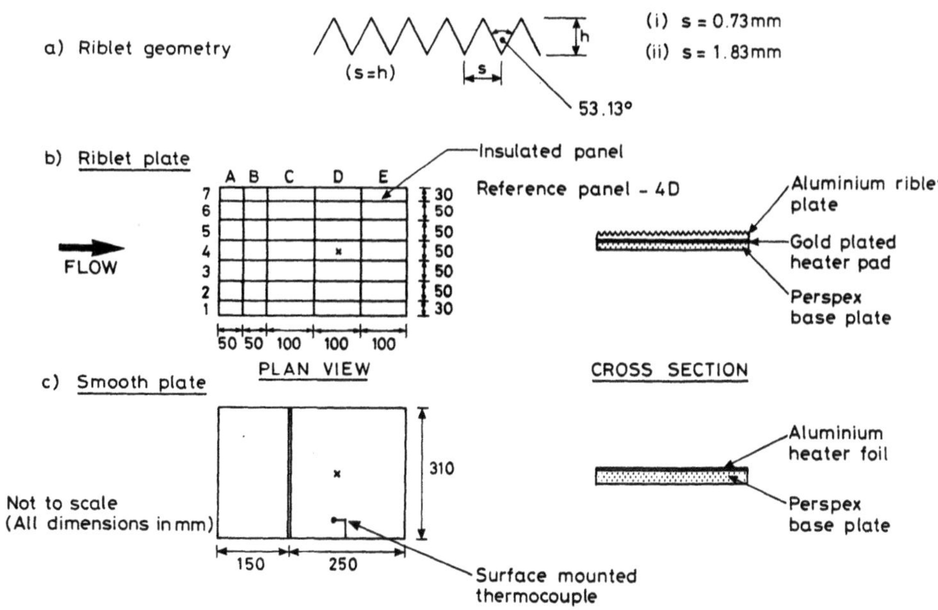

Figure 3 Details of test plates.

Due to the comparatively high power requirement for the low-resistance heater foil and the power input restrictions imposed by the 13 amp. power supply, two separately heated foil sections were used on the smooth test plate (see Figure 3c). To ensure identical power input per unit area of these heated foils the same potential difference was applied across the terminals of the two sections. Since the aluminium heater foil is very thin (of the order of 0.1 mm) lateral heat conduction for the smooth plate is considered to be negligible.

Electrical power is supplied to the heater pad (or foil) via two copper conducting strips running along the edges of the plate. The conducting strips, which act as terminals, are secured in position by 20 mm x 20 mm U-channels, as shown in the Figure 2, for a uniform electrical contact along the plate.

2.3 Temperature Measurement Techniques

The surface temperature of the heated plates was measured using both T-type thermocouples and thermochromic liquid crystals sprayed on the surface. For the riblet plate the thermocouples were embedded in the aluminium at the centre of each centre panel. For the smooth plate one thin thermocouple was surface-mounted at an off-centre location (see Figure 3c) to avoid disturbance to the flow near the centre reference point. For each surface, the temperature was measured near the centre of the plate marked x in Figure 3. The surfaces of all the test plates were sprayed matt black before applying the 10 μm thick layer of encapsulated liquid crystals. A calibration of the liquid crystals against the thermocouples was made with a narrow-band colour of yellow at 30.4 °C. Consequently the surface temperature for each test plate was set to 30.4 °C using this colour, which gave a temperature difference of approximately 12 °C relative to the ambient temperature, which was also measured with a thermocouple. Note that the spectrum displayed by the liquid crystals on the smooth test plate was used to visually check the continuity of the temperature distribution across the boundary of the two separately heated sections of this plate.

2.4 The Traverse Mechanism

Hot-wire anemometry was used for the flow survey over the test plates. A single-wire probe was traversed using a 750 mm long Unislide traverse gear bolted to one side of the tunnel with its lead screw normal to the side wall (see Figure 1). The majority of the hot-wire measurements were made at the mid-chord position of the test plate at x = 2000 mm, where x is measured from the leading edge of the development plate. A limited number of measurements were also made just downstream of the trip rod at x = 40 mm and the probe was calibrated ahead of the leading edge (at X = 175 mm). These three streamwise positions of the test section (3, 2 and 1 respectively, see Figure 1) were accessed through three vertical slots in the tunnel side wall. The slots also enabled the probe to be traversed in the spanwise (Z-) direction. Traverses normal to the plate surface (Y-direction) were automatically controlled by a Zenith micro-computer via a Digiplan drive unit. The positioning accuracy of the traverse mechanism was of the order of 50 μm per 300 mm of traverse with a step resolution of 10 μm. The accuracy was improved by approaching a point always from the same direction to reduce backlash error. The mounting of the traverse gear on the side of the tunnel was such that the probe could be accurately positioned manually in five discrete spanwise positions covering approximately the central 40% of the span. All unused access holes and gaps were sealed with adhesive tape.

3. Velocity and Turbulence Measurements

The main objectives of the flow survey were firstly to ascertain the flap deflection angle required to avoid any flow separations at the leading edge of the plate and secondly to investigate the spanwise uniformity of the flow at the mid-chord position of the test plate. We have also documented the mean velocity and turbulence profiles together with the basic flow parameters before the heat-transfer measurements over the test plates. This part of the experiments entailed measurements of mean flow and turbulence over the unheated smooth test plate only at the nominal freestream velocities of U_∞ = 2.25, 4.5, 6.75 and 9.0 m/s. A DISA 55M constant temperature anemometer was used with a DISA 55P05 boundary layer type single-wire probe. By using this type of boundary layer probe with its offset prongs and by pitching the probe support shaft by 4° towards the plate enabled measurements as close as 0.3 mm from the surface. All the hot-wire data were digitally linearized before data processing.

3.1 Flow near the Leading Edge

In setting up the test plate for the present experiments, we have tried to avoid the flow separation at the leading edge by adjusting the angle of the trailing-edge flap attached to the end of the test plate. The stagnation point on the test plate can be moved around the leading edge with a change of the flap angle, effectively by changing the strength of the circulation around the test plate. Our strategy was to move the stagnation point to the working side of the leading edge and to keep it there most of the time. Since the stagnation point tends to move about, it is important that it is located slightly downstream of the leading edge on this side of the test plate. The flap angle has to be, therefore, positive (towards the working surface); it was set to $\beta = +6°$ throughout the present investigation. With this flap angle, there was no indication of flow separation at the leading edge of the test plate.

3.2 Spanwise Uniformity of the Flow

The spanwise uniformity of the flow was investigated at the nominal freestream velocity of 4.5 m/s by measuring the mean velocity and turbulence intensity profiles at five spanwise positions (Z = -65, -35, -5, +25 and +55 mm), see Table 1. At each of the other three freestream velocities only the traverse at the central position (Z = -5 mm) was made. Plots of mean velocity (U) and turbulence intensity (u'), normalized by the local freestream velocity, plotted against the normalised distance (y/Θ) are shown in Figure 4 for the freestream velocity of 4.5 m/s. The variation of integral parameters such as displacement thickness (δ^*) and momentum thickness (Θ), computed from these profiles, as a function of spanwise position Z and freestream velocity U_∞ are given in Table 1. The data for U_∞ = 4.5 m/s covering the central 40% of the plate span indicate

Z(mm)	U_∞(m/s) nominal	U_∞(m/s) actual	δ^*(mm)	θ(mm)	u^*(m/s)	Re_θ
-65	4.5	4.47	5.66	3.55	0.196	988
-35	4.5	4.32	6.11	3.87	0.189	1042
-5	4.5	4.19	8.00	5.23	0.176	1365
+25	4.5	4.41	7.57	5.05	0.189	1388
+55	4.5	4.47	7.93	5.46	0.191	1521
-5	2.25	2.00	5.87	5.86	0.093	730
-5	6.75	6.07	6.39	4.11	0.249	1554
-5	9.0	8.09	5.61	3.63	0.326	1830

Table 1 Boundary layer integral parameters

a substantial lack of uniformity in the integral parameters. For example, the momentum thickness changes by about 40% over this span. However, the variation in momentum thickness is less than 10% across the upper half (+ve Z) of this central region indicating that the core exists but is shifted upwards away from the geometric centre of the tunnel. Also, the reasonably good collapse of the profiles in Figure 4 suggests a good measure of similarity of the mean flow and turbulence structure across the span in spite of the non-uniformity in the integral parameters as mentioned.

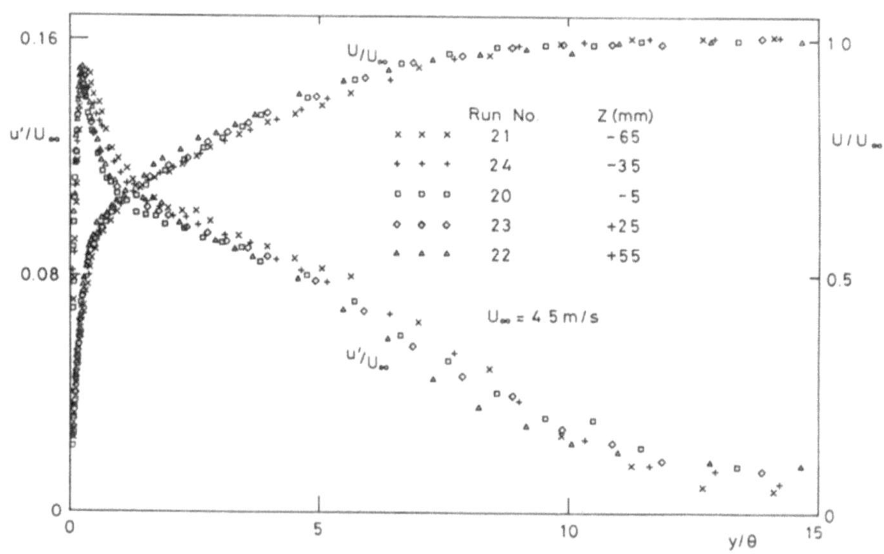

Figure 4 Mean velocity and turbulence intensity profiles.

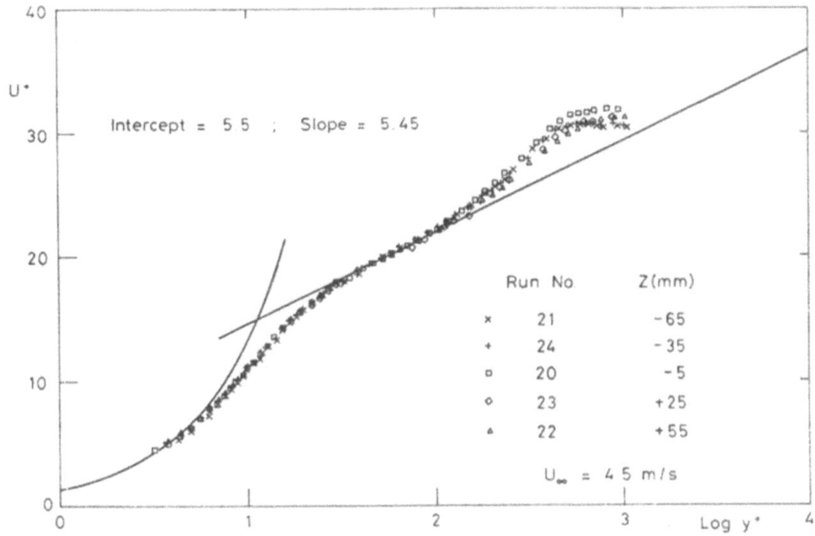

Figure 5 Log-law plot of mean velocity profiles.

3.3 Near-Wall Velocity Profiles

Plots of the mean velocity profiles in log-law format given by

$$U^+ = 5.5 \log y^+ + 5.45 \tag{1}$$

are shown in Figure 5 for the freestream velocity of 4.5 m/s, where $U^+ = U/u^*$, $y^+ = y.u^*/\nu$ and ν is the kinematic viscosity. The friction velocity u^* was obtained using the Clauser plot from the best fit to the log-law in the range $50 < y^+ < 160$. The relatively short log-law region is indicative of the comparatively low Reynolds numbers (Re_θ = 900 to 2700) of these experiments. Table 1 shows the variations of u^* and the peak turbulence intensity u'_m as a function of Z and U_∞. Here, unlike the asymmetric spanwise variation of the integral data, these near-wall parameters for U_∞ = 4.5 m/s indicate a symmetric variation about the centreline. Also the spanwise uniformity of the near-wall parameters is more acceptable, with the centreline values of u^* and u_m' only 10% and 5% lower than the values at the edge of the central region (at 20% span). Details of the near-wall region ($y^+ < 100$) are shown in Figure 6 in terms of u'/U_∞ versus y^+, where the non-dimensional height y^+ is an appropriate length scale for this region. It is encouraging to note that for the study of the heat-transfer characteristics of riblets, two-dimensionality of the near-wall region of the turbulent boundary layer is of primary importance, whilst the outer layer may be of secondary importance.

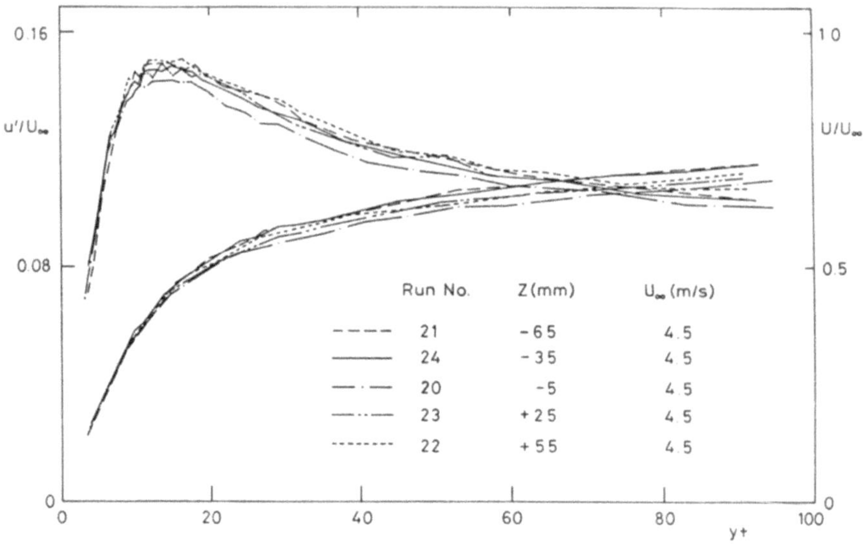

Figure 6 Near-wall mean velocity and turbulence intensity profiles.

3.4 Measurements of Freestream Turbulence

A limited number of single-wire measurements were made to obtain the turbulence intensities with the wire at position 1 (X = 175 mm, Y = 135 mm, Z = -5 mm). Both the DC and the AC signals from the wire were recorded at each of the four freestream velocities with the trailing edge flap angle set at +6°. Table 2 summarises the turbulence intensities at four different tunnel speeds of 2.25, 4.5, 6.75 and 9.0 m/s. As one can see from the table, the turbulence intensity is smaller at higher tunnel speed, reducing its percentage value from 0.74 at the lowest speed of 2.25 m/s to 0.35 at the highest speed of 9.0 m/s.

U_∞ (m/s)	u'/U_∞ (%)
2.25	0.74
4.5	0.53
6.75	0.43
9.0	0.35

Table 2 Turbulence intensities in the wind tunnel

4. Heat-Transfer Measurements

The heat-transfer coefficient, h, defined as

$$h = \frac{\dot{q}}{A \cdot \Delta T} \quad \textbf{(2)}$$

where $\dot{q}$ is the heat flux, A is the surface area of the heated plate and ΔT is the temperature difference between the plate and the ambient air, was measured for each test plate at each of the four freestream velocities. The heat flux, which was kept constant over the test plate, is the power input to the heater pad given by

$$\dot{q} = I \cdot V \quad \textbf{(3)}$$

where I is the current and V is the voltage supplied to the heater pad. The temperature difference

$$\Delta T = T_w - T_a \quad \textbf{(4)}$$

was measured using a combination of thermocouples and liquid crystals. At each test configuration the tunnel speed, the plate temperature (T_w) and the ambient temperature (T_a) were allowed to attain equilibrium conditions before I, V, T_w and T_a were recorded a minimum of four times. This procedure usually took approximately one hour and the repeatability for the measurement of h was within $\pm 1\%$ of the mean value.

Over the spanwise extent of the heated test plate, a temperature variation exists due to spanwise change in mean flow and turbulence profiles, particularly near the edges of the plate. In addition there is a streamwise temperature change due to the developing thermal boundary layer over the test plate. Preliminary tests with one of the riblet plates indicated a temperature variation over the spanwise as well as streamwise extent of the plate of up to 2 °C. The tests, however, confirmed the effectiveness of the insulated panels in minimizing lateral heat conduction. That is to say the colours displayed by the liquid crystals on adjacent panels were not continuous across the boundaries. A good thermal diffusivity of the aluminium panels was also confirmed which resulted in a uniform colour display over each panel, particularly along station D (see Figure 3b). Thus the electrical power supplied became a uniform convective heat flux from these panels. The centre of panel 4D at x = 2050 mm was selected as the reference point for the heat-transfer measurements.

In general, the measured h is the total heat-transfer coefficient with two main components, i.e.

$$h = h_{rad} + h_{conv} \quad \textbf{(5)}$$

where h_{rad} and h_{conv} represent the components of heat loss due to radiation and forced convection respectively. The heat loss due to free convection was negligible compared with these two components for the present experimental conditions. Under a two-dimensional turbulent boundary layer, the convective heat-transfer coefficient is a function of flow speed and becomes zero as the flow speed becomes zero. Therefore, the radiative heat-transfer coefficient can be determined by extrapolating the measured 'total' heat-transfer coefficient to zero speed. In this comparative study of the convective heat-transfer enhancement by riblets, an estimate of h_{rad} is needed to quantify any differences in h_{conv} relative to the smooth surface, at low speeds in particular.

The heat-transfer part of the experiments was carried out in two parts. In the first part h_{rad} was estimated by extrapolation as described above, and in the second part an attempt was made to measure h_{conv} directly by nullifying the radiative part of the heat-transfer by employing a 'radiation shield'.

4.1 Measurements without Radiation Shield

A dimensionless coefficient of heat-transfer is the Nusselt number, Nu, which is defined by

$$Nu = \frac{h \cdot x}{k} \tag{6}$$

where h is the heat-transfer coefficient of the test surface and k is the thermal conductivity of the fluid. Here x is the streamwise length of the geometric boundary layer measured from the trip rod near the leading edge of the unheated development plate, although the virtual start of the boundary layer is slightly upstream of this position. The solid lines in Figure 7 show the variation of the Nusselt number as a function of Reynolds number for each of the three test plates (without radiation shield), where the Reynolds number, Re, is defined as

$$Re = \frac{U_\infty \cdot x}{\nu} \tag{7}$$

The approximately linear increase in the Nusselt number with increase in the Reynolds number is in reasonably good agreement with the empirical heat-transfer relation (see Kays & Crawford [5], for example) for a turbulent boundary layer over a smooth surface, i.e.

$$Nu = 0.0287 \cdot Pr^{0.6} \cdot Re^{0.8} \cdot \left[1 - \left(\frac{\xi}{x} \right)^{\frac{9}{10}} \right]^{-\frac{1}{9}} \tag{8}$$

where the Prandtl number Pr is a constant dependent only on the properties of the fluid (in this case, air). The effects of delayed start of the thermal boundary layer are

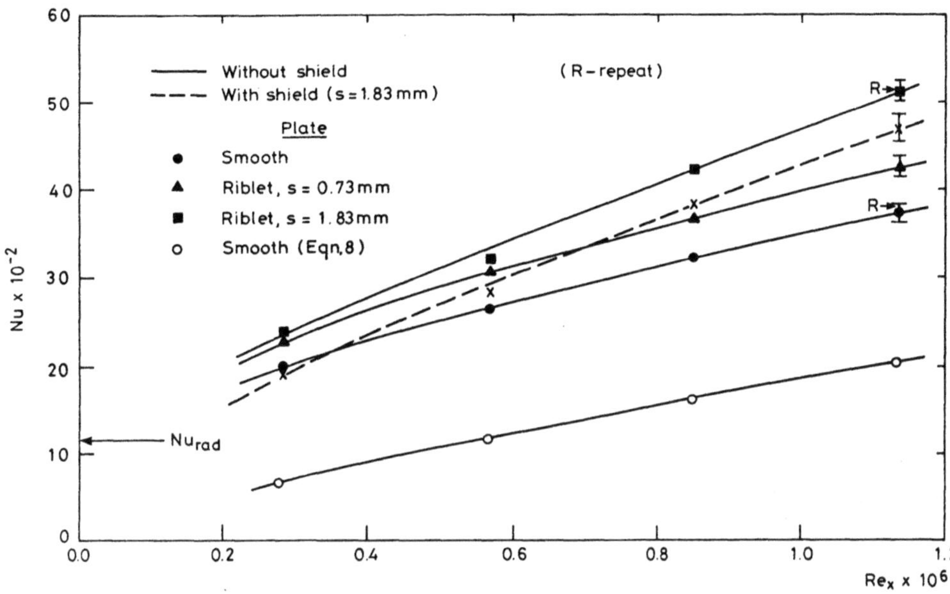

Figure 7 Nusselt number versus Reynolds number with and without radiation shield.

incorporated in the factor within the square bracket in equation 8, where ξ is the distance between the origins of the geometric and thermal boundary layers. The values predicted by this equation for the smooth plate are also given in Figure 7, which give a reasonable comparison with the measured data if we account for the heat loss due to radiation. Each data point in Figure 7 represents the average of at least 4 measured values as described above with an error band indicating 95% confidence limit. This figure indicates a significant increase in heat-transfer due to riblets for all the riblet tests carried out in this investigation.

4.2 Measurements with Radiation Shield

Thermal (infrared) radiation is a characteristic of the heated body and depends only on the temperature of the test plate and the ambient temperature and is given by

$$h_{rad} = \sigma \, . \, [\frac{T_w^4 - T_a^4}{T_w - T_a}] \tag{9}$$

where σ is the Stefan-Boltzmann constant of radiation, see reference [6] for example. The thermal radiation from the test plate will, therefore, become zero if we set T_a equal to T_w. The aim here was to measure the convective part of the heat-transfer coefficient h_{conv} whilst trying to nullify the radiative part of the heat loss, h_{rad}. The radiation shield was formed by attaching gold plated heater pads to the floor, ceiling and side wall of the tunnel. Figure 8 shows the positions of these heater pads relative to the test plate. To

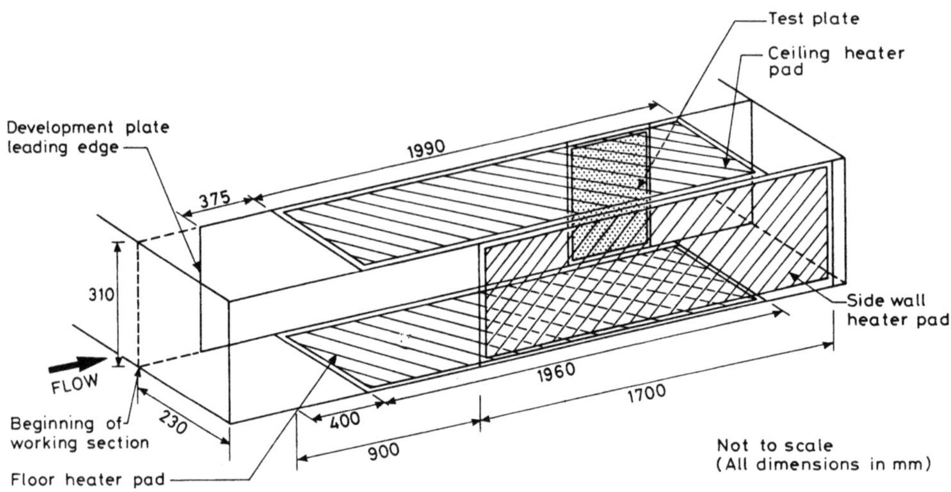

Figure 8 Schematic view of tunnel heater positions.

keep the same radiation level of the test plate throughout the experiments, the surface temperature of the test plate T_w was set to the value corresponding to an appropriate constant value of h_{rad} for a given ambient temperature T_a. The temperature of the radiation shield was set equal to this surface temperature.

With the radiation shield installed in the tunnel, the plate with the larger riblets (s = 1.83 mm) was tested. As a check on the repeatability of the measurements, the heat-transfer coefficient was first measured again without heating the shield, at the highest velocity (U_∞ = 9.0 m/s) only, before employing the tunnel heater. The variation of Nu_{rad} with Reynolds number for this riblet plate both with and without the radiation shield can be seen in Figure 7. Firstly, the excellent repeatabilty of the measurements with the unheated shield gives a measure of confidence in our heat-transfer data and secondly, the heated radiation shield has the effect of shifting the intercept towards zero. The approximately 35% reduction in the intercept, although encouraging, is somewhat lower than might be expected and may indicate some radiated heat loss from the test plate to the un-shielded area, and possibly some residual heat loss due to conduction.

As above the repeatability test for the smooth plate at U_∞ = 9.0 m/s with the unheated radiation shield was excellent, (see Figure 7). However, due to the differences in the fabrication of the smooth plate compared to the riblet plate (i.e. no thermocouple at the reference point and the use of two separate heater foils) and also because the required T_w was in general different from the operating temperature of the liquid crystals (30.4 °C), it was not possible to repeat the tests with the heated radiation shield for the smooth plate. Nevertheless, the consistency of the data in Figure 7 again gives a measure of confidence in the heat-transfer data.

4.3 Heat-Transfer Enhancement by Riblets

As shown in Figure 7, the magnitude of the increase in the heat-transfer coefficient depends on the riblet size and the Reynolds number. It is also shown that the increase relative to the smooth surface is greater for the plate with the larger riblets and the difference further increases at higher Reynolds numbers. On the other hand, the difference between the Nusselt numbers for the three plates diminishes as the Reynolds number decreases and there is some indication that the three curves in Figure 7 (without shield) converge to a single point as the Reynolds number tends to zero. Assuming that this intercept is the same for each plate, which implies identical heat loss due to radiation Nu_{rad}, it is possible to estimate the increase in the convective heat-transfer of the riblet surface relative to the smooth surface as a function of non-dimensional riblet spacing s^+, where

$$s^+ = \frac{s \cdot u^*}{\nu} . \tag{10}$$

Figure 9 shows the change in convective heat-transfer due to riblets

$$\frac{Nu_{riblet} - Nu_{smooth}}{Nu_{smooth} - Nu_{rad}} \tag{11}$$

as a function of s^+. Not surprisingly there is some scatter in the data at low s^+, i.e. at low Reynolds numbers, where the percentage change in convective heat-transfer is very sensitive to any experimental error, since Nu_{rad} is a significant fraction of total Nu there

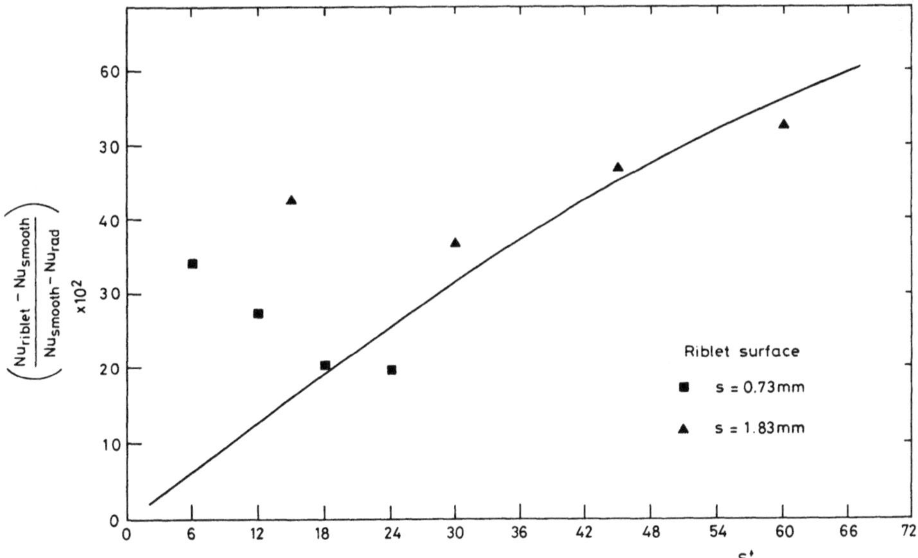

Figure 9 Percentage increase in convective heat-transfer versus dimensionless riblet height.

(see Figure 7). Thus in Figure 9 it is possible to draw a curve through the more reliable data, corresponding to the higher Reynolds numbers, with the knowledge that any curve must pass through the origin. Again, as might be expected, the heat-transfer increases with increase in riblet size (in terms of wall units). However, the important conclusion that may be drawn from this figure is that there seems to be a significant increase in the heat-transfer coefficient for triangular riblets even within the drag reduction regime of $s^+ \leq 30$. Walsh and Weinstein [1] noted that the increase in heat-transfer obtained using large longitudinal ribs as cooling fins in heat exchangers is always accompanied by a greater increase in drag. It appears from the present work that the magnitude of the convective heat-transfer enhancement can be as much as 30% without incurring a penalty of increasing the drag if we chose riblets of $s^+ \approx 30$.

5. Concluding Remarks

A comparative study of the heat-transfer characteristics of the smooth and riblet surface in a two-dimensional boundary layer has been carried out, both with and without compensation for thermal radiation. The results indicate the following differences in the characteristics.

1. Relative to the smooth surface, there seems to be an increase in the heat-transfer coefficient for the riblet surface within their drag reduction regime ($s^+ \leq 30$) and beyond it ($s^+ > 30$).
2. The magnitude of the convective heat-transfer enhancement by riblets can be as much as 30% without incurring a penalty of increasing the drag at $s^+ \approx 30$.
3. The radiative heat loss is a significant part of the total heat-transfer coefficient at low Reynolds numbers, which may be reduced or eliminated by heating the surrounding walls to the same temperature as the test surface.

All the present experiments were carried out in a wind tunnel using air as the working fluid. The results from these experiments should, therefore, be interpreted carefully when an application of riblets is considered for heat-transfer enhancement in water. It is quite a different situation from the drag reduction characteristics of riblets, where the results will be valid for any (Newtonian) fluids including air, water or oil. This is due to the fact that the thermal boundary layer plays an important role in the heat-transfer characteristics of riblets and the Prandtl number (Pr) of water (Pr = 7.0) is different from that of air (Pr = 0.7). In other words, the relative thickness of the thermal boundary layer (δ_T) to the geometric boundary layer thickness (δ_M) is different in water from that in air. This can easily be understood as the Prandtl number is a ratio of the viscous diffusivity (ν) to the thermal diffusivity (α):

$$Pr = \frac{\nu}{\alpha} . \tag{12}$$

Since the thickness of the boundary layer is proportional to the square root of the diffusivity, the ratio of thermal boundary layer thicknesses of air and water will be the square root of the ratio of the Prandtl numbers, which is about 3.2, assuming that the Reynolds number is kept constant. This suggests that the relative size of thermal boundary layer to the geometric thickness (δ_T/δ_M) in water will be nearly one-third of that in air. An important implication of this is that the heat-transfer coefficient of riblets will be greatly increased in water to an equivalent level where the physical size of riblets in air is increased by 3.2 times.

This argument is certainly correct for *laminar* flows, so that the heat-transfer of longitudinal grooves in laminar boundary layers would be very much more efficient when used in water than in air. This would not strictly apply for *turbulent* boundary layers in which riblets can be used for drag reduction. In this situation, we have to use the *turbulent* Prandtl numbers instead of (molecular) Prandtl numbers. Since the turbulent Prandtl number of water will not be much greater than that of air, improvement in heat-transfer efficiency in turbulent boundary layers in water will not be as great as in laminar boundary layers illustrated above. It is, however, worthwhile to investigate the heat-transfer characteristics of riblets for water applications. It must be noted that the turbulent Prandtl number over a riblet surface may be quite different from that over a smooth surface because of the modified turbulence structure in the near-wall region near the riblets [2].

Acknowledgements
We would like to thank Dr. Peter Ireland and Dr. Zolan Wang of Oxford University for their assistance in setting up these experiments and for the many useful discussions relating to the heat-transfer tests. This work was supported by the Ministry of Defence UK, Department of Trade and Industry UK and BMT Fluid Mechanics Ltd.

References

1. Walsh, M.J. and Weinstein, L.M. "Drag and Heat-Transfer Characteristics of Small Longitudinally Ribbed Surfaces"; AIAA Journal, Vol. 17, No. 7, p770, 1979.
2. Choi, K-S. "Near-wall structure of a turbulent boundary layer with riblets"; J Fluid Mech., Vol. 208, pp. 417-458, 1989.
3. Walsh, M.J. "Drag characteristics of V-groove and transverse curvature riblets"; In *Viscous Drag Reduction* (ed G. R. Hough), AIAA, 1980.
4. Walsh, M.J. and Lindeman, A.M. "Optimisation and application of riblets for turbulent drag reduction"; AIAA Paper 84-0347, 1984.
5. Kays, W.M. and Crawford, M.E. "Convective Heat and Mass Transfer"; 2nd Ed., McGraw-Hill Book Co., 1980.
6. "Handbook of Engineering Fundamentals"; ed. Eshbach, O.W., 2nd ed., Wiley Engineering Handbook Series, 1961.

Performances of internal manipulators in subsonic three-dimensional flows

E. COUSTOLS

ONERA/CERT
Toulouse, France

K.-S. Choi (ed.), Recent Developments in Turbulence Management, 43–64.

Abstract
The present paper summarizes one of the latest experimental studies, undertaken at ONERA/CERT in subsonic flows. It deals with the efficiency of **internal** manipulators, commonly named by many researchers **riblets**, in three-dimensional boundary layer flows. The flow developing on both sides of an ONERA D section aerofoil, set at 22.5° angle of sweep, has been manipulated using different riblet vinyl adhesive films. These latter were covering almost 85% of the chord length. Total drag variations were estimated from Pitot tube surveys, performed at about one chord length behind the aerofoil trailing edge. Two chord Reynolds numbers have been considered : 2.65 10^5 and 4.25 10^5. On both sides of the aerofoil, grooves have been aligned with the free-stream flow direction. Skin-friction drag decreases of up to 5-6% have been recorded when the dimensionless rib height, scaled with wall units, is less than 15-20.

1 Introduction

Past studies of the detailed structure of turbulent boundary layers have shown that the flow structure very near the wall exhibits a pattern of alternating low- and high-speed streaks, with the periodic break-up of the low-speed streaks responsible for almost 70% of the production of turbulent energy. Thus, the wall region is dominated by a sequence of eddy motions that are collectively called the "bursting" phenomenon [25], [26]. Capitalizing upon the rapid accumulation of knowledge about the eddy structures of the near-wall flow region, there have been few successful efforts to favourably alter or control them. However, several studies have examined the potential for modification of this turbulence structure using passive surface modifications. As quoted by Johansen and Smith [24], "generally, the motivation of these tudies has been to develop means for reduction of surface drag by turbulence control". One of the most effective and relatively easy to manufacture drag reducing modifications are **riblets**, which are micro-grooves aligned with the free-stream flow direction.

The mechanisms involved in such a drag reducing process are still under investigation in some laboratories. Indeed, one could think that those ribs might increase the spanwise streak spacing and then decrease the burst intensity [5],restrict the spanwise motion of the longitudinal vortices [9], "act as a nucleation site causing a focusing of low-speed streaks over their peaks" as quoted by Gad-el-Hak and Blackwelder [20], increase the viscous sublayer thickness [11], ...etc.

Because of the micro-geometry of these grooves, measurements just above or inside the ribs are very difficult. To our knowledge, the only attempts to measure within triangular grooves have been performed by Vukoslavcevic et al [30], and later by Benhalilou et al [6], while Hooshmand et al [23], or Coustols et al [12], looked at the flow modification in the close vicinity above the crests plane. These investigations revealed that the wall shear is increased near the peak of the riblet but substantially reduced within the valley, so that a nett drag reduction ensues. It has been surmised that the effect of grooves is to retard the flow in the valley, thus creating a viscosity dominated region where the local skin friction is greatly reduced. Thus, viscous effects might play the leading part in this near-wall flow manipulation and the resulting outcome on the Reynolds stress components be a consequence.

Numerous studies performed during the last decade or so, have shown that manipulation of turbulent boundary layers using internal manipulators ("riblets") is one of the most promising methods of reducing drag in turbulent flows. Furthermore, they confirmed some of the earliest and most important results obtained by Walsh [31], and Walsh and colleagues [32], [33], [34]. Drag reductions by as much as 8% in skin-friction has been reported with grooves, the height and spacing of which are of the order of 15 wall units, i.e $15\nu/u_\tau$.

Several sets of experiments have been carried out at ONERA/CERT since mid-1986. They were performed in zero- as well as moderate- pressure gradient conditions, in either low subsonic or transonic speeds [13] to [17]. Through either momentum balance technique or wake surveys, they confirmed that such internal devices could provide nett friction drag reductions. In order to go closer to flight applications, the effect of small streamwise grooves on a one thirty-eighth scale Airbus-type fuselage and on a complete one-eleventh scale Airbus A320 model was respectively checked at the F2-ONERA/Le Fauga wind tunnel [12] and in the S1-ONERA/Modane wind tunnel [19]. Total drag variations were measured through appropriate internal drag balances, instead of afore-mentioned laboratory measurements. Those last experiments showed that important drag reductions could be achieved at cruise conditions. Indeed, maximum nett decreases of 1.6% in Cd were obtained at M_∞=0.7 and at cruise level, with negligible changes in ΔCd over the explored Cl range : 0.1-0.6 [19]. It has been expected that those results would be easily applied to practical flight test conditions for which the fuselage Reynolds number is only increased in a ratio close to 5. Let us point out that recent flight tests carried out by Airbus Industrie and its partners confirm these findings.

However, in case of future applications of such internal passive devices to subsonic transport aircraft onto, not only the fuselage, but also the wings, fin and horizontal tail, it is necessary to analyse the behaviour of "riblets" in three-dimensional boundary layers.

The influence of yaw angle, between grooves and infinite free-stream flow direction has been essentially investigated by several researchers for two-dimensional boundary layers [14], [15], [19], [27] or [29]. Those experiments showed up that performances of grooved surfaces were unaffected by misalignment up to 15°. However, in three-dimensional flows, the effect is completely different since the velocity vector varies very rapidly close to the wall, over distances the height of which is close to the ribs depth.

To our knowledge, very few experimental work has been completed as regards turbulent manipulation of three-dimensional boundary layers. Indeed, "riblets" have been applied to :

- the hull of a one-third scale model of an Americas cup yacht [8] ;
- the upper and lower sides of the wings of a one-eleventh scale A320 aeroplane model [19] ;
- the upper surface of the wing of a T-33 aeroplane [27] ; although the angle of sweep is almost 0°, the tapering of the wing induces three-dimensional effects ;
- the "MOBY-D" high-speed buoyancy propelled vehicle [10].

Thus, an experimental study has been undertaken at CERT ; "riblets" have been applied on both sides of an ONERA D section aerofoil, set at at an angle of sweep of 22.5°. The performances of three models, having the same aspect ratio, have been determined through total drag variations estimated from Pitot tubes surveys in the aerofoil wake, one chord length behind the trailing edge.

2 Test facility - Configuration

Experiments have been performed in a wind tunnel at low subsonic free-stream speeds. The cross section is rectangular : 30cm high, 40cm wide and approximately 120cm long. The upper and lower side walls of the test section diverge slightly to ensure practically a zero-pressure gradient flow. Static pressure probe measurements carried out along the centre-line of the test section indicate a weak favourable pressure gradient along the whole length of the test section [14].

The turbulence level in the external flow is roughly constant and of the order of 0.25% for outer flow velocity range : 18-36 ms^{-1} ; preliminary surveys made in the potential flow through a four-wire probe allowed to check the two-dimensionality of the flow [14].

2.1 Experimental set-up

The chosen aerofoil is an ONERA D one, the chord length of which, c, is 200mm. The dimensions of this aerofoil are rather important versus those of the test section ; thereby, one would have to worry about the "blocking" effect, and then to take into account the downstream free-stream velocity instead of the upstream one. The aerofoil is steadily held on one lateral side wall through a set-up which allows, for a given angle of sweep, to fix the aerofoil along the test section centre-line and to adjust it at a given angle of attack.

The ONERA D aerofoil is set at an angle of attack of 0°, and at an angle of sweep, ϕ of 22.5° ; that latter was chosen because it is close to that of an Airbus-type wing, i.e of a usual transport-aircraft wing. Of course, modifications of the angle of attack induce different favourable and/or adverse pressure gradients on both sides of the aerofoil. That was not the aim of the present study since such an effect, on groove performances, had already been study in subsonic [14], [15] and transonic flows [19].

Two mean free-stream velocities have been considered : 20.0ms^{-1} and 32.2ms^{-1}, which lead to chord Reynolds numbers ,R_c close to respectively 2.65 10^5 and 4.25 10^5. The transition is tripped on the upper as well as lower side of the aerofoil at about 7.5% chord length from the leading edge, with a cylindrical wire (d=0.8mm) ; this diameter has been chosen in order to satisfy the usual displacement thickness criterion (2 to 3 times greater). Compared to carborundum band, this tripping device allows to get rid of possible variations in time of the induced over-thickenning of the boundary layer.

2.2 "Riblet" models

Different models, made from thin vinyl sheets having an adhesive backed film, have been applied between x/c=0.15 and 1.00, on the upper as well as lower sides of the aerofoil. On the other hand, forwards to the leading edge a smooth vinyl sheet, the thickness of which is $\simeq$ 100μm, has been stuck. Thus, when the groove surfaces are applied upon the aerofoil, the bottom of the valleys is approximately mounted flush with the adjacent upstream smooth part, since the thickness of the self-adhesive bed of the "riblet" film is varying between 90 and 130μm. As regards the "reference" configuration, the aerofoil was entirely covered with smooth film ; of course, the tripping wire is set upon the film (figure 1). Symmetrical V-shaped films have been supplied by the 3M Company. Their aspect ratio, s/h, is constant and equal to 1; s and h denote respectively the spacing and the height of the ribs. Three depths have been considered : h=0.152mm, 0.076mm and 0.051mm.

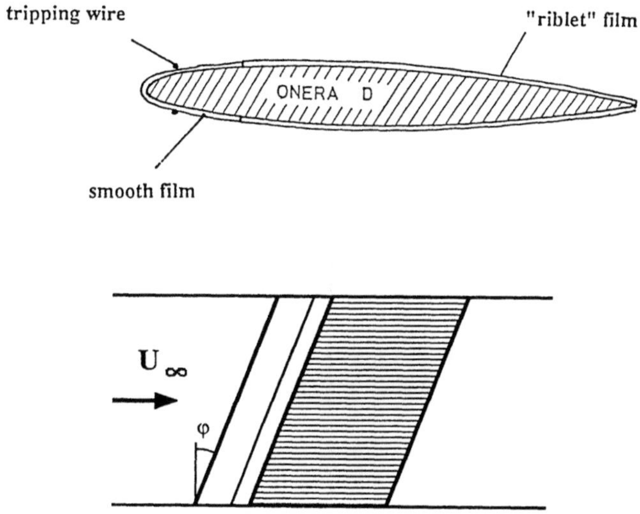

Figure 1: ONERA D aerofoil covered with "riblet" film.

The grooves are approximately aligned with the infinite free-stream flow direction, instead of being set perpendicular to the aerofoil leading edge. This choice was deliberate, since this study does not deal with the effect of angle of yaw on "riblet" performance but with the behaviour of these devices in three-dimensional flows. It is worth noting here that on the one-eleventh scale model as well as on the A320 prototype, the ribbed surfaces were applied in such a manner. One must recognize it is easier to know the infinite free-stream flow direction, i.e fuselage axis, than the external streamline which is close to the former for wings application. Finally, the films covered the entire wetted area, in the spanwise direction, from one side wall of the test section to the other (figure 1).

3 Reference configuration

For the reference configuration, the total drag coefficient of the aerofoil, entirely covered with a smooth vinyl sheet, has been determined through wake surveys, at about one chord behind the trailing edge, along the centre-line of the test section. It was decided to manipulate both sides of the symmetrical ONERA D section aerofoil in order to increase the possibility of detecting small differences on the wake momentum thickness.

3.1 Pressure distribution

The pressure distributions along the aerofoil have been obtained with the help of a Girard-Guienne probe (G-G), previously calibrated. That probe is equipped with a knee-joint set with a return-spring ; this device allows the probe to follow rather precisely the exact shape of the surface, except close to the trailing edge and in the region of high curvature at the leading edge.

On figure 2, are plotted the pressure distributions, on the upper side of the symmetrical aerofoil, for an angle of attack of 0°, in a diagram -Kp/$\cos^2\phi$ versus the longitudinal abscissa. The static pressure distributions are almost independent of the free-stream velocity, U_∞, either 20.0ms^{-1} or 32.2ms^{-1} [18]. Then, this observation suggests a similar "blocking" effect of the ONERA D section aerofoil within the test section. Furthermore, when taking into account the presence of the upper and lower sides of the test section, a slight difference could be observed on the pressure distribution, figure 3. Also, the reference dynamic pressure could be evaluated from either the upstream or downstream free-stream velocity. The difference between these two velocities is not negligible (2.1%).

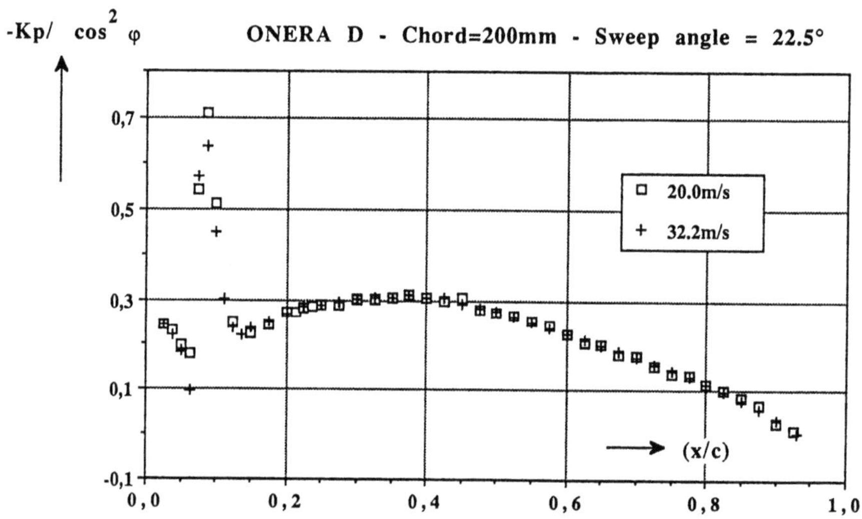

Figure 2: Pressure distributions on the ONERA D aerofoil.

An inviscid two-dimensional calculation, based upon a singularity method in confined atmosphere, has been run ; the results are compared with the experimental pressure distributions. However, the thickness of the smooth vinyl sheet covering the aerofoil is not taken into account. Nevertheless, the agreement between theoretical and experimental pressure evolutions is rather good, except in the leading edge and tripping wire vicinities where measurements with the G-G probe were very critical, and also in the trailing edge area where viscous effects are rather important. Moreover, the thickness of the trailing edge is significant, slightly greater than 0.2mm.

One can observe a velocity peak close to the leading edge of the aerofoil. The flow is smoothly accelerating towards x/c=0.35-0.40, which corresponds to the location of the maximum cross-section, then decelerating towards the trailing edge (figure 3). The effect induced by the tripping wire, chosen not only for fixing the transition but also for increasing the boundary layer thickness, is clearly visible on the pressure distributions.

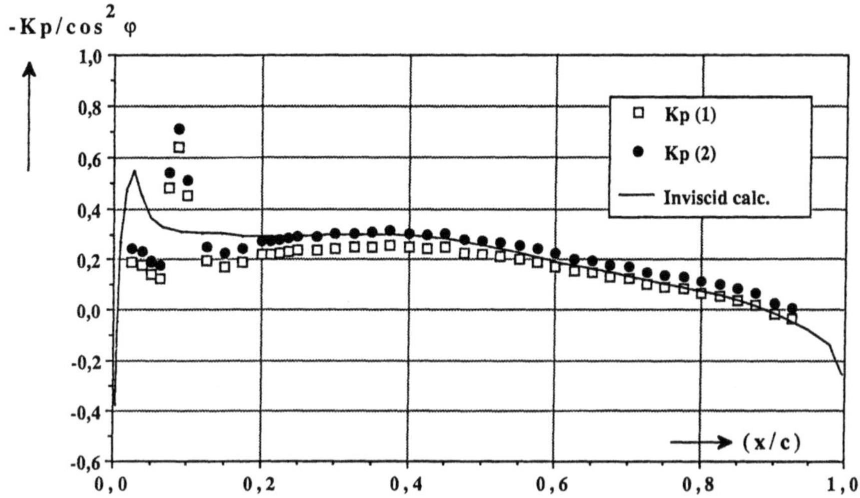

(1) - Estimation of Kp with downstream free-stream velocity (+2%)
(2) - Estimation of Kp with upstream free-stream velocity

Figure 3: Pressure distributions on the ONERA D aerofoil.

3.2 Mean velocity measurements in the aerofoil wake

For a free-stream velocity of 20.0ms^{-1}, a dozen of surveys have been performed downstream of the trailing edge of the ONERA D aerofoil, down to x/c $\simeq$ 2. Note that x is evaluated along a direction normal to the aerofoil leading edge. Of course, because of the angle of sweep, streamlines are deflected within the boundary layer and the aerofoil wake, generating cross-flow. Those secondary motions are rather important close to the trailing edge, but can be neglected at one chord length behind the trailing edge ; as it will be seen later on, the streamwise section corresponding to x/c=1.92 will be used as a reference in order to judge of "riblet" performances in terms of drag reduction.

Mean velocity profiles have been obtained using a Pitot-static probe, the axis of which has been aligned with the infinite free-stream flow direction. Then, results are certainly wrong over the first 20% of chord length behind the aerofoil ; in fact, accurate measurements would require a directional probe aligned with the mean local vector. As a consequence, this study of the wake has to be considered as a qualitative approach and analyse in such a way. However, estimates of drag variations carried out at about x/c=1.92, where the flow comes back to its upstream two-dimensional state, are reliable.

Measurements in the wake showed that the flow is accelerating from the trailing edge down to x/c=1.75 ; further away, the downstream external velocity remains uniform and is 2.1% greater than the upstream external velocity. Pitot-probe surveys confirm that the wake is strictly symmetrical along the test-section centre-line and that it does not spread out too much over one chord length. Integrating mean velocity profiles provide the momentum thickness ; its streamwise evolution is plotted on figure 4. Just behind the trailing edge, over a distance close to $70\Theta_s$, Θ decreases because of the favourable pressure gradient and then, reaches an asymptotic value Θ_s. According to such a streamwise evolution, the efficiency of the different "riblet" models will be checked at the last measurement section, by comparing the momentum thicknesses deduced from Pitot-probe surveys with and without grooved surfaces ; that location is at 200mm from the aerofoil trailing edge along the test-section centre-line, i.e corresponds to x/c=1.92.

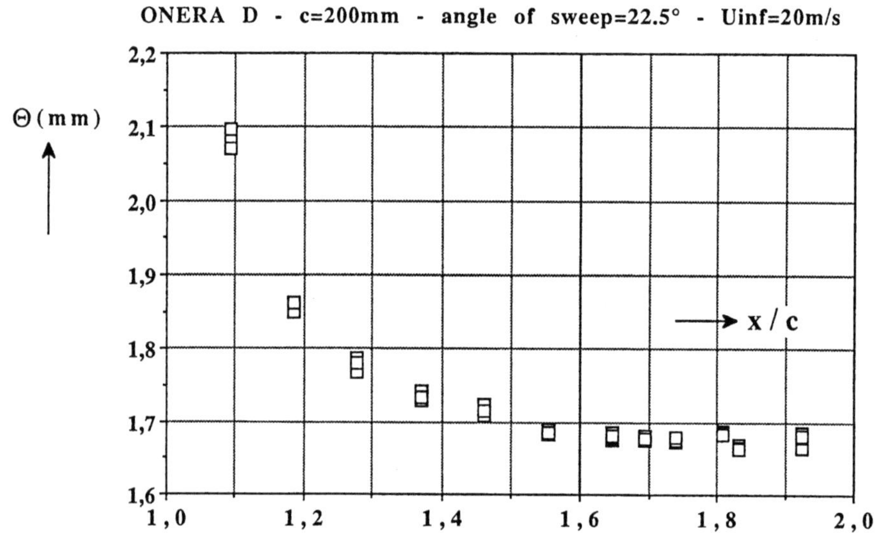

Figure 4: Evolution of streamwise momentum thickness within the wake.

On figure 5 is plotted the streamwise growth of the mean longitudinal velocity U_c, measured along the wake axis ; these two quantities have been reduced with the friction velocity U_τ evaluated at the trailing edge through a boundary layer calculation. The results are compared at first with those from Chevray and Kovasnay [7] and Hebbar [22] for symmetrical turbulent wakes behind respectively flat plate and aerofoil, and secondly with the empirical laws suggested by :

- Andreopoulos and Bradshaw [2] : $U_c/U_\tau = 2.02\ \mathrm{Ln}(xU_\tau/\nu) + 0.7$;
- Alber [1] : $U_c/U_\tau = 1/\chi\ [\ \mathrm{Ln}\ g(xU_\tau/\nu) - \gamma\ ] + B$; where g satisfies :

$g(x)\ [\mathrm{Ln}\ g(x) - 1\ ] = \chi^2\ x$; with χ=0.418, B=5.5 and the Euler constant γ=0.5772.

It is interesting to notice that in the near-wake region (defined as $x < 50\ \Theta_s$, i.e. $x/c < 1.45$) as well as in the intermediate wake region ($x > 50\ \Theta_s$) the centre-line velocity increases logarithmically with downstream distance. This result was obtained when considering the inner variables of the turbulent boundary layer taken at the trailing edge. However, the constants in the logarithmic relationship for the present data are different from those for the flat-plate data [7] or the arofoil data [22] ; then, that might be attributed to the presence of a favourable pressure gradient in the near-wake region.

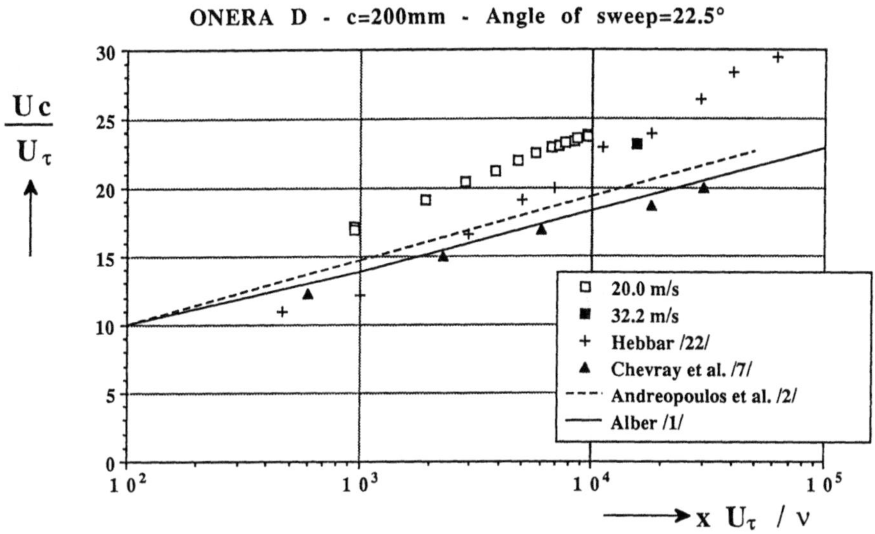

Figure 5: Evolution of the mean streamwise velocity along the wake.

3.3 Drag measurements

Performances of different "riblet" models have been determined by variations of total drag coefficient estimated by integration of the velocity profile at x/c=1.92. Thus, if the chosen reference length is the aerofoil chord (c=200mm), the total drag coefficient is $Cx = 2\ \Theta_s/c$.

For each chord Reynolds number, i.e each free-stream velocity, several wake surveys have been performed (figure 6). Velocity profiles have been obtained through one hundred points, the spacing of which is equal to 0.3mm. Although there is no static pressure gradient within the wake, there is a slight difference in the momentum thickness, depending on the choice of the upper and lower borders of the wake. Then, for a given survey, maximum variation of Θ_s is close to 1% (resp. 1.8%) for U_∞=20.0ms^{-1} (resp. 32.2ms^{-1}). Calculating the average as well as the standard deviation (σ) gave :

- $U_\infty = 20.0\text{ms}^{-1}$; $\Theta_s = 1.674\text{mm}$; $\sigma = 0.58\%$;
- $U_\infty = 32.2\text{ms}^{-1}$; $\Theta_s = 1.555\text{mm}$; $\sigma = 0.83\%$.

As these measurements of total drag forces, through wake surveys, are very exacting, maximum care has been taken in order to minimize uncertainties and then, to ensure a good consistancy of presented results.

3.4 Boundary layer calculations

Knowing the experimental pressure distributions on the ONERA D aerofoil, boundary layer calculations were performed in order to :
- extrapolate the momentum thickness in the wake at x/c=1.92 and check it with the experimental value ;
- estimate the contribution of friction component to the total measured drag ;
- know the streamwise variation of the rib height scaled with respect to the inner variables of the turbulent boundary layer.

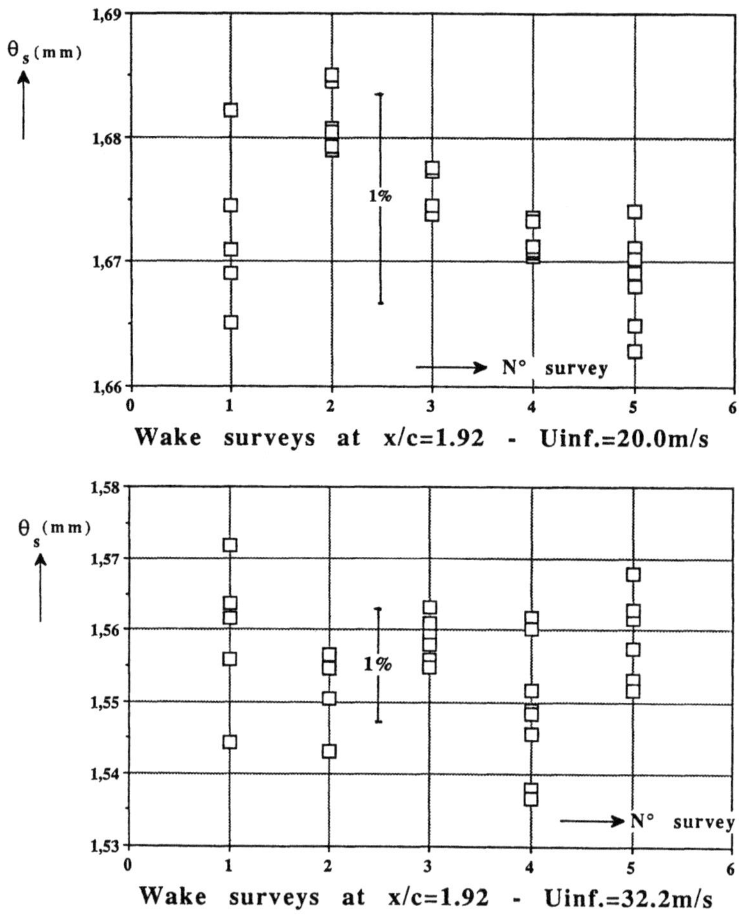

Figure 6: "Reference" configuration (angle of sweep : 22.5°).

In the leading edge vicinity, as well as in the tripping device neighbourhood, pressure distributions provided by inviscid calculations have been considered. However, from x/c=0.20 down to x/c=0.90, experimental static pressure data have been used. Close to the aerofoil trailing edge, it has been necessary to perform extrapolation ; it was decided to follow a behaviour similar to the one given through inviscid calculations.

The code developed at CERT solves the local boundary layer equations for three-dimensional incompressible flows. Transition was fixed at x/c=7.5%, at the location of the tripping wire. The extent of the transition region was calculated using empirical relationships [4].

On figure 7, for U_∞=20.0ms^{-1}, the calculated external and wall streamlines are plotted versus x ; α denotes the angle between the external streamline and the direction normal to the leading edge, while β_0 refers to as the angle between the wall streamline and the external one. Over the future manipulated length with "riblets" ($0.15 < x/c < 1.00$), α is almost constant and close to 20°. On the other hand, if β_0 is very weak over the first sixty percents of the chord length it then increases rapidly due to the adverse pressure gradient ; in the trailing edge vicinity, the turbulent boundary layer is almost separated. The angle $\beta_0+\alpha-\phi$ illustrates the yaw angle between the wall streamlines and the infinite free-stream flow direction ; this angle is less than 10° up to x/c=0.90, thus giving proof of suggested alignment of longitudinal grooves with the free-stream direction.

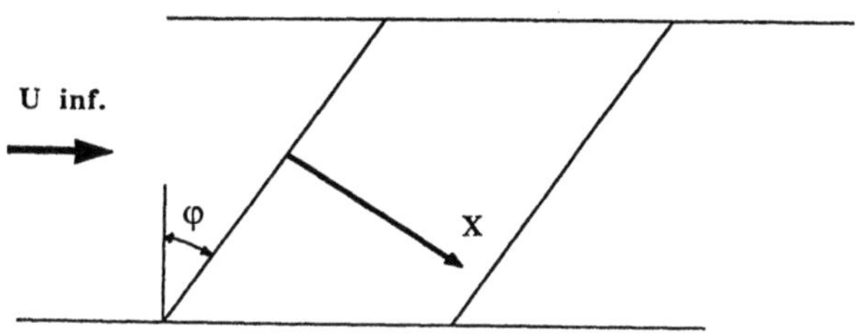

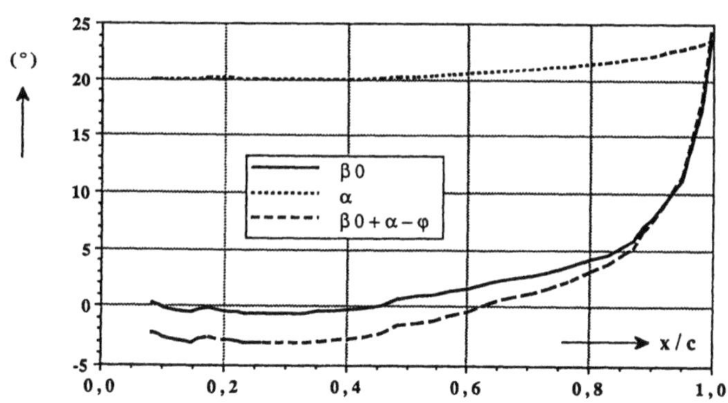

Figure 7: Deviations of wall and external streamlines.

In order to quantify the intensity of the pressure gradient, the parameter $\beta - \beta = -(\delta/u_\tau) \times (dUe/dx)$, where δ denotes the physical boundary layer thickness and u_τ the friction velocity – has been computed through boundary layer code. As β varies with the streamwise abscissa, especially on the rear part of the aerofoil, its average value along the manipulated length (i.e between 15% and 100% of chord length) has been considered. This average value varies slightly with the Reynolds number, in turbulent regime. Then, $\overline{\beta}$ =0.0288 (for U_∞=20.0ms^{-1}) and 0.0282 for U_∞=32.2ms^{-1}, [18]. As these two values are rather weak, three-dimensional effects would be more important than pressure gradient effects.

From the boundary layer code, it is also possible to compute the streamwise evolution of the non-dimensionalised rib height parameter, h_w^+ on the upper side of the aerofoil ; $h_w^+ = (h/\nu_w) \cdot \sqrt{\tau_w/\rho_w}$. However, over the manipulated length, its streamwise variation is smooth enough so that it is possible to define an average value $\overline{h_w^+}$. Then, for h=1μm, $\overline{h_w^+}$ = 0.055 (resp. 0.088) for U_∞ = 20.0ms^{-1} (resp. 32.2ms^{-1}). The three riblet films used give a range of riblet heights from $\overline{h_w^+}$=2.8 to 13.4 (figure 8). According to previous results obtained in two-dimensional subsonic flows, with or without pressure gradient, this parameter range has provided nett skin-friction drag reductions,as reported in [14] or [34], for instance.

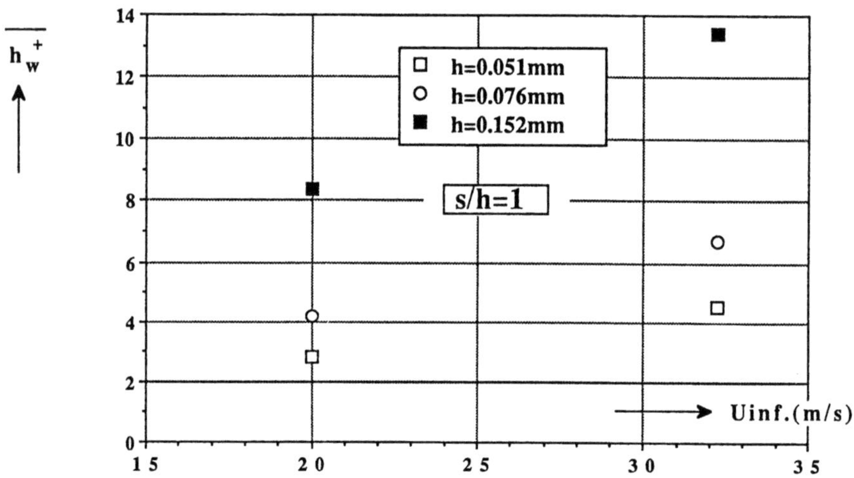

Figure 8: Average riblet height evaluated along the manipulated surface.

Based upon the continuity of wake and boundary layers developing on both sides of the aerofoil, and using integral momentum equation, Squire and Young connected the momentum thickness in the wake, Θ_s, with the one taken at the trailing edge. This relationship is valid in two-dimensional flows. Along that present aerofoil, set at 22.5° angle of sweep, the external streamlines are almost aligned with the infinite free-stream direction (figure 7). It would then be possible to apply that specific relationship, by using the external velocity as well as the integral thicknesses, evaluated in the co-ordinates system fitted to the external streamline. Because of the symmetry of the aerofoil, if $\Theta_{T.E}$ denotes the momentum thickness calculated at the trailing edge of the upper side, the relationship would be read :

$\Theta_s = 2\ \Theta_{T.E}\ (Ue_{T.E}\ /\ U_\infty)^{(H_{T.E}+5)/2}$
where $H_{T.E}$ is the trailing edge shape parameter, U_∞ the infinite free-stream velocity and $Ue_{T.E}$ the potential flow velocity. That external velocity could not be either measured with the probe or estimated from inviscid calculations because of viscous effects ; therefore, it was extrapolated from experimental pressure measurements performed upstream of that location. The total drag coefficient could then be computed ($Cd_c = 2\ \Theta_s/c$) and compared with the experimental value, Cd_e, obtained through wake surveys. A difference of almost 45% has been recorded between Cd_c and Cd_e. This is surprising since that method from Squire and Young usually provides nice agreement, especially if the transition onset is consistent with the experimental location.

That difference could be explained since transition was tripped using a wire, the diameter of which was rather important (d=0.8mm). Then, that induces inevitably the problem of over-thickening due to roughness drag, which is not of course taken into account in the boundary layer code. Using earlier work, performed at CERT [3], one could evaluate the step in momentum thickness at the transition location. Indeed, $\Delta\Theta \simeq 0.5\ d\ Cd\ (U_d/U_e)^2$, where U_d refers to as the laminar velocity taken at the height y=d and Cd is the drag coefficient of the trip. At the device location, d/δ_1 is equal to 4-4.5, which implies that U_d=U_e. From several sets of experiments Cd was estimated to be 0.4 [3]. Thus, one ended up with a value of $\Delta\Theta$ at the device location close to 0.16mm ; that over-thickenning is not negligible since it represents almost 21% of half Θ_s at U_∞=32.2ms^{-1}. Consequently, boundary layer computations were performed again, by moving artificially upstream the transition onset in order to take into account this over-thickenning at x/c=7.5%. Furthermore, because the thickness of the trailing edge of the aerofoil is significant (twice the thickness of the smooth vinyl film, that is to say t=0.20mm), the residual afterbody drag has also been computed ; its value is independent of the chord Reynolds number : (t/c) $Cp_{T.E.}$, i.e. 1.07 10^{-4}. It represents almost 1% of the total drag coefficient. Then, when dealing with all the terms included in the drag balance, the comparisons gave :
- for U_∞=20.0ms^{-1} ; 100 Cd_c=1.6006 and 100 Cd_e=1.6737 ;
- for U_∞=32.2ms^{-1} ; 100 Cd_c=1.5414 and 100 Cd_e=1.5552.

The agreement is rather good between the measured and the computed total drag coefficient (maximum difference : 4.6%). So, one could be confident especially in the way considered for estimating the total drag forces, through wake surveys.

Then, by integrating the local shear stress along the curvilinear abscissa, one ended up with the computed skin-friction drag coefficient, Cd_f. Comparing that value with the measured total drag coefficient revealed that the contribution of the friction part was rather important and was slightly dependent upon the chord Reynolds number : 47% (resp. 49%) for R_c=2.65 10^5 (resp. 4.25 10^5).

4 Performances of "riblet" models

Let us recall that for the reference configuration, the total drag coefficient of the aerofoil, covered entirely with smooth vinyl sheet, has been estimated through wake surveys performed at a streamwise abscissa located at 200mm downstream the trailing edge, in the free-stream direction, i.e. x/c=1.92.

Grooved surfaces have been applied between x/c=0.15 and 1.00, on the upper as well as lower sides of the aerofoil. Data, recorded from wake surveys, are plotted for the two infinite free-stream velocities and given for two triangular symmetrical V-shaped models : h=0.051mm (figure 9) and 0.152mm (figure 10).

For every "riblet" model, at least 7 to 8 wake surveys have been performed in order to minimize experimental uncertainties and so, to increase the accuracy of the average momentum thickness. Generally speaking, the largest scatter has been recorded for the highest free-stream velocity, but not greater than 1%. One can observe that Θ_s is reduced whichever groove depth is used ; this observation remains valid for either free-stream velocity, 20.0 or 32.2ms^{-1}.

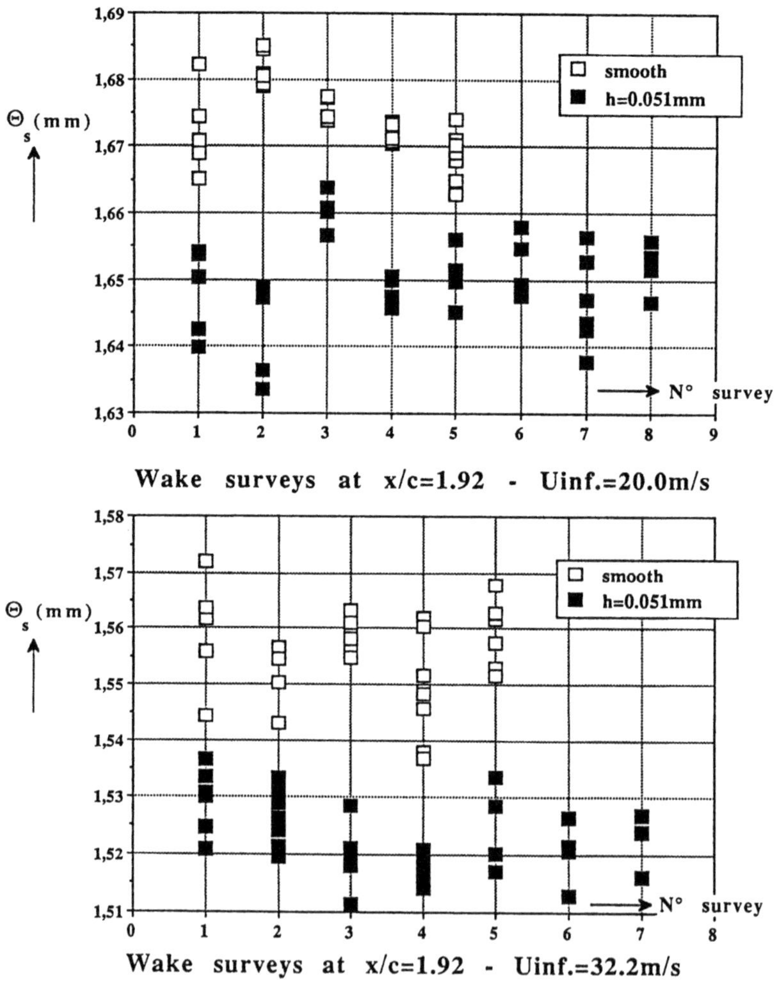

Figure 9: Efficiency of riblet : s=h=0.051mm on total drag force.

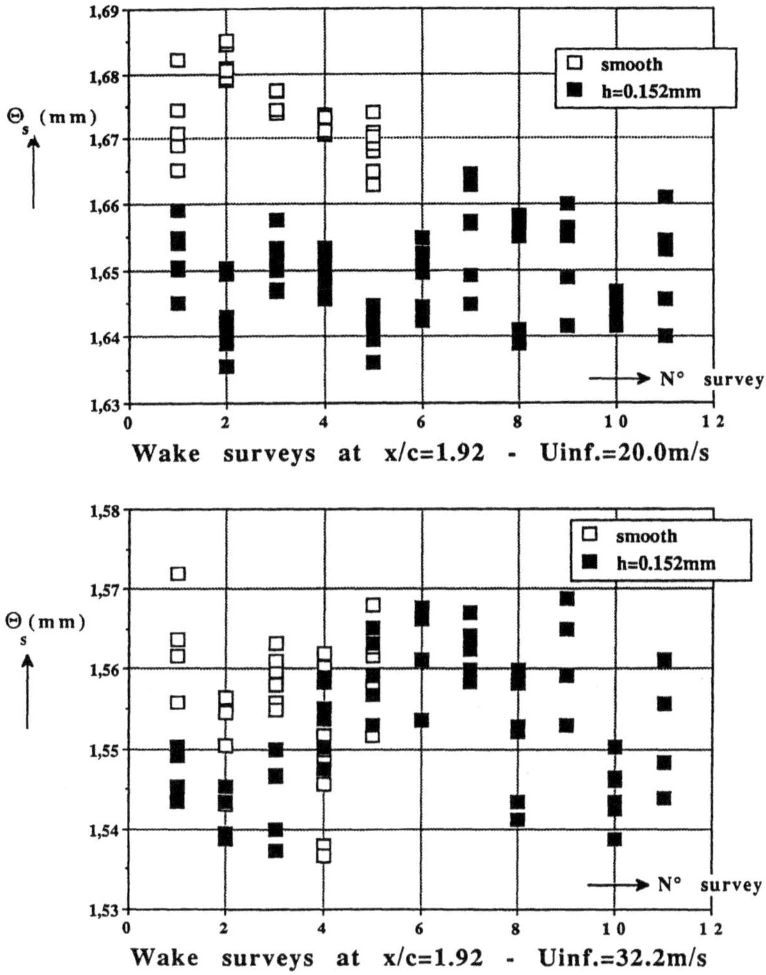

Figure 10: Efficiency of riblet : s=h=0.152mm on total drag force.

The variations of the total drag coefficient have been deduced from the following relationship :
$\Delta Cd/Cd = (Cd_{riblet} - Cd_{smooth})/Cd_{smooth} = (\Theta_{s\ riblet} - \Theta_{s\ smooth})/\ \Theta_{s\ smooth}$.

For the three "riblet" models and the two chord Reynolds numbers, results are plotted on figure 11 versus $\overline{h_w^+}$. On this diagram, vertical lines refer to as the experimental uncertainty concerning the measurement of the wake momentum thickness. When that parameter is less than 9, nett drag reductions up to 3-3.5% have been recorded. As regards the "intermediate" model (h=0.076mm), the total drag decreases seem to be really under-estimated for U_∞=20.0ms^{-1}. This weaker performance might come from the model itself, since, firstly, such an observation could be also pointed out for $32.2ms^{-1}$, and secondly the scatter over the number of surveys is even less important than for the other models [18]. The thickness of the adhesive self-backing film is maybe greater than 0.1mm, inducing a forward step drag penalty ; let us mention again the sensitivity of drag reductions to cross-section uniformity, surface finish,...

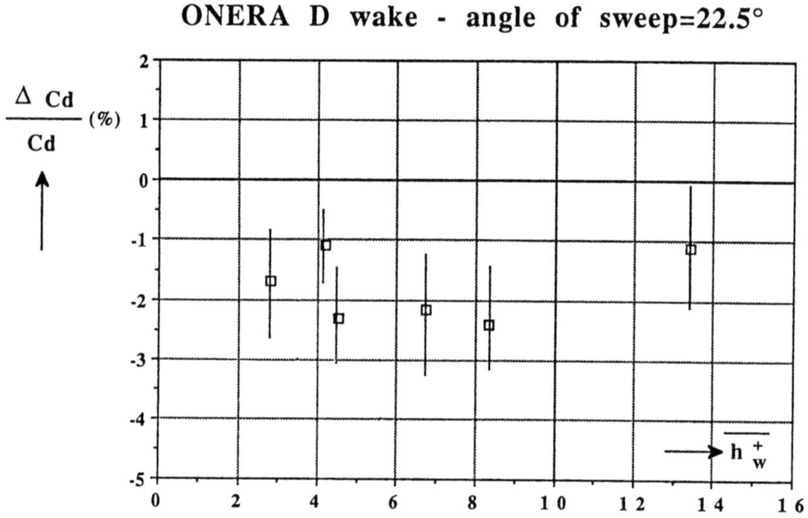

Figure 11: Variations of total drag coefficient.

In order to estimate as carefully as possible the effect of streamwise grooves not only on the total drag coefficient but also on the skin-friction drag coefficient, a couple of features have to be mentionned :

- the aerofoil is not entirely covered with "riblet" film ; an estimate of the ratio manipulated area / total wetted area is close to 85% ;
- the contribution of the friction part in the total drag balance represents between 47% and 49% ; this slight difference is attributed to the small variation in the chord Reynolds number ;
- in this specific set of experiments, the grooved surfaces were applied onto the trailing edge of the ONERA D aerofoil. As a consequence, depending upon the considered vinyl sheet (smooth or "riblet" film), the thickness of the aerofoil base-line varies from 0.2mm (reference case) up to 0.484mm (h=0.152mm). Then, compared to the smooth configuration, varying the groove depth from 0.051mm up to 0.152mm increases the afterbody drag from 41% to 142% ; its maximum value represents almost 1.7% of the calculated total drag coefficient. As small variations of the total drag coefficient are being sought, corrections due to these afterbody drag penalties have been taken into account.

In future, it would be better to keep the thickness of the trailing edge constant, so that the only recorded changes in drag forces would be those caused by skin-friction decreases or increases. Of course, that point is very useful for small models tested in wind tunnels, but would be useless for either larger models of transport aeroplane or aircraft applications.

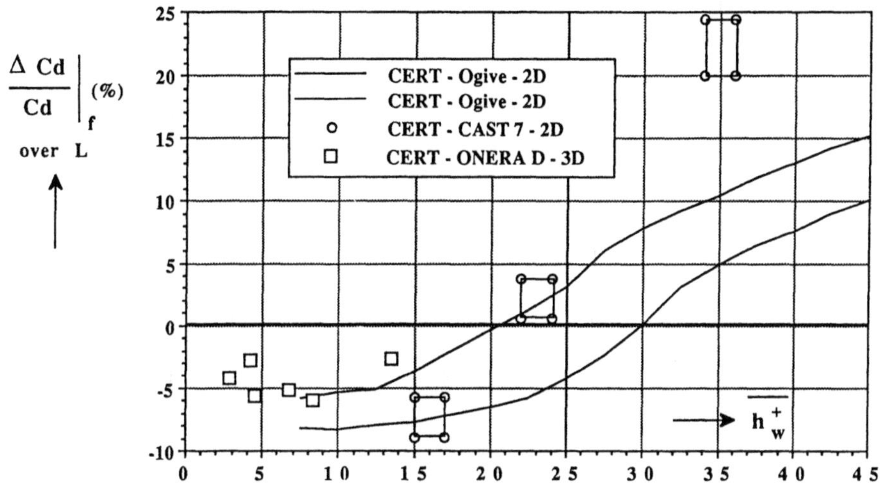

Figure 12: Variations of friction drag coefficient over the manipulated length.

Then, for the three tested "riblet" models, the variations of the friction drag coefficient over the manipulated length L, are plotted versus $\overline{h_w^+}$ on figure 12. Results are also compared with those recorded under transonic conditions, in the T2-wind tunnel at CERT, when manipulating turbulent boundary layers developing over :
- a cylinder-type body (ogive) under zero-pressure gradient conditions ; measurements were made using an internal drag balance [14] or [19].
- a CAST 7 section aerofoil under adverse and/or favourable pressure gradients ; in that case, "riblet" efficiency had been checked through wake surveys [19].

Although the contribution of the friction component to the total drag might vary in the ratio 1 to 2, there seems to be a good agreement between data recorded in two- and three-dimensional boundary layers. Indeed, results obtained on the ONERA D would extend, for the lowest values of $\overline{h_w^+}$, those completed on the ogive. It seems fair to guess that ΔCd/Cd would return to 0 as $\overline{h_w^+}$ vanishes, since there is no evidence, even through computations, there exist an asymptotic "drag reduction" value... The data band labeled "CERT-Ogive-2D" represents a band which fits all the data which were obtained with four symmetrical V-shaped grooves. Let us add that these results were consistent with other experiments performed either at Cambridge or RAE Bedford, as mentionned by Savill [28]. The data band labelled "CERT-CAST7-2D" refers to manipulation of both sides of the aerofoil, with three different symmetrical V-shaped ribs ; only data for the largest groove (h=0.051mm) did not fall within the "CERT-Ogive-2D" band. This could be attributed to a bad quality of the model surface finish, to a lack of cross-sectional uniformity, and also but not only to the fact that the afterbody drag penalty had not been taken into account, ... [19].

However, one might argue that the recorded drag decreases are somewhat smaller than those registered for pure two-dimensional boundary layers. This difference could come from the presence of adverse pressure gradients on the rear part of the aerofoil, which would increase the deviation of wall streamlines (figure 7). All the benefit obtained through internal manipulation of the boundary layer over almost 75% of the chord length could be offset over the last 10% of the chord length, where the layer is very close to separation.

The effect of misalignment between grooves and the infinite free-stream direction had been investigated at transonic speeds on the ogive. In that case, the model axis was maintained aligned with the direction of the free-stream flow direction ; on the other hand, the grooves were set at a given angle to the cylinder symmetry line [14] or [19]. The assumption was made that no induced three-dimensional or helicoïdal effect existed. Data are plotted on figure 13 for a single "riblet" model, s=h=0.023mm and two angles of yaw 10° and 20°. These results had confirmed previous experiments in either low-speed or transonic experiments ; the higher the angle of yaw is, the weaker the skin-friction drag reductions are and the smaller the zero drag reduction crossover point is.

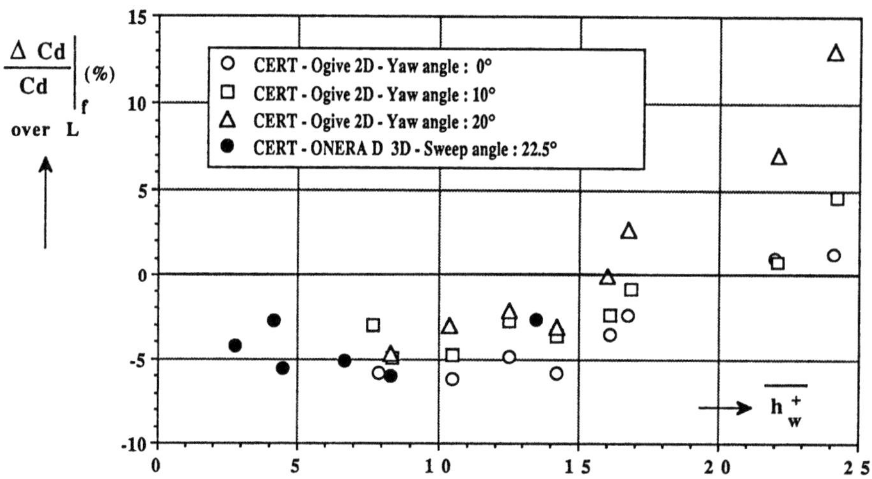

Figure 13: Effects of sweep angle in 3D and yaw angle in 2D flows.

When comparing those results to the present data, except for the highest value of $\overline{h_w^+}$ close to 13.5, there is no obvious evidence that drag reduction points recorded in three-dimensional flows (ONERA D set at 22.5° angle of sweep) fall within those obtained when setting grooves at 20° angle of yaw to the cylinder symmetry line. Thus, the behaviour of grooves in three-dimensional boundary layers is completely different to that observed by giving some yaw in two-dimensional boundary layers. The fact that the velocity vector varies very rapidly close to the wall, might prevent the tendancy of the flow, driven by the grooves, to revert to a two-dimensional state.

Walsh and Anders summarized and correlated all the available "riblet" film data for two-dimensional turbulent boundary layers under subsonic or transonic conditions [34]. They also considered results obtained by Gaudet [21] at a Mach number of 1.25. As shown on figure 14, one can point out that significant "riblet" data had been recorded to "firmly establish their drag reduction performance", as quoted by Walsh and Anders [34]. Of course, there seems to emerge different data band, but one has to be aware that these data have very often been obtained with various techniques for estimating the drag variations : laboratory measurements (changes in momentum thicknesses, wake surveys, ...) or direct drag measurements through internal balances. Nevertheless, data recorded by manipulating three-dimensional turbulent boundary layers (filled symbols) did not fit too badly in this diagram. There might be a shift of the "Low-speed film data band" towards the lowest values of $\overline{h_w^+}$; as a consequence, the uper limit of the reduction domain (zero-drag reduction crossover point) would be smaller for this set of experiments.

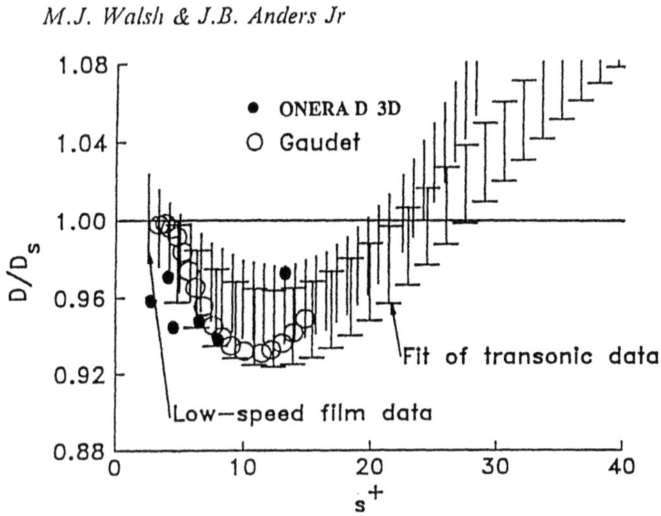

Figure 14: Synthesis of drag data (from Walsh & Anders /34/).

5 Concluding remarks

Following the great amount of experimental work concerning manipulation of turbulent boundary layers by passive devices such as "riblets", in two-dimensional flows, world-wide data established their potential for skin-friction drag reduction. Although details of the mechanisms involved are not firmly understood, the concept is rather close to industrial application. In view of this, the behaviour of grooved surfaces in three-dimensional flows was a necessary step if it is planned to apply such devices to the majority of transport aircraft, including wings, fin, horizontal tail and maybe rear part of the fuselage, in order to increase the percentage of manipulated area, i.e. area covered with "riblet" film.

So, an experimental study was conducted in the Aerothermodynamics Department of CERT, on an ONERA D section aerofoil, set at an angle of attack of 0° and an angle of sweep of 22.5° within the lateral walls of a low-speed wind tunnel. Grooved surfaces (vinyl films) have been applied on both sides of the aerofoil between 15% and 100% of the chord length. "Riblet" performances have been determined through drag variations estimated from wake surveys performed approximately one chord length behind the trailing edge. As discussed previously, it would have been better to keep constant the thickness of the trailing edge, so that the afterbody drag penalty would not have been dependent upon the ribbed model. Of course, such an advice would be useful when dealing with small models in wind tunnels, and useless for flight applications.

Drag reductions were recorded for each of the three tested models over the range, $\overline{h_w^+}$: 2.5-13.5. Maximum decreases up to 5-6% have been obtained ; these maxima are somewhat weaker than those generally recalled when manipulating two-dimensional turbulent boundary layers. That could be attributed not only to three-dimensional effects, but also to the fact that grooves were applied on the last 10% chord length of the aerofoil where strong adverse pressure gradients induced important wall streamlines deviations since the layer is close to separation. Nevertheless, the "riblet" data have been recorded when the grooves are approximately aligned with the infinite free-stream flow direction. It is likely that if the crests had been set in a direction normal to the leading edge of the aerofoil, either small decreases or even increases would have been measured, since the deviation to the wall streamlines and *a fortiori* to the external streamlines be too important.

These results obtained in low-speed wind tunnel are consistent with those recorded in the S1-wind tunnel of Modane, at a realistically higher Reynolds number [19]. Let us recall that these experiments were carried out on a one-eleventh scale Airbus A320 model, in collaboration with Aérospatiale and Aerodynamics Division of ONERA, Chatillon. Among others, two configurations had been considered : a wing-body configuration with "riblets" set only upon the fuselage (Case A) and a whole wing-body configuration covered with "riblets" (Case B). The corresponding percentages of wetted areas covered with longitudinal grooves were 47% for Case A and 66% for Case B. For a Mach number at infinity of 0.7 and at cruise conditions, adding "riblets" on the wings and fairings yielded a drag coefficient Cd lower than the one measured in Case A ; indeed, Cd_{CaseB}-$Cd_{CaseA}$$\simeq$1 10^{-4}. An estimate of the average nett skin-friction drag reduction, evaluated over the manipulated surface, was close to 3.3%. This gain was certainly under-estimated since the groove depth had not been optimized on the wings (same geometry as that used on the fuselage) and also because of boundary layer tripping necessary for this set of wind-tunnel tests ; as a consequence, the leading edge of the "riblet" model was rather far downstream on the upper as well as lower side of the wings. Nevertheless, the benefit of covering three-dimensional parts of the model with "riblets" had been observed, though the percentage of wing wetted area was rather small at 57%.

Since the Reynolds number based upon the fuselage length was close to 40 10^6, it was expected that these results would be easily applied to practical flight test conditions, where the Reynolds number is only increased by a factor 5. Several French and European newspapers, reported the success of the flight test programm carried out by Airbus Industrie and its partners on the prototype Airbus A320 N°1. Let us just add that the results measured during these flights, were in complete agreement with the promising S1-wind tunnel results ...

Acknowledgements
This experimental study was financially supported by the "Service Technique des Programmes Aéronautiques" (Grant N° 89 95 009-51). Special thanks are due to F. Marentic from 3M-USA and A. Delachanal from 3M-France, for providing us with all of the riblet material.

References

[1] Alber I.E. : AIAA Journal, Vol. 18, pp. 1044-1051, (1980)
[2] Andreopoulos J., Bradshaw P. : J. Fluid Mech., Vol. 100, pp. 639-668, (1980)
[3] Arnal D. : AGARD Report N° 709, pp.2.1-2.71, (1984)
[4] Arnal D. : AGARD Report N° 741, pp.4.1-4.34, (1986)
[5] Bacher E.V., Smith C.R. : AIAA Journal, Vol. 24, N°8, pp. 1382-1385, (1986)
[6] Benhalilou M., Anselmet F., Liandrat J., Fulachier L. : 8th Symp. on Turb. Shear Flows, Münich (1991)
[7] Chevray R., Kovasnay L.S.G. : AIAA Journal, Vol. 7, pp. 1641-1643, (1969)
[8] Choi K.S., Pearcey H.H., Savill A.M. : Int. Conf. on Turbulent Drag Reduction by Passive Means, London (1987)
[9] Choi K.S. : J. Fluid Mech., Vol. 208, pp. 417-458, (1989)
[10] Choi K.S. : Aeronautical Journal, March 1990
[11] Clark D.G. : In Turb. Control by Pass. Means, ed. Coustols E., Kluwer Acad. Press., pp. 79-96, (1990)
[12] Coustols E., Gleyzes C., Schmitt V., Berrue P. : 24ième Colloque AAAF Poitiers - France (1987)
[13] Coustols E., Cousteix J. : 16th ICAS Congress, Jerusalem (1988)
[14] Coustols E. : AIAA Paper 89-0963 (1989)
[15] Coustols E., Cousteix J. : 2nd IUTAM Symp., Zürich (1989)
[16] Coustols E. : 4th Int. Conf. on Drag Reduction, Davos (1989)
[17] Coustols E., Cousteix J. : 7th Symp. on Turb. Shear Flows, Stanford (1989)
[18] Coustols E. : CERT Internal Technical Report (March 1990)
[19] Coustols E., Schmitt V. : In Turb. Control by Pass. Means, ed. Coustols E., Kluwer Acad. Press., pp. 123-140, (1990)
[20] Gad-el-Hak M., Blackwelder R.F. : AIAA Paper 87-0358 (1987)
[21] Gaudet L. : Applied Scientific Research, Vol. 46, pp. 245-254, (1989)
[22] Hebbar K.S. : Exp. in Fluids, Vol. 4, pp. 214-222, (1986)
[23] Hooshmand D., Youngs R., Wallace J.M. : AIAA Paper 83-0230, (1983)
[24] Johansen J.B., Smith C.R. : AIAA Journal, Vol. 24, No. 7, (1986)
[25] Kim H.T., Kline S.J., Reynolds W.C : J. Fluid Mech., Vol. 50, Pt. 1, (1971)
[26] Kline S.J., Reynolds W.C., Schraub F.A., Runstadler P.W. : J. Fluid Mech., Vol. 30, Pt. 4, (1967)
[27] McLean J.D., George-Falvy D.N., Sullivan P.P. : Int. Conf. on Turbulent Drag Reduction by Passive Means, London (1987)
[28] Savill A.M. : 2nd IUTAM Symp., Zürich (1989)
[29] Squire L.C., Savill A.M. : Int. Conf. on Turbulent Drag Reduction by Passive Means, London (1987)
[30] Vukoslavcevic P., Wallace J.M., Balint J.L. : Int. Conf. on Turbulent Drag Reduction by Passive Means, London (1987)

[31] Walsh M.J. : AIAA Paper 82-0169 (1982)
[32] Walsh M.J., Lindeman A.M. : AIAA Paper 84-0347, (1984)
[33] Walsh M.J., Sellers III W.L., McGinley C.B. : AIAA Paper 88-2554, (1988)
[34] Walsh M.J., Anders Jr. J.B. : Applied Scientific Research, Vol. 46, pp. 255-262, (1989)

High resolution conformal mesh computations for V, U or L groove riblets in laminar and turbulent boundary layers

L. DJENIDI, L. C. SQUIRE* and A. M. SAVILL*

University of Newcastle
Newcastle, NSW, Australia

* *University of Cambridge*
Cambridge, U.K.

K.-S. Choi (ed.), Recent Developments in Turbulence Management, 65–92.

Abstract
This paper presents results from a series of numerical studies of various geometry riblets in both laminar and turbulent external flows, which complement results already reported by colleagues at IMST, ONERA-CERT, and UMIST, for V and L-groove riblets in laminar internal and external flows . Although these earlier studies suggested that a 2-3% reduction could be achieved in the latter case, questions remained as to the sensitivity of this finding to further grid refinement, particularly as no drag reduction was indicated for fully developed internal flows. New, higher resolution, CRAY 2 V-groove riblet computations, employing both orthogonal cartesian and conformal grids, have tended to confirm an optimum 2.5% drag reduction for riblets scaling on the boundary layer thickness (rather than wall units as for turbulent flow), and computed mean velocity profiles agreed well with LDA experimental data. Conformal mesh computations for semi-circular U, L, and intermediate form, riblets indicated a similar drag reduction in laminar flow, but also revealed important differences in the spanwise distribution of Cf across these riblets compared to the V-groove type. Subsequent turbulent computations for V and L-groove riblets, employing a conformally transformed mixing length, resulted in mean velocity profiles which again compared well with the limited available turbulent flow data. Parametric computations revealed that the maximum drag reduction was considerably overpredicted for the V-groove riblets, but the optimum performance of the L-groove riblets and the variation in drag reduction with h+ were rather better predicted. Any discrepancies could be associated with secondary motions set-up inside the riblets, since these may be responsible for additional momentum transport, and hence higher Cf, but were not accounted for in the modelling.

1. Introduction

The desire to be able to control the coherent structure of a turbulent boundary layer in order to reduce the drag of vehicles, regardless of the fluid they are moving in, has attracted the attention of the researchers for the last decade. Passive drag reduction devices are of particular interest since such devices do not require any energy supply, and of all the passive techniques studied thus far, a change of wall geometry through the use of longitudinal grooves (so-called "riblets") appears to be one of the most attractive e.g. see Thomas (1985), Bushnell (1985). A general consensus has now been reached that this technique, when applied in turbulent boundary layers, can reduce skin-friction drag by about 7 % under a variety of flow conditions, [e.g. see various authors (1987) and (1989)]. However, to date the manner in which riblets act has not yet been fully established. One interpretation is that the drag reduction comes primarily from the establishment of low speed viscous regions within the grooves as suggested by Gallagher and Thomas (1984). Several experimental studies have been carried out to check on this idea [e.g. by Wilkinson et al. (1987); Djenidi (1989); Djenidi et al. (1989 a, b); Launder and Li (1989); Liandrat et al. (1990)], and it appears from these that viscous effects may well play the main role in the drag reduction due to riblets, despite a large wetted area augmentation.

Although there have been many detailed studies of the flow over riblets, further parametric investigations are needed. Indeed, the shapes that the grooves can have are unlimited and therefore it is impossible to carry out experiments for every conceivable configuration. One reason why numerical studies have been conducted is that these may provide further understanding of the drag reduction mechanisms, but in addition they can be used to help define the optimum shapes of the grooves. An early computational study - which seems to have been the first one on this topic - has been carried out by Khan [1985]. Both laminar and turbulent regimes were simulated on V grooves. Although the drag reduction observed for turbulent flow conditions agreed with experiment, it is fair to say this study was incomplete, and indeed there must be some concern regarding the low mesh resolution employed. Subsequent numerical studies have concentrated on laminar boundary layer computations for V and L grooves and these have indicated that even then the skin friction drag is not increased on riblet surfaces despite their larger wetted areas. Instead a drag reduction of about 2.5 % is suggested, but this value is strongly dependent upon the grid refinement; especially that close to the crest vicinity. Parallel numerical studies for internal laminar flows with L-grooves by Launder and Li (1989) have shown a drag increase. However, the different nature of external and internal flows prevents any direct comparison being made between these findings.

In previous laminar boundary layer studies, for the sake of simplicity, only V and L shapes have been considered; using a cartesian grid. In fact, using such a cartesian mesh precludes calculations of other groove shapes, such as U grooves for instance, although these are of equal practical interest. A more appropriate mesh for riblet computations would be curvilinear. An orthogonal curvilinear mesh is particularly appropriate, and is easily generated for riblet problems. Such an approach has advantages in simplifying the application of boundary conditions and naturally increasing the grid resolution in regions where gradients are often highest (i.e. crest vicinity). Such conformal mapping was used by Doormaal et al. [1981] to predict natural convection in non-rectangular enclosures. More recently Bechert et al. [1986] have studied several groove cross-sections using a similar approach. In using their analytical model they have considered riblets immersed in a viscous Couette type shear flow. By neglecting the convective term of the boundary layer equations they solved the Laplace equation in two dimensions for the velocity U. The same conformal transformations have been adopted for the present work, where they have been introduced into the code originally developed for a cartesian mesh by de Saint Victor (1987) at ONERA-CERT and used subsequently by Djenidi at I.M.S.T. For turbulent flow computations an eddy viscosity based on a similarly transformed mixing length scheme was adopted.

The main reason for choosing this approach was that it offered the simplest possible extension to the laminar flow case, and required the minimum number of assumptions regarding the flow behaviour with the riblets; the mixing length scale being transformed in an identical manner to any other (mesh) length scale. The only assumption made was that the Van Driest damping along the normals to the riblet surface was the same as in the original untransformed plane boundary layer flow.

It was clear from the outset that such a model approach was likely to overpredict the effectiveness of the riblets since it ascribes all of their influence to changes in the viscous near wall region and takes no account of any turbulence driven secondary motions initiated by the change in wall geometry. However it was felt that such simple idealised computations might shed some light on the differences between laminar and turbulent boundary layer application as well as the relative importance of different proposed riblet drag reduction mechanisms. At the same time various higher level closure models have been evaluated by researchers at U.M.I.S.T. as part of the same SERC funded research project and these will be reported separately elsewhere.

2. Numerical Approach

Previous calculations on cartesian grids by Djenidi (1989), Djenidi et al. (1989a, b) have indicated that drag reduction may be achieved in laminar flow on a riblet-wall. However, there remained some doubt as to whether this finding was independent of the number of grid points. This is why we began the present study by considering this question first. Because exactly the same numerical code was used, we were able to compare directly the results obtained, in both curvilinear and cartesian co-ordinates, for laminar flow before proceeding to turbulent flow computations.

2.1 Riblet geometries and associated grids

The principal groove shapes considered in this study are depicted in Figure (1). (The flow direction is perpendicular to the $y - z$ plane illustrated). For each we used orthogonal curvilinear meshes as illustrated in Figure 1(b - c). The mesh used to compute the flow over the V-groove was derived using a Schwarz-Cristoffel transformation from the original plane wall cartesian grid (Figure 1a). Similar grids have been derived using this procedure to calculate the flow in cascade rows of turbomachines and the precise technical details of the transformation technique may be found in the paper of Bechert et al. [1986]. A series of numerical experiments established that the predicted flow fields were not particularly sensitive to the grid spacing. This is partly because of a natural mesh refinement around the peak. The particular numbers of grid points used for the final computations represented a compromise between short computational time and the fine precision of the results. It is believed that the predictions of percentage drag change are numerically accurate to ± 1 %. Whereas for the earlier IMST studies, and our own initial grid refinement test, computations were initiated on a smooth surface for the conformal mesh computations. The leading edge of the riblet was also the entrance of the calculation domain. It appeared, however, that this was not a source of difficulty, provided the riblet height was relatively small in comparison with the boundary layer thickness. In fact, as we shall see, after a short 'transition' region the behaviour of the flow rapidly approaches equilibrium.

2.2. Mathematical Formulation

The equations embodying the mathematical model can be written in any of several co-ordinate systems. In the present study the problem is solved using an orthogonal curvilinear system. The details of derivation of the conservation equations in general orthogonal curvilinear co-ordinates may be found in Nash and Paton (1972).

Because the riblets are immersed in the viscous sublayer of the turbulent boundary layer (in the laminar case the riblets influence only the linear part of the Blasius profile) we assume the boundary layer equations are also satisfied on the riblet surface. Therefore, with these considerations, we can now write the equations of the motion where $\frac{\partial P}{\partial x} = 0$ and $\frac{\partial}{\partial x} << \frac{\partial}{\partial y}$ to obtain:

$$\frac{\partial}{\partial x}(h_2 h_2 U) + \frac{\partial}{\partial \beta}(h_1 h_2 V) + \frac{\partial}{\partial \alpha}(h_2 h_3 W) = 0 \tag{1}$$

$$\frac{U}{h_1}\frac{\partial U}{\partial x} + \frac{V}{h_2}\frac{\partial U}{\partial \beta} + \frac{W}{h_3}\frac{\partial U}{\partial \alpha} = \nu \frac{1}{h_1 h_2 h_3}\left[\frac{\partial}{\partial \beta}\left(\frac{h_3 h_1}{h_2}\frac{\partial U}{\partial \beta}\right) + \frac{\partial}{\partial \alpha}\left(\frac{h_2 h_1}{h_3}\frac{\partial U}{\partial \alpha}\right)\right] \tag{2}$$

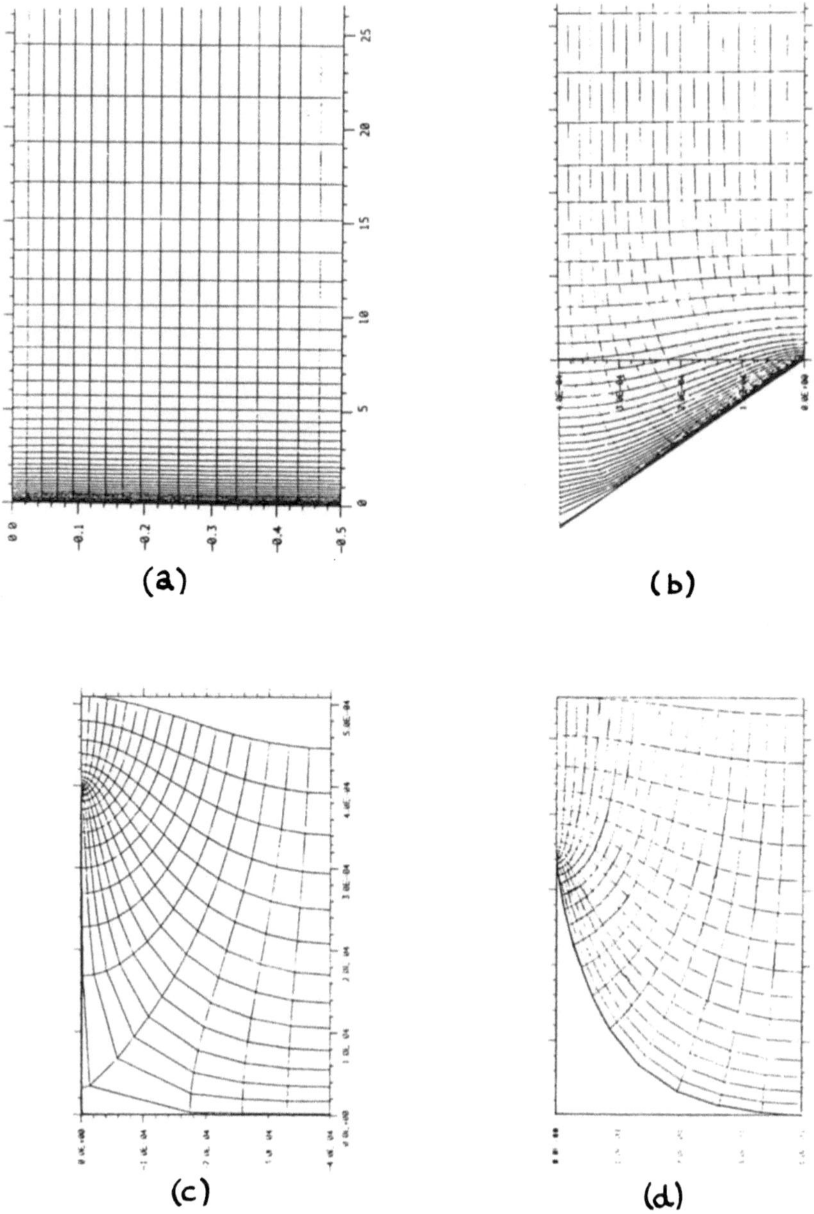

Fig.1. Examples of ortho-curvilinear meshes derived from conformal mapping.
(a) Original flat plate Cartesian grid (b) V-groove
(c) L-groove (d) U-groove

$$-\frac{1}{h_2}\frac{\partial\overline{uv}}{\partial\beta}-\frac{1}{h_3}\frac{\partial\overline{uw}}{\partial\alpha}-\frac{\overline{uv}}{h_2h_3}\frac{\partial h_2}{\partial\beta}-\frac{\overline{uw}}{h_2h_3}\frac{\partial h_3}{\partial\alpha} \tag{3}$$

where U, V, W, u, v and w are the components of the mean velocity and fluctuations of the velocity in the x, β and α directions respectively, and h_1, h_2, and h_3, (known as the material coefficients or metrics) are in general functions of x, β, α. The last two terms of the right number are introduced to take account of the curvature of the surface. Because the x co-ordinate is untransformed and the transformation in (y, z)-plane is conformal we have

$$h_1 = 1,\ \ h_2 = h_3 = h. \tag{4}$$

By using the mass-equation and regrouping terms we can write the momentum equation in its semi-conservative form,

$$\frac{\partial h^2UU}{\partial x}+\frac{\partial hUV}{\partial\beta}+\frac{\partial hUW}{\partial\alpha}=\nu\left(\frac{\partial^2U}{\partial\beta}-\frac{\partial^2U}{\partial\alpha^2}\right)-\frac{\partial(h\overline{uv})}{\partial\beta}+\frac{\partial(h\overline{uw})}{\partial\alpha} \tag{5}$$

It is important to note that simply by choosing $h = 1$ we obtain the equations of motion in the cartesian co-ordinate system.

However, the equations contain five unknown U, V, W, $\overline{uv}$ and $\overline{uw}$. As regards the $\overline{uv}$ and $\overline{uw}$ components we shall see later on how we can close the problem. There remains one unknown, either V or W. If we choose V to be solved by the mass equation we need to find one more equation to solve W. Of course, one solution would consist of writing the W-momentum equation, but by doing this we would complicate the resolution and increase the calculation time. Fortunately another approach, based on the previous numerical and experimental studies of Djenidi (1989) and de St. Victor (1987) leads to a reasonable result without any such penalties and is consequently much simpler and faster.

Locally the cartesian and curvilinear co-ordinates are as illustrated in Figure 2.

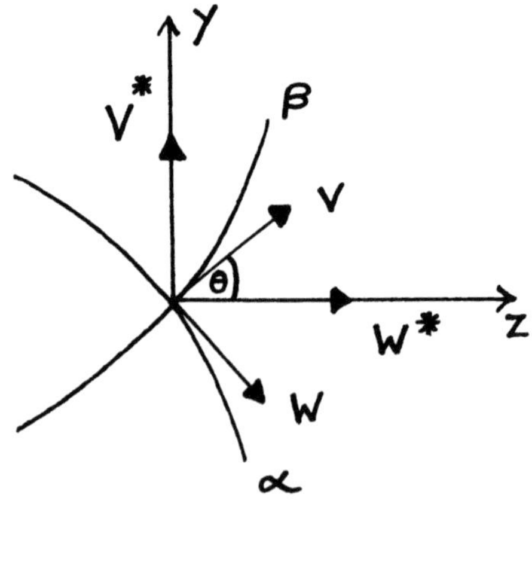

Fig. 2 Local cartesian and curvilinear co-ordinate directions

where V^* and W^* are the components of the velocity in the cartesian co-ordinate directions y and z, respectively.

By geometrical construction we have

$$W^* = V cos\Theta + W sin\ \Theta \tag{6}$$

Previous numerical and experimental studies in laminar flow by de St. Victor (1989) and Djenidi (1989) have shown that W^* is zero, and there are no secondary motions (counter rotating eddies) within the grooves in that case. Therefore we have

$$W = -V\frac{cos\Theta}{sin\Theta} \tag{7}$$

which is the third equation we need to solve for the equations of the motion.

It should be noted that at this point we chose to continue to use this equation even in the turbulent flow case. Of course physically W may not be zero then because of turbulence driven secondary motion within the groove. However this choice is consistent with the very simple turbulence model (a mixing length eddy viscosity approach) we have adopted.

At first sight there would appear no reason for using such a simple closure scheme, other than the rapidity and ease with which one can then 'solve' the problem. However we shall see the results are informative even with such a simple model. The Reynolds stresses $\overline{uv}$ and $\overline{uw}$ take the following analytical expressions

$$\overline{uv} = -\nu_t\frac{\partial U}{h\partial\alpha}, \quad \overline{uw} = -\nu_t\frac{\partial U}{h\partial\alpha} \tag{8}$$

where ν_t is the eddy viscosity and is related to the mixing length, ℓ, by

$$\nu_t = F^2\ell^2\sqrt{\left(\frac{\partial U}{h\partial\beta}\right)^2 + \left(\frac{\partial U}{h\partial\alpha}\right)^2} \tag{9}$$

We chose to use the mixing length expression proposed by Michel et al. (1969) which is applicable everywhere in the boundary layer thickness.

$$\ell = 0.085\delta tanh\left(\frac{\kappa y}{0.085\delta}\right) \tag{10}$$

where δ is the boundary layer thickness and κ the von Karman constant ($\kappa \simeq 0.41$).

The F in the eddy viscosity expression is the damping function which takes account of the wall effect and was first introduced by Van Driest [1955].

$$F = 1 - exp\left(\frac{-y^+}{26}\right) \tag{11}$$

Before going further let us re-state the set of the equation of the motion:

$$\frac{\partial h^2 U}{\partial x} + \frac{\partial hV}{\partial\beta} + \frac{\partial hW}{\partial\alpha} = 0 \tag{12}$$

$$\frac{\partial h^2 UU}{\partial x}+\frac{\partial hUV}{\partial \beta}+\frac{\partial hUW}{\partial \alpha}=\frac{\partial}{\partial \beta}\left[(\nu+\nu_t)\frac{\partial U}{\partial \beta}\right]+\frac{\partial}{\partial \alpha}\left[(\nu+\nu_t)\frac{\partial U}{\partial \alpha}\right] \tag{13}$$

$$W=-V\frac{cos\Theta}{sin\Theta} \tag{14}$$

(Again it is important to note that we only need to assign $\nu_t = 0$ to recover the laminar flow equations.)

2.3 Boundary and Initial conditions

On the wall the non slip conditions applied were $U = V = W = 0$, while at the upper boundary we have $\frac{\partial U}{\partial y} = 0$. In order to simulate symmetrical conditions on both lateral sides we impose $\frac{\partial U}{\partial z} = 0$.

To start the calculation an initial velocity field was required. For the laminar flow case this was provided by a solution of the two dimensional Blasius equation. For the turbulent case we used a theoretical formulation to calculate the U-profile at the entrance to the computational domain and the corresponding transverse velocity V, was then obtained from the mass equation.

2.4 Numerical Procedure

The code being basically the same as the one used at I.M.S.T. (Marseille), (only the coefficients were altered because of the introduction of the scale factors) the numerical procedure was also the same [see Djenidi (1989), Djenidi et al. (1989a, b) for details]. The equations were solved using an X-marching method, where a finite volume scheme was used for discretisation at each step in the X direction. The discretised system of equations was solved with the Modified Strongly Implicit algorithm (M.S.I.) of Schneider and Zedan (1981). Further details about the resolution technique are provided by Djenidi (1989).

3. Results

3.1. Laminar flow

(i) Calculations using a cartesian mesh

Before presenting the new high resolution results obtained using the CRAY 2 at the E.P.F. Lausanne ERCOFTAC Pilot Centre, it is appropriate to summarise the previous results obtained in the study carried out at I.M.S.T. Since one of the main aims of this paper is to answer the questions arising from the I.M.S.T. laminar flow studies. (In particular: Does the number of grid points influence the calculation of the skin friction with a cartesian mesh? And, is the skin friction on the ribleted wall actually reduced in comparison to that of a smooth flat plate?)

Figure 3 shows the main results stemming from I.M.S.T. studies which included both experimental and numerical investigations. Their initial experimental measurements showed that the mean skin friction over triangular riblets is apparently not different from the one on smooth plate (Figure 3.a).

$$\left.\frac{\partial U}{\partial y}\right)_o riblet = \left.\frac{\partial U}{\partial y}\right)_o smooth. \tag{15}$$

(Although it should be noted that the spanwise dimension of the hot-wire probe used was 4mm while the distance between two crests, s, was only 0.7mm). Furthermore

(a) Spanwise averaged velocity profiles for smooth wall (o) and riblet wall (•): s=0.7mm, h=0.3mm, Ue=14.9cm/s, δ=8mm

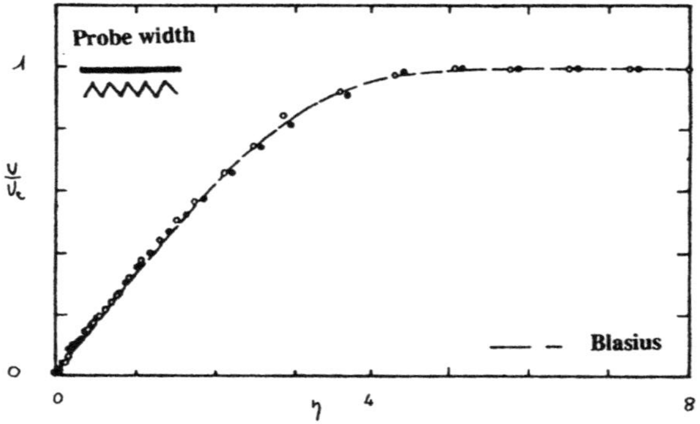

(c) Velocity profiles at selected spanwise locations: s=5mm, h=4.3mm, Ue=15.5cm/s

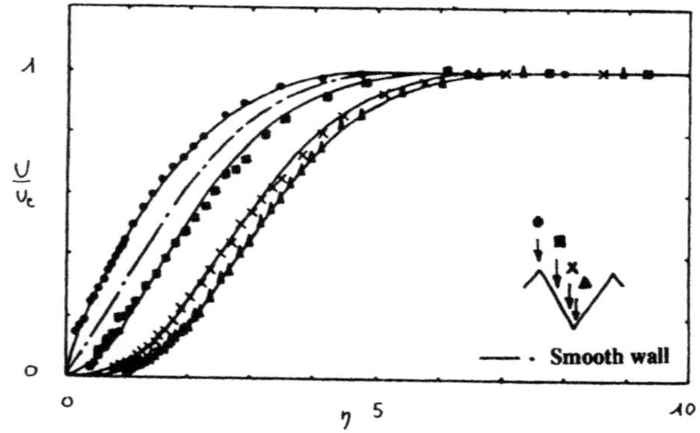

(b) Comparison of streawise (•) and spanwise (o) velocity components at mid-riblet location: s=5mm, h=4.3mm, Ue=7.5cm/s

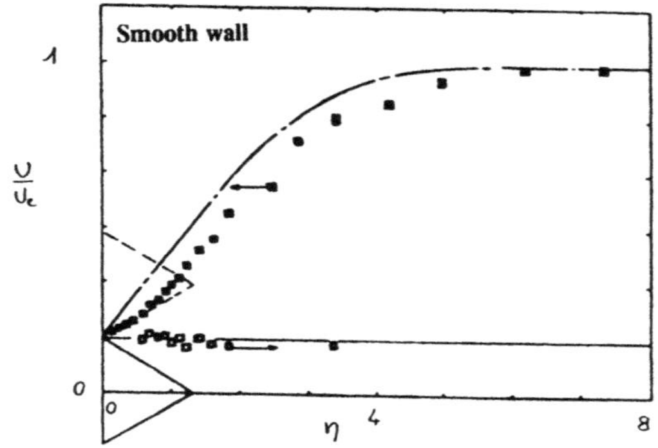

(d) Drag reduction calculations with two alternative integration methods: • : Newton-Cotes technique, o : Trapezoidal rule

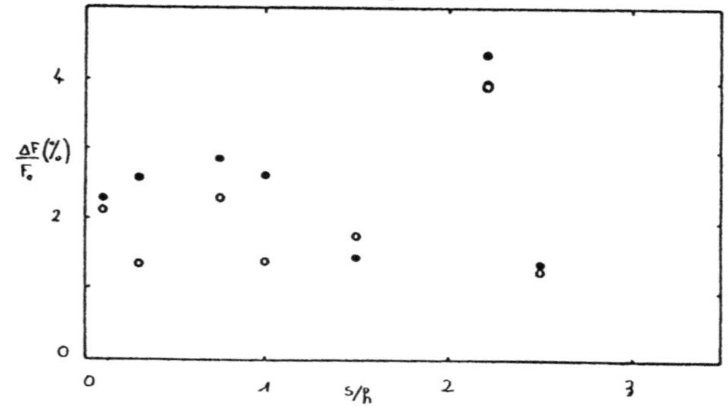

Fig.3. Summary of results from the IMST studies: Comparsion of LDA measurements (Symbols as on figures) with low-resolution Cartesian mesh computations.

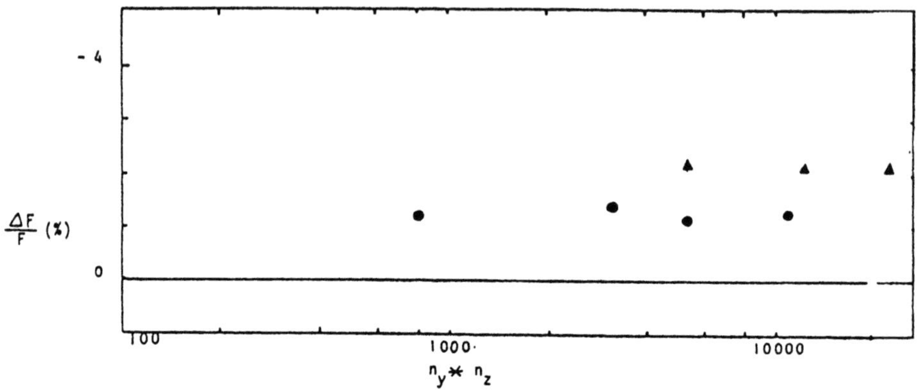

Fig.4. Influence of the number of grid points on the skin friction estimate: ●: h=1mm, s=1.5mm (IMST 1989), ▲: h=1mm, s=1.5mm (EPFL 1990)

subsequent measurements using LDA techniques, showed that the spanwise component of the velocity, W, is equal to zero, (or at least negligible) inside the grooves as well as outside (see Figure 3-b). (For these latter measurements, a backscatter technique was used to reduce the effective probe volume to $1.32 \times 0.12 \times 0.12\text{mm}^3$ compared to the height and width of the groove which were 4.3mm and 5mm respectively).

This finding is of great significance for the numerical approach, since it allows the above mentioned simplification of the equations of motion and is supported by initial 3-D computations performed by de St. Victor (1987). In the next graph (Figure 3-c) both experimental and numerical velocity profiles, calculated with W assumed equal to zero, are plotted. A remarkable agreement is observed between the experimental profiles and the predicted ones. The last graph (Figure 3-d) where $-\Delta F = F_{riblet} - F_o$ and F_o is the friction on a flat plate with a span equivalent to s, (this being the distance between two crests of a groove), shows perhaps the most important results of the study. Not only is the skin friction on riblets not increased, compared to the smooth surface as found in laminar internal flows, but it is apparently reduced by approximately 2.5 %.

Because this result is of potentially great importance, especially to the understanding of the mechanism of the turbulent drag reduction by riblets, one needs to be sure about its validity. In particular one has to ensure that the skin friction calculation is independent of the grid, and above all the number of grid points. Therefore, a systematic study of the influence of the number of grid points on the computed friction reduction was carried out, in which the grid resolution varied from 41×19 up to 115×95. The results are presented in Figure 3d where the black circles represent results taken from the I.M.S.T. study for a non-optimum $h = \frac{2}{3}s$ V-groove geometry. They indicate that the drag reduction was not very sensitive to the number of grid points, but, for reasons of time and available computational resources it was unfortunately not possible to pursue this point further. The present authors have been able to repeat and then extend the original I.M.S.T. calculations using the CRAY 2 at the E.P.F. Lausanne, PC. Our additional finer resolution Cartesian grid results were all obtained for the optimum shape of V-groove, $s/h = 1$, indicated by the I.M.S.T. study. These computations confirmed that the drag reduction of about 2.5 % is indeed independent of the number of grid points - see Figure 4. (It should be noted that the last point on the Figure represents a total of 20898 points on a 161×129 mesh, for which computations required 3400s CPU on the CRAY2.)

We concluded these initial calculations by checking the asymptotic behaviour of the flow in the streamwise direction. Because the computational domain consisted of two regions in the streamwise direction for the Cartesian grid calculations (a smooth flat plate section of 25 cm length followed by a triangular riblets section of the same length), we investigated two kinds of 'transition' from smooth - to - riblet surfaces: first with a constant Δx step throughout the whole domain, and then with a variable Δx step from the leading edge of riblet. The results reported in Figure 5 show that an asymptotic behaviour was reached at sufficiently large whatever the Δx step employed through the transition region from smooth to riblet surfaces. A similar asymptotic behaviour, independent of Re, has been found by Coustols [private communication], using basically the same code at ONERA - CERT Toulouse. This confirms that the apparent drag reduction is not simply an artifact due to the changes in wall geometry altering the virtual origin of the flow.

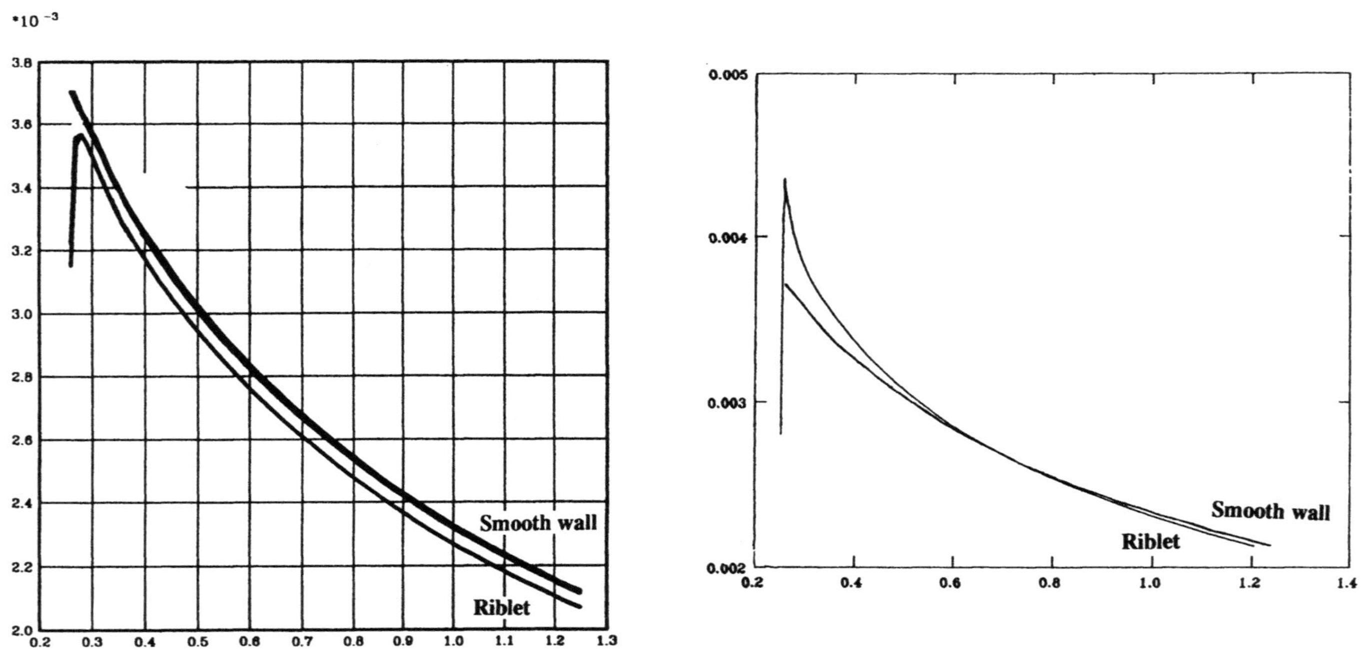

Fig.5 Streamwise evolution of the skin friction coefficient: effect of the Δx step
(a) Constant Δx (b) Variable Δx

(ii) Calculations using the ortho-curvilinear grid

In order to test the conformal grid calculations we first compared the velocity profile results produced with ones obtained from similar resolution cartesian grid calculations. Two sizes of V-grooves were computed. The first test, with $h/delta \simeq 0.49$ was performed for the same experimental conditions as in the IMST studies (see Figure 3). The results presented in Figure 6 show that the conformal mesh computations do not in fact agree particularly well with the cartesian grid ones (and accordingly also with experimental measurements). However we believe that this discrepancy has nothing to do with the resolution technique. Indeed, in a second test computation for smaller riblets, with $h/\delta = 0.1$ (again close to the optimum value found in the IMST numerical studies), the agreement was perfect between conformal mesh and cartesian grid calculation (see Figure 7). So it seems that the explanation for this discrepancy can be found by looking at the particular conditions of the calculation in the first test. In particular it would appear that, because of the manner in which the initial velocity profiles were imposed coupled with the relatively short development length prior to the flow moving onto the riblet surface and the rather large ratio h/δ, the flow had probably not reached an equilibrium state at the station where comparisons were made.

Further parametric computations suggested that the conformal mesh computations could predict correctly the laminar flow over a ribleted wall. It should be noted however, that the conformal description is rather different from the one used by cartesian grid, especially in the vicinity of the wall (right above the crest and valley, both of these descriptions are of course identical) since the distances y and β from the wall are not the same. It is also important to note that the gradient $\left.\frac{\partial U}{h\partial\beta}\right)_{\beta=a}$ at the wall represents the skin friction whatever the location in the z-direction. This is not the case in the cartesian representation with $\left.\frac{\partial U}{\partial y}\right)_{y=0}$, but the latter does allow direct comparison with experimental data.

From a physical point of view it is of interest to ask what is happening to the behaviour of the flow when one replaces a flat plate by a ribleted wall. Let us therefore go back to the conformal mapping. In this we have transformed a cartesian grid where the $iso - \beta$ lines (horizontal lines) also represent, to within a small coefficient, the iso-velocities on a flat plate. By the transformation we obtained a curvilinear mesh (c.f. Figure 1 a-b) which is in fact, the original mesh plotted in cartesian system. It is then very instructive to see how the velocity behaves along the β-lines. In Figure 8(a) we plot β-lines while on Figure 8(b) we plot the velocity along these β-lines. The conclusion is obvious: the β-lines do actually represent, again to within a small coefficient, the iso-velocities. This observation shows that the behaviour of the flow, especially in the vicinity of the ribleted wall, is strongly governed by the viscosity of the fluid. In other words, it means that the convective terms of the equations of motion are essentially negligible within the grooves (a result actually proved earlier by Djenidi (1989)). It is worth pointing out that our results are therefore akin to those of Bechert et al. (1986) who, in carrying out their analytical study, assumed that since the riblets were immersed within the inner layer of a turbulent boundary layer, the convective terms would be negligible. Accordingly they then solved the Laplace equation for the streamwise component of the velocity and obtained the iso-velocities in a YZ plane over the grooves by using exactly the same conformal transformations, as we have been using to draw the meshes and to transform a flow over a smooth flat plate into a flow over riblets.

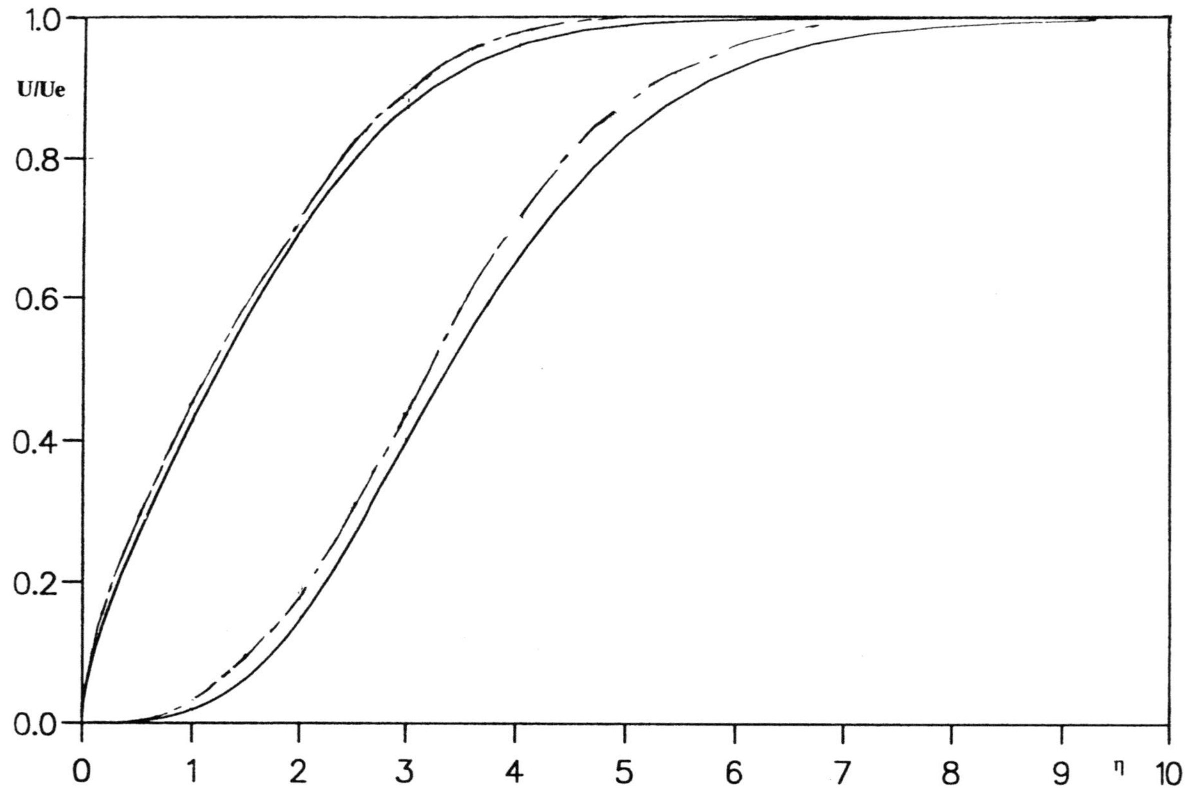

Fig.6. Comparison of Cartesian mesh predictions of velocity profiles at peak and valley (groove centre-line) for V-groove riblet: s=5mm, h=4.3mm, h/δ=0.5, Ue=15cm/s: ____ : EPFL 1990, __ _ : IMST 1989

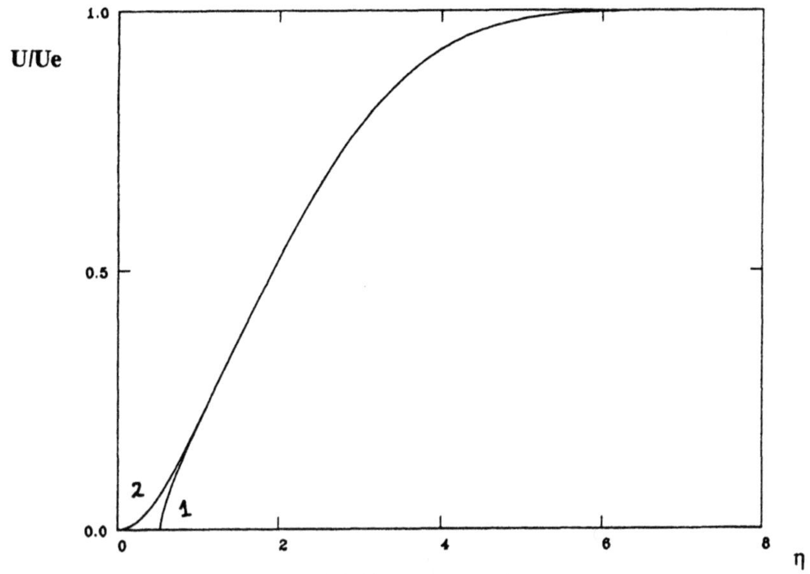

(a) Cartesian mesh

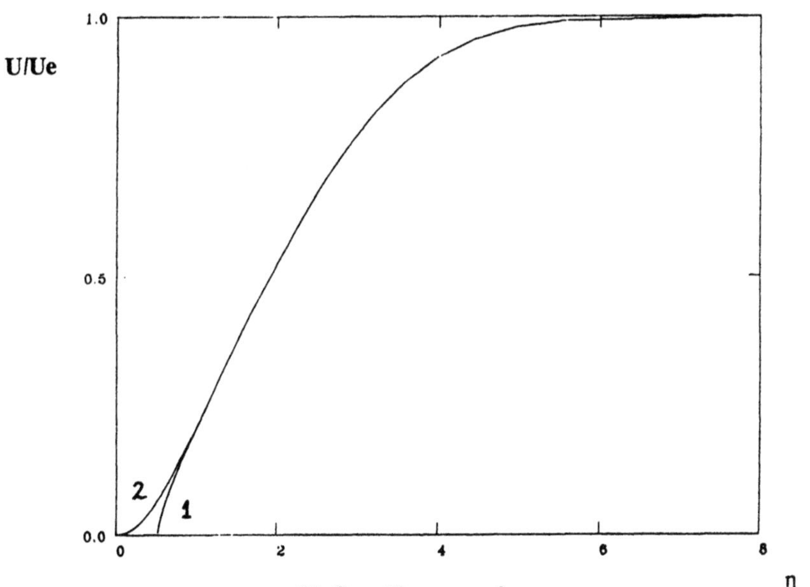

(b) Curvilinear mesh

Fig.7. Predicted velocity profiles over peak[1] and valley[2] of V-groove: s=1mm, h=1mm, Ue=6cm/s

Figure 8 also indicates how the skin friction reacts to the spanwise modification of the wall. Around the crest, the selected lines plotted on the Figure indicate a relatively strong skin friction, while one can easily anticipate a correspondingly low friction in the valley. This is actually observed on Figure 9, where the spanwise distributions of skin friction coefficient along the surface for four shapes of groove under the same flow conditions, are presented, (note: the friction coefficient scale is a logarithmic one). The riblet shapes are V, L, LU (L with a rounded corner) and a semi-circular U-groove. The friction distributions on L and LU grooves have quite the same shape and differ from those on the V and U-grooves: they are characterised by the presence of a minimum value of the friction due to the corner. Perhaps surprisingly a similar minimum friction is also found for the semi-circle U-groove, while for the V-groove this minimum is shifted to the riblet valley centre line. It is remarkable that, at least in this particular case where $s = 2h$, the skin frictions on the L, LU and semi-circular U grooves take the same values at the half-riblet location in the valley.

At the riblet crest, the evaluation of the local C_f is problematic. Preliminary calculations correctly showed that the value of the skin friction on the peak did not reach a limiting value as the grid refinement was increasing around the crest, even with preferential β refinement. This is because using conformal mapping a numerical problem is created by the presence of a singularity. However, it is possible to get around this mathematical difficulty in the following way: provided there is sufficient refinement around the crest, one can assume that this skin friction at the peak is not very different from that at the closest spanwise grid location. The authors therefore believe that an extrapolation of the peak neighbouring points can quite correctly approximate the friction at the peak. In fact tests showed that by using a three degree Lagrange's polynomial extrapolation procedure, the continuity of the normal direction at the wall was preserved right up to the riblet peak. We therefore made use of such a technique to calculate the overall skin friction coefficient for the four shape grooves already described above. The results presented in Figure 10, are expressed in terms of the ratio between the overall riblet Cf and that of an equivalent smooth flat plate (Cf_o) versus h/δ where δ is the boundary layer thickness. Cartesian mesh parametric computational results from Liandrat et al. (1990) are also reported. Although one might have some reservations regarding the appropriateness of the extrapolation technique, our results are consistent with the previous IMST findings (which used two alternative procedures for evaluating the total surface drag of a whole riblet cross-section) confirming at least that there is no drag increase on the ribleted walls, and that the drag results are independent of the coordinate system of the grid.

3.2 Turbulent flow

3.2.1 Calculations using an ortho-curvilinear grid

In an earlier part of this paper we said that the use of the mixing length - eddy viscosity approach was dictated partly by its simplicity but that other considerations were taken into account in making this choice of model for turbulent flow. It is perhaps necessary to explain these other considerations in a little more detail. Our own calculations in laminar flow revealed that working within this particular curvilinear system allowed us to have an exact bi-dimensional flow ($W = 0, \frac{\partial}{\partial z} = 0$) such that the iso-$\beta$ are also iso-U, and the calculations of Bechert et al. (1986) led to the same result when they considered only the inner part of the turbulent flow. This is one reason why we thought that using a mixing length, which gives good results for a bi-dimensional flow over a flat plate, might also be appropriate with the conformally transformed grid. The second reason was linked to the nature of the

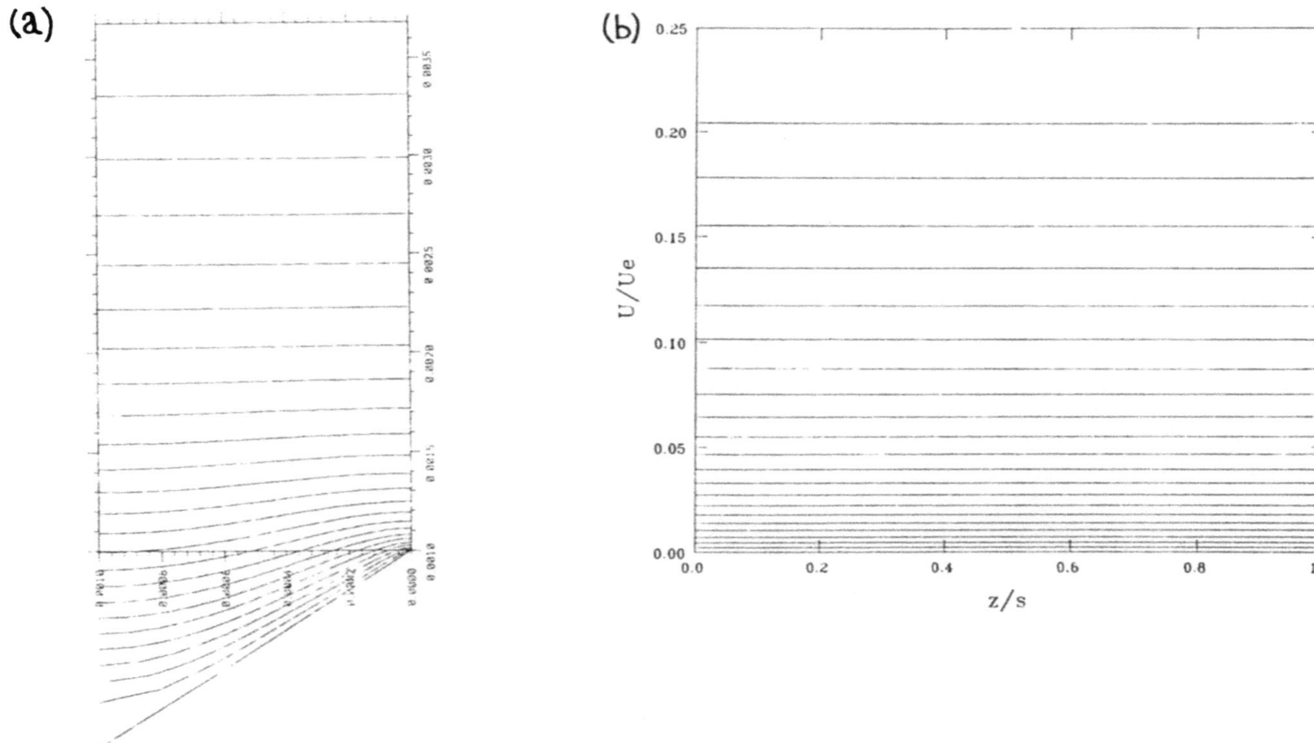

Fig.8.(a) Iso-β lines and (b) Iso-velocity lines along β for V-groove: s=2mm, h=1mm

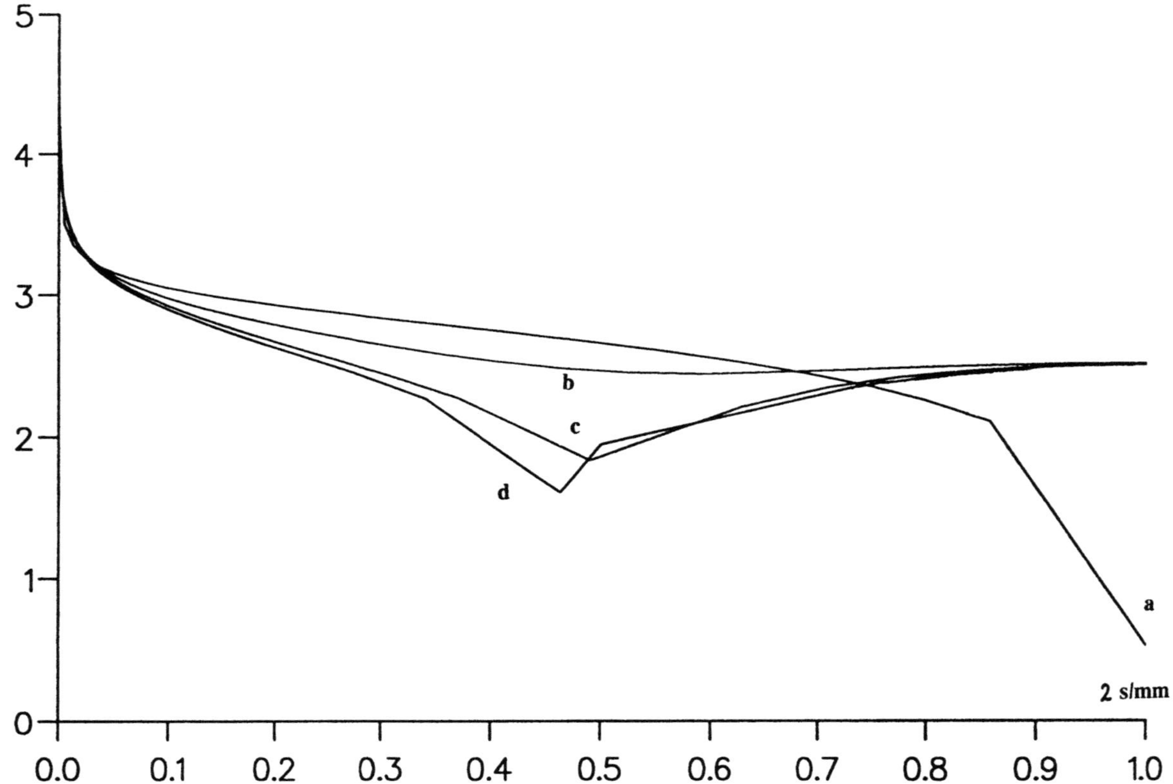

Fig.9. Spanwise Cf distributions across groove surfaces: s=1mm, h=0.5mm, Ue=10cm/s, δ=6mm

(a) V-groove, (b) U-groove, (c)L-groove with rounded corner (LU) & (d) L- groove

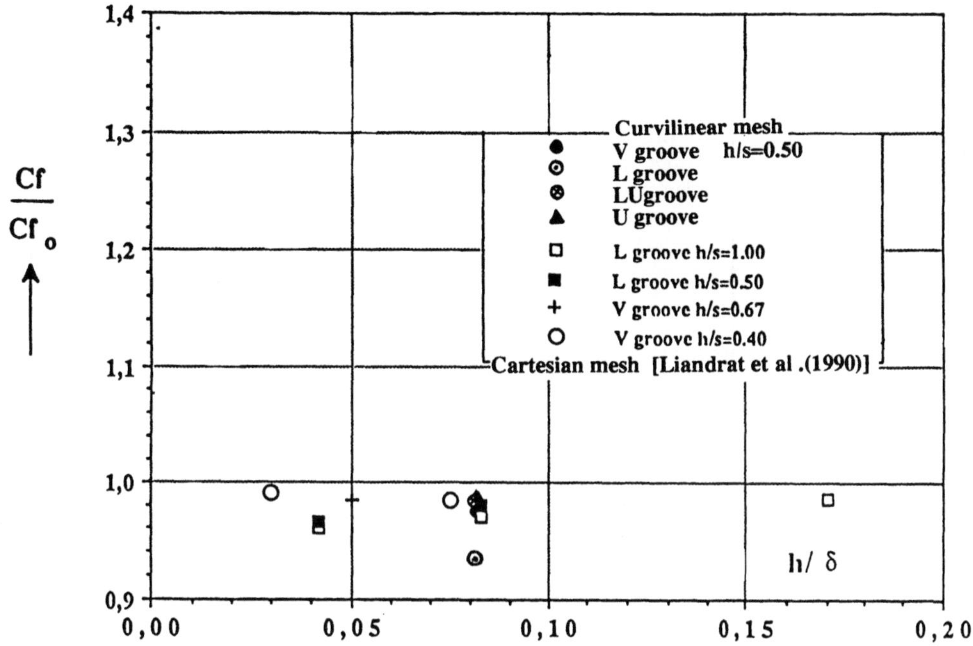

Fig .10. Variation of relative skin friction drag with groove shape

mixing length which can be regarded as having a vertical direction, in the case of a smooth flat plate, parallel to the β direction. Thus following conformal transformation the mixing length, as defined on a smooth flat plate, is curved on the riblet surface, just as the β-direction is curved in the physical grid, and remains parallel to the latter. Such a transformation therefore allows one to conserve the original idea of a mixing length and implicitly takes the effect of the valley, and the peak of the groove into account; since moving from a crest to a valley along an iso-β line, the mixing length undergoes a different variation according to the groove shape. The slope of the curve $\partial \ell / \partial y$ at the wall (ℓ being the mixing length), increases when moving towards the crest (narrowing of the iso-β lines) and decreases when moving towards a valley.

Unfortunately very few experiments have so far been conducted on turbulent boundary layers over a ribleted wall in which any detailed measurements have been made within the grooves. Indeed, to our knowledge only two such studies have been carried out. We shall therefore compare our numerical results with the results stemming from these two experiments.

The first such study was performed by Vukoslancevic et al. [1987], who used V-grooves of large physical size (s = 10mm, h = 5mm) in a very low speed wind-tunnel in order to carry out hot-wire measurements above and inside the grooves. Because of the low Reynolds number the boundary layer was tripped and the measurement section was 7m downstream of the trip. The mean flow velocity was $U_e = 1.23m/s$, $\delta = 13.55cm$, and $R_\Theta \simeq 1000$.

The second study is still in progress at I.M.S.T. Here also the riblets have a V-shape and are of similar scale (s = 7mm, h = 3.5 mm). But the experiments are being conducted in a water tunnel using a Laser Doppler Anemometry system (see Djenidi (1989) for further details). The mean flow velocity is U_e = 0.091 m/s, $\delta \simeq$ 35mm and $R_\Theta \simeq 297$, and, once again the turbulent boundary layer is tripped. The experimental and predicted mean velocity profiles plotted in terms of the normalized velocity $U^+ \left(= \frac{U}{u_\tau}\right)$ as a function of the non- dimensional distance to the wall $y^+ \left(= \frac{y u_\tau}{\nu}\right)$ for smooth and riblet walls in both studies are presented in Figure 11. For flow over the smooth flat plates the predicted profiles fit quite well with the experimental data, except in the outer part of the boundary layer, in both cases. Any discrepancies between measurement and prediction observed in this region may be explained by the fact that the experimental turbulent boundary layers were tripped whereas the numerical ones were not. In particular it is not surprising to find that the wake law is not very well satisfied in the outer part of the boundary layer since this is known to be the case when when Re_Θ is small. Furthermore in the IMST experiments one would not expect to find any logarithmic law region at such low Re if the flow had not been tripped.

Consider now the results for ribleted walls also shown on Figure 11. Presenting the results in wall co-ordinates (U^+ *versus* y^+) allows easy comparison between experimental and numerical velocity profiles over a single groove, but because of the nature of the conformal mesh, especially in the close vicinity of the wall, we can best compare the experimental and calculated profiles above peaks and valleys. It is fair to say that the predicted velocity profiles are quite comparable with the experimental ones for each of the flow cases studied; in particular the shapes of the predicted profiles are rather close to the experimental ones, showing that the behaviour of the boundary layer over the riblets is reasonably well predicted. Indeed the main discrepancy between measurements and calculations appears between the

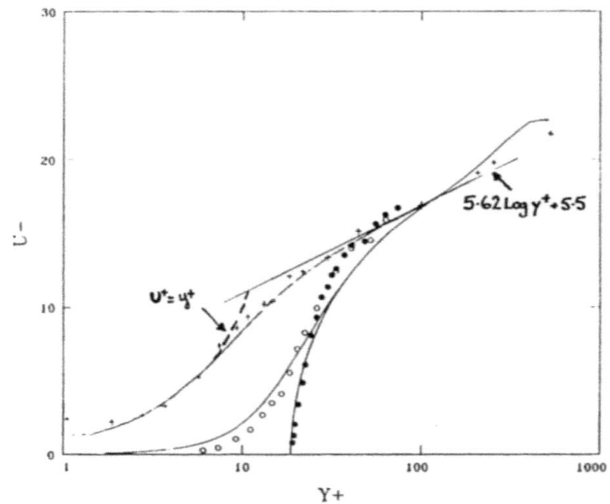

(a) Wind-tunnel experiments of Vukoslancevic etal.(1987): Reθ=1000

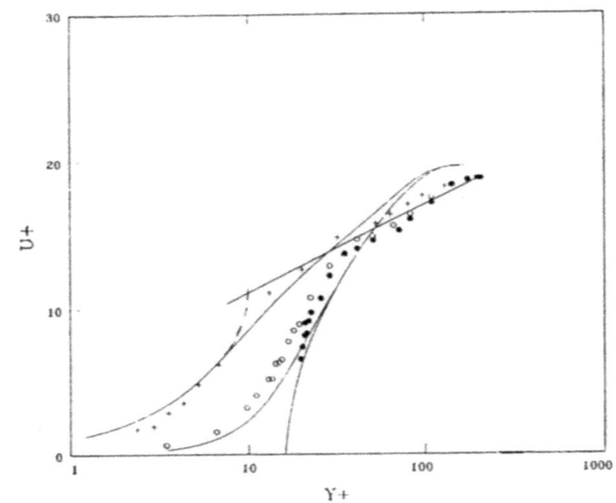

(b) Water-tunnel experiments of Djenidi et al.(1989): Reθ=297

Fig.11. Comparison of turbulent flow computations with experimental data for a V-groove riblet wall (+: Smooth wall, •: Riblet peak, o: Riblet valley)

y location where the profile above a peak is identical to that over the valley one and the boundary layer edge. It would appear therefore that the calculations do not accurately predict the 'blending' from the wall vicinity to the upper region, but, bearing in mind the fact that R_θ is low and the initial conditions are not ideal, one might expect such a discrepancy. Some further comparison between other experiments and computations are clearly required , but to date very few accurate measurements have been made which provide sufficient information to do this.

Let us therefore examine the predicted skin friction. Bearing in mind the fact that our turbulence model is not particularly sophisticated, we have only performed skin friction calculations to provide a qualitative guide to the capabilities of different riblet geometries. The skin friction for two kinds of groove shapes have been investigated: an aspect ratio one ($h = s$) , V-groove and a semi-circular $h = \frac{1}{2}s$ U-groove. Both are compared with c_f of a smooth flat plate with a spanwise extent equal to s. The results are presented in Figure 12, where the dashed area represents the scatter of experimental data from early parametric studies conducted by Walsh [1980] who measured the the difference between the drag on ribleted and smooth flat walls experimentally using a direct drag balance. It is clear that the predicted maximum "drag" reduction is not very close to experiment. Instead, the V-groove calculations overestimate the reduction by a factor of at least 3, but they do reproduce the correct trend in variation of drag with h^+. In contrast the U-groove computations show approximately the correct magnitude of reduction, but spread over far too wide a range of h^+. (Note however that the shape of the drag reduction curve predicted for U-grooves is quite similar to some experimental findings obtained by Wilkinson and Lazos [1988] for L-grooves, and as we have seen laminar flow predictions for L, U and LU were similar). Overall then the agreement is not so bad and the over-prediction of the V-groove optimum performance could be attributed to the fact that the present model cannot simulate the secondary flows which are known to occur within the grooves, and which may well produce additional momentum transport, thereby increasing the actual riblet Cf to the level measured experimentally.

4. General discussion

We shall concentrate the discussion on the turbulent flow predictions over riblets. The laminar boundary layer results, have already demonstrated that there are no numerical problems in that case, and that there is very good agreement between calculations and measurements. This does not of course mean that laminar flow over riblets is not of interest. On the contrary the laminar experimental and numerical studies may be of considerable practical significance and have helped to shed some further light on the turbulent drag mechanism: in particular it now seems clear that viscous effects must play a major rôle in the skin friction drag reduction process. One problem with modelling turbulent boundary layer flow over riblets is that despite the very large number of studies which have been undertaken there are very few experiments which can be used to provide detailed test case data for numerical studies. Very recently (while this manuscript was in press) Choi et al. (1991) have reported some Direct Simulations of riblets in both fully developed laminar channel flow, which confirms a drag increase in such internal flow, and turbulent boundary layer flow, although not yet at sufficiently high resolution for optimised drag reducing riblets. However some Direct Simulations of a square duct turbulent flow performed by Gavrilakis (1989) may also, surprisingly, provide a useful data set. The reason is that, in the near vicinity of any of the four corners of the duct, the flow appears typical of a 90° V-groove, since the influence of the flow in the centre of the duct and the other three corners can then largely be ignored. In order to

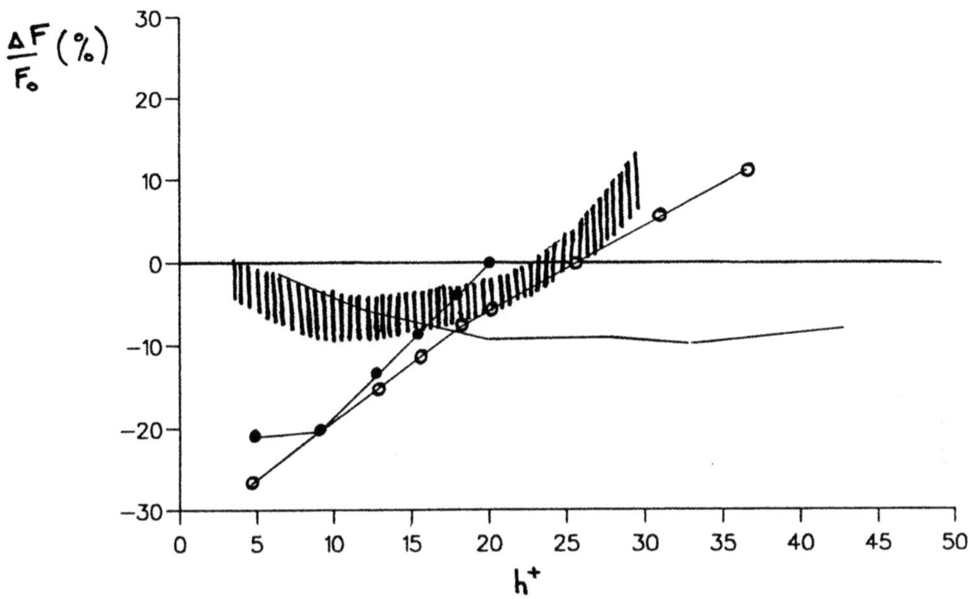

Fig.12. Turbulent drag reduction predictions compared with experimental findings: s=2h, Ue=12m/s: V-groove riblets: ●: h=0.25mm, o: h=0.5mm; U-groove riblets: ______: h=0.5mm

test this idea, we have plotted in Figure 13 the long-time mean flow velocity profiles obtained from low Reynolds number square duct Direct Simulations compared with experimental velocity profiles of the I.M.S.T. 90° V-groove experiment. One can see that the velocity profile along the duct wall median is similar to the smooth flat plate one and the velocity profile along the duct corner bisector does indeed show the same characteristic as the riblet valley bisector profile. Of course in the corner of the square duct there are no crests, but this is not a serious inconvenience if the comparison between riblet and square duct corner flow concentrates on what happens *inside* the groove. The very extensive square duct Direct Simulation data base could therefore provide far more information regarding mixing length, Reynolds stresses, turbulent kinetic energy, and even dissipation profiles and higher order moments such as pressure strain etc., than can be extracted from experiments, and may be used to assist more refined modelling of turbulent flow over riblets.

Of course, great care should be taken before attempting modelling based on such data since they show that secondary motions (counter rotating vortices) occur in each corner of the square duct corner, but this does not mean simply because a 90° V-groove has a similar corner geometry that such secondary motions are generated in each of the riblet grooves. Several other factors must be taken into account. For example the optimum size of riblets is so small ($h^+ < y^+ = 20$) that the riblet is largely immersed within the viscous sublayer of the turbulent boundary layer, and therefore it is unclear whether the Reynolds stresses will be sufficiently strong inside the grooves to generate such secondary flows. It may be that time dependent interactions between the groove and the turbulent flow structure or between flows in and above adjacent grooves are involved in this process. In addition the flow inside the groove might well behave differently depending on the exact groove shape.

It is also not a priori obvious that such secondary motion might not aid the viscous effect, and thus increases a net drag reduction, as was suggested by Khan [1985], instead of preventing it from fully playing its role and thus generate less drag reduction, as indicated by the difference observed between experimental measurements and our own predicted results for V-groove riblets. It may in fact be possible that secondary motions act in both ways since the tendency is to homogenize the friction along a non-planar surface, as shown by ALy et al. (1978).

Conclusions

The main aims of this present numerical study, were: **(a)** to essentially complete previous laminar computations performed elsewhere; **(b)** to develop a computational code with some freedom in the choice of the groove shapes which could be considered; and finally **(c)** to take a first step towards predicting a turbulent boundary layer over riblets using curvilinear co-ordinates. The results presented in this paper also provide some indirect information regarding the turbulent drag reduction mechanisms. It appears that the laminar skin friction over suitably scaled ribleted wall in external boundary layer flow is not increased, as found for laminar internal flows, but is actually reduced by ~ 2.5 % in comparison to an equivalent smooth flat plate. This is despite the fact that, as in internal laminar and turbulent flows, the wetted area is increased by up to 40 % in the optimum ribleted wall case. Calculations using either cartesian coordinates with a very large number of grid points or curvilinear co-ordinates showed that this result is independent of the mesh. Since this laminar drag reduction is due to a viscous effect, it is reasonable to think that this effect should also play a significant rôle in turbulent drag reduction. However, in the latter case, the viscous effect is not the only phenomenon present. Instead, secondary motions may then occur inside the grooves and our

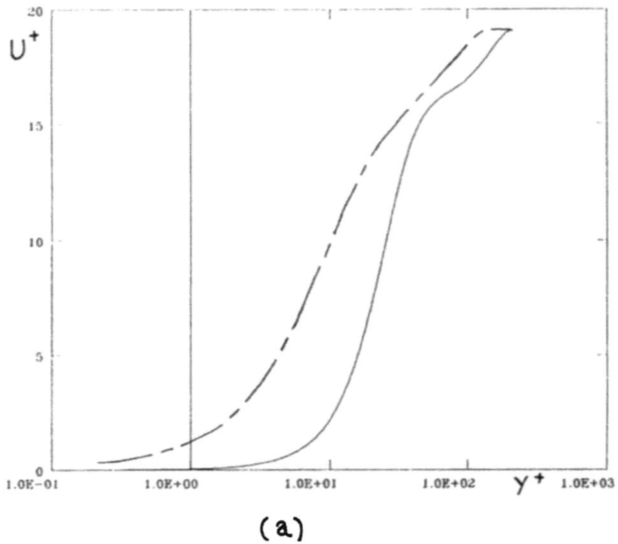

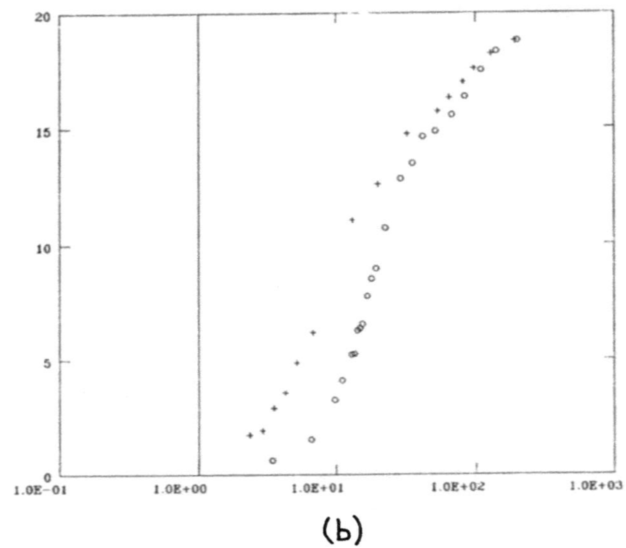

Fig.13. Comparsion of velocity profiles from (a) Square Duct Flow Direct Simulations (___ _ : wall bisector, _____ : corner bisector), (b) turbulent boundary layer computations for a right-angle V-groove riblet (+: Smooth wall, o: Riblet valley)

results suggest these resist the viscous effect, even though the nett drag reduction is then ~ 7 %. This needs to be proved however and we therefore think that future experiments should take care to determine the presence or absence of such secondary flow inside different groove shapes and to define their effects on the turbulent drag reduction. As regards future modelling, although a mixing length - eddy viscosity closure implemented in curvilinear co-ordinates has given rather encouraging results more refined model computations are now required which can handle the stress anisotropy associated with turbulence driven secondary motions. Such modelling is presently being conducted at U.M.I.S.T., using higher order closure low-Re models, and alternative mesh transformations, which avoid any singularity at the riblet peak.

Acknowledgements.

This work was supported under a joint Cambridge/UMIST/SERC Research grant GR/E 7703.9.

We are very grateful to our research colleagues at I.M.S.T., Marseille (Drs. L. Fulachier, F. Anselmet and J. Liandrat) at ONERA/CERT, Toulouse (Dr. E. Coustols) and also at C.E.A.T. Poitiers (Dr. C. Tenaud) in France for their close collaboration in this project. Special thanks are also due to Dr. I. L. Ryhming for providing additional support for travel to and from, the EPFL ERCOFTAC Pilot Centre in Lausanne, Switzerland and for the Cray 2 facilities made available through this Centre.

References

A. M. M. Aly, A. C. Trupp and A. D. Gerrard: Measurements and Prediction of Fully Developed Turbulent Flow in an Equilateral Triangular Duct. J. Fluid Mech. Vol. 85, pp. 57-83, 1978.

D. W. Bechert, M. Bartenwerfer, G. Hoppe and W. E. Reif. Drag Reduction Mechanisms Derived from Shark Skin. ICAS-86 - 1.8.3, 1986.

D. M. Bushnell: Turbulent drag reduction for external flows. AGARD Rep. 723 (1985), 5.1 - 5.26.

H. Choi, P. Moin, J. Kim: Turbulence Control in Wall-Bounded Flows using Direct Simulations. Abstracts of AFOSR Turbulence Structure and Control Conference, Columbus, Ohio, 1991.

L. Djenidi, J. Liandrat, F. Anselmet and L. Fulachier: "Mechanism Involved in a Turbulent Boundary Layer over Riblets". Proc. 3rd European Turbulence Conference. H. Fernholz and H. E. Fiedler (ed.)Springer Verlag, 1989

L. Djenidi: Contribution a l'etude de Couches Limites sur Parois Rainurées - Ph.D. dissertation, I.M.S.T. 1989.

L. Djenidi, J. Liandrat, F. Anselmet and L. Fulachier: Numerical and Experimental Investigation of the Laminar Boundary Layer over Riblets. Applied Scientific Research, Vol. 46, 1989, pp. 263-270.

J. P. van Doormaal, G. D. Raithby and A. B. Strong. Prediction of Natural Convection in Nonrectanglar Enclosures Using Orthogonal Curvilinear Coordinates. Numerical Heat Transfer, Vol. 4, 1981, pp. 21-38.

J. Gallagher and A. S. W. Thomas: Turbulent Boundary Layer Characteristics over Stream-wise Grooves. AIAA Paper 84-2185, 1984.

S. P. Gavrilakis: Numerical Simulation of the Turbulent Flow through a Duct of Square Cross Section. ERCOFTAC Bulletin II, p. 12 and Supercomputing Review 1, p. 14 (1989).

M. M. S. Khan. A Numerical Investigation of the Drag Reduction by Riblets Surface. AIAA Paper, 86 -1127, 1985.

B. E. Launder and S.-P. Li: A numerical study of riblet effects on laminar flow through a plane channel. Applied Scientific Research, Vol. 46, pp. 271-280, 1989.

J. Liandrat, E. Coustols, L. Djenidi, F. Anselmet, X. de Saint Victor, F. Fioc, R. F. Fulachier. Effect of Riblets on either Fully Developed Boundary Layers or Internal Flows in Laminar Regime. In Turbulence Control by Passive Means. (Ed. E. Coustols; Kluwer Academic Publishers). p.141. 1990.

R. Michel, C. Quemard and R. Durant Application d'un schéma de longueur de mélange a l'étude des couches limites d'équilibre. Note interne ONERA No. 154, (1969).

J. F. Nash and V. C. Patel. Three-dimensional Turbulent Boundary Layers. SBC Technical Books, 1972.

X. de Saint Victor: Resolution des equations de N.S. appliqué à l'étude de l'ecoulement laminaire dans des riblets. Rep. Tech. OA No. 1815025 AYD, ONERA-CERT (1987), pp. 1-40.

G. E. Schneider and M. Zedan: A Modified Strongly Implicit Procedure for the Numerical Solution of Field Problems. Num. Heat Transfer 4, (1981) pp.1-19.

A. S. W. Thomas: Aircraft drag reduction technology. A summary. AGARD Rep. 723 (1985), pp. 1.1 - 1.20.

E. R. Van Driest. On Turbulent Flow Near a Wall. J. Atms Sci., Vol. 23, 1955.

Various authors: Proceedings Int. Conf. on Turbulent Drag Reduction by Passive Means, Royal Aeronautical Society, Volumes 1 and 2, 1987.

Various authors: Proceeding Drag Reduction 89 (Ed. R. H. J. Sellin and R. T. Moses, Ellis-Horwood, 1989.

P. Vukoslancevic, J. M. Wallace and J. L. Balint. On the Mechanism of Viscous Drag Reduction Using Streamwise Aligned Riblets: a review with new results. Proc. R. Ae.S. Int. Conf. on Turbulent Drag Reduction by Passive Means; The Royal Aeronautical Society Volume 2 p, 1987.

M. S. Walsh, Direct Characteristics of V-groove and Transverse Curvature Riblets. Viscous drag reduction - Progress in Astronautics and Aeronautics. Vol. 72 - Gary R., Hough, editor, 1980.

S. P. Wilkinson, and B. S. Lazos. Turbulent Viscous Drag Reduction with Twin-Element Riblets. AIAA J. 26, p.496, 1988.

S. P. Wilkinson, J. B. Anders, B. S. Lazos, D. M. Bushnell: Turbulent Drag Reduction at NASA Langley: Progress and Plans (Int. Conf. on Turbulent Drag Reduction by Passive Means.) The Royal Aeronautical Society, 1987.

Coherent structures over a smooth and a triangular riblet drag reducing surface

A. D. SCHWARZ-VAN MANEN, R. HOOGSTEEN, J. C. STOUTHART, K. KRISHNA PRASAD and F. T. M. NIEUWSTADT*

Eindhoven University of Technology
Eindhoven, the Netherlands

* *Delft University of Technology*
Delft, the Netherlands

K.-S. Choi (ed.), Recent Developments in Turbulence Management, 93–112.

Abstract

This paper reports measurements on the behavior of ejection- and sweep-like motions of fluid in the near–wall region of a turbulent boundary layer in a water channel. The measurements were performed over a smooth as well as a drag–reducing grooved surface. The results of the signal analysis are shown to be consistent with the flow visualization studies of Blokland and Krishna Prasad (1984, 1986) and the experimental work of Choi (1989). It is further shown that the grooved surface manifests a modest level of manipulating capability over the specific motions associated with ejections and sweeps.

1. Introduction

Drag reduction by microgrooves in the streamwise direction has been confirmed by many measurements in the past ten years. However, the mechanism responsible for this behavior is still unclear. The drag reduction by passive means so far is quite small, about 7%. An understanding of the mechanism of drag reduction can perhaps lead to a development of more effective drag reducing surfaces. The main problem is that there is still no accepted theory that explain the observed behavior of turbulent flows over smooth surfaces. Thus it is mandatory in any investigation on drag reduction to include measurements on the turbulent flow over a smooth surface.

Various heuristic arguments have been proposed to explain the mechanism of drag reduction by means of longitudinal micro grooves. In this work we restrict our attention to one argument based on the hypothesis that the bursting process, which is assumed to contribute to a significant proportion of turbulent production, near the grooved surfaces is less vigorous than near a smooth surface.

The present work is a continuation of an earlier investigation (Schwarz–van Manen *et al.* 1989). Fresh experiments were carried out with longer data series at higher sampling frequencies. In addition the experiments covered more heights than earlier and included the sweeps apart from ejections. Several statistical properties of the different quantities that are used to describe ejections and sweeps are examined. In addition we present a few properties of the conditional averaged signatures obtained from the fluctuating signals. Finally we include some comparisons with earlier studies.

2. Experimental set–up

All the measurements were conducted in a fully developed turbulent boundary layer in a two–dimensional low–speed water channel. The measurement section of the water channel is 0.3 m wide, 0.3 m high and has a length of 7 m (fig.1). The main stream velocity can be adjusted between 0 and 0.4 m/s . The pressure gradient created by the side wall boundary layers was small as inferred from the LDA measurements. All the measurements were performed above a surface, mounted 0.158 m below the water surface (fig.1b). The upstream surface in the channel is made of fixed glass with a sharp leading edge. A tripping wire is placed at 0.6 m downstream of the leading edge of the glass surface. The replaceable test surfaces are of Erthaliet.

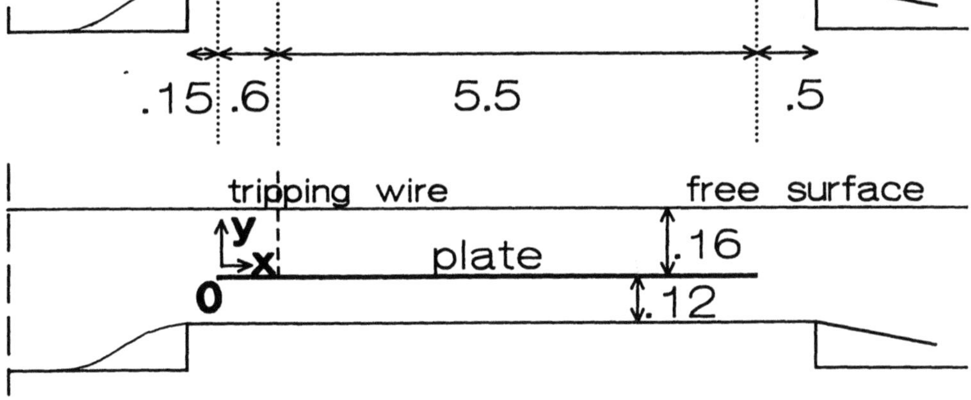

Fig.1. Waterchannel with the dimensions in meters.

δ	$9.6 \cdot 10^{-2}$ m	l^+	$11.2 \cdot 10^{-5}$ m
Re_x	$8.0 \cdot 10^5$	Re_θ	1760
t^+	$12.5 \cdot 10^{-3}$ s	U_0	$20.0 \cdot 10^{-2}$ m/s
u^*	$8.9 \cdot 10^{-3}$ m/s		
sampling frequency : 512 Hz		duration data series : 900 s.	
y^+	7, 10, 15, 22, 25, 27, 30, 33, 35, 50, 55, 60, 65, 70, 100, 200, 300		

Table 1. The Experimental conditions

The measurement system is described in detail by Pulles (1988). The primary element of this system is the two–component Laser Doppler Anemometer, which samples alternately the $(\boldsymbol{u} + \boldsymbol{v})$ component and the $(\boldsymbol{u} - \boldsymbol{v})$ component, where $\boldsymbol{u}$ is the time dependent streamwise velocity and $\boldsymbol{v}$ the time dependent velocity normal to the surface. The measurements were performed over two surfaces, a smooth surface and a longitudinal grooved surface (fig.2), where $y^+ = 0$ was choosen at the top of the grooves. In table 1 the experimental conditions of our measurements are given.

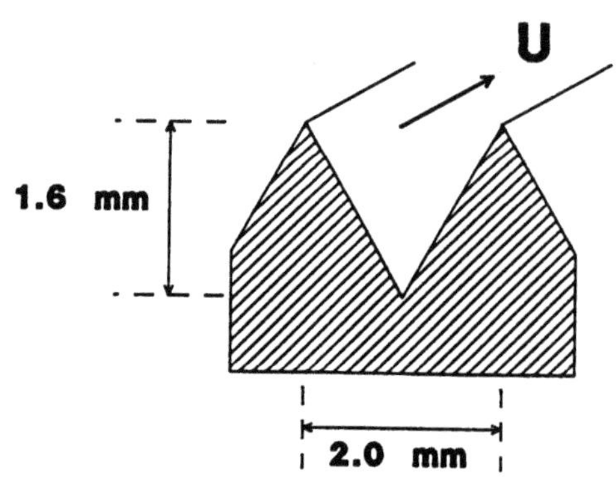

Fig.2. The geometry of the grooved surface.

3. Data analysis method

The velocity components, $(\boldsymbol{u} + \boldsymbol{v})$ and $(\boldsymbol{u} - \boldsymbol{v})$, are alternately sampled with a time interval $\Delta t = 0.15\ t^+$. The $\boldsymbol{u}$ and $\boldsymbol{v}$ components of the velocity field are obtained by adding and subtracting respectively the successive samples and placing the results at the middle of the time interval; this is allowed because the time interval between two successive samples is small. The velocities are decomposed into a time –fluctuating component (u or v) and a time – average component (U or V). The time fluctuating components were analyzed by the quadrant technique. This technique was first introduced by Lu and Willmarth (1973) and was further developed by Comte–Bellot *et al.* (1978). Their contribution to the technique was to introduce three threshold levels (fig.3), which can all be calculated from the data–set, to detect the intermittent events. The hole size (a concept introduced by Lu and Willmarth) is calculated as the average value of the measurements in the particular quadrant that is being investigated:

$$H_{\mathrm{q}} = \frac{\Sigma\ (uv)_{\mathrm{q}}}{N_{\mathrm{q}}\quad u_{\mathrm{rms}}\quad v_{\mathrm{rms}}} \tag{1}$$

where H_{q} is the average value of all the samples in quadrant q , N_{q} is the number of samples in that quadrant and u_{rms} and v_{rms} are the root mean squares of u and v respectively.

The two other threshold levels are based on the fact that u and v have to be simultaneously relatively large, to eliminate signals that qualify according to $|(uv)_{\mathrm{q}}| > |H_{\mathrm{q}}|$ but obey the inequalities $|u/u_{\mathrm{rms}}| >> |v/v_{\mathrm{rms}}|$ or $|v/v_{\mathrm{rms}}| >> |u/u_{\mathrm{rms}}|$. These threshold levels h_{u} and h_{v} of the specific quadrant can be

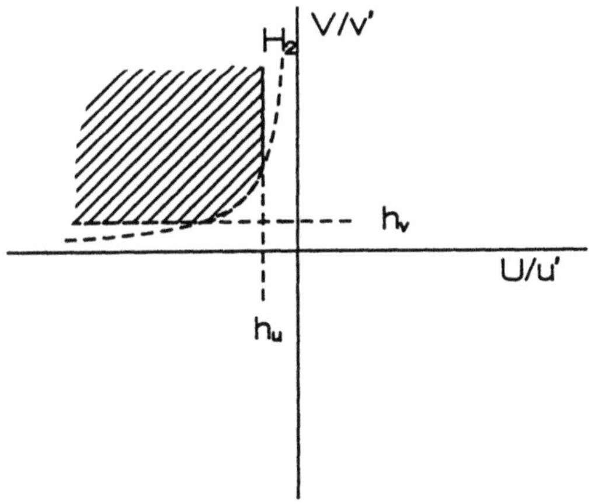

Fig.3 The quadrant analysis technique shown for the second quadrant.

determined after H_q is obtained for that quadrant. This method was suggested by Comte–Bellot for all four quadrants.

Bogard and Tiederman (1986) improved this theory for the second quadrant. They performed detailed single point measurements simultaneously with flow visualization studies. To analyze the point measurements they used several existing detection techniques to determine the burst frequency. They compared the frequencies so obtained with the frequency obtained from the flow visualization study, and for every detection technique they calculated the probability of detecting the "wrong" burst and the probability of detecting the "right" burst. The technique that came out best was the quadrant analysis technique when employed with an additional constraint. This constraint was based on the fact that the flow visualization studies indicated that the breakup of a streak involved a single ejection or multiple ejections, that are closely grouped together. This feature was also observed by Corino and Brodkey (1969).

The analysis identified a critical time, τ_{max}, which is defined as follows. When the period between two successive ejections is smaller than τ_{max} these ejections are taken to belong to the same group. On the other hand when this period is greater

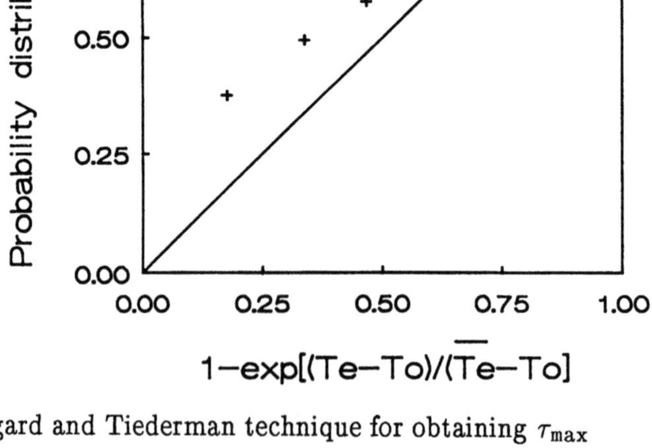

Fig. 4. The Bogard and Tiederman technique for obtaining τ_{max}

then τ_{max} the two successive ejections are deemed to belong to different groups. Figure 4 shows the method of obtaining τ_{max} . The probability distribution of the period of the intermittent events (T_s) is compared with the exponential distribution of T_s. The continuous line represents a perfect exponential distribution; the markers represent the T_s distribution of a single point measurement. τ_{max} is given by the point at which the two distributions intersect.
Other methods of obtaining τ_{max} were tried (Hoogsteen 1990). The results indicated that the above mentioned method produced the most consistent results; this was also confirmed by Luchik and Tiederman (1987). Bogard and Tiederman used this technique only for the second quadrant; in this study we have used this technique for the fourth quadrant as well.

4. Results

The primary purpose of the present investigation was to delineate differences in coherent structures, if any, between smooth and grooved surfaces. Generally stated no differences can de observed for $y^+ > 100$. Similar results have been noticed by Grass (1971), Pulles (1988) and Choi (1989). Most striking differences were observed at the top of the viscous layer and in the buffer layer. The differences appeared in both the intensity and frequency of occurrence of the coherent structures. In the following sections we will discuss the effect of the threshold level, H_q (which really determines the strength of intermittent events), the period between the occurrence of ejections (second quadrant events) and sweeps (fourth quadrant events) and the conditional averages of these events.
From now on the bold lines in the figures will represent the results of the grooved surface and the normal light lines will represent the smooth surface, unless mentioned otherwise.

4.1. Threshold levels H_q

The quadrant analysis assigns specific types of physical events to the points located in the four quadrants of the $u - v$ plane. Thus points in the first quadrant represent the outward interactions, the second quadrant the ejections, the third quadrant the inward interactions and finally the fourth quadrant the sweeps. The threshold level, H_q, whose primary purpose is to separate noise from "real" events, varies for each quadrant and can also be expected to be a function of height over the surface.

Comte–Bellot identified the following values for H_q:

$$H_1 \simeq 0.4\ ;\quad H_2 \simeq -1.0\ ;\quad H_3 \simeq 0.4 \text{ and } H_4 \simeq -0.7$$

The essential differences between our experiments and theirs are: their experiments were in a fully developed turbulent pipe flow (air) and the present experiments were in a fully developed turbulent boundary layer (water); their experiments were for $260 \lesssim y^+ \lesssim 3000$, and our experiments were for $7 \lesssim y^+ \lesssim 300$; Re_θ of Comte–Bellot's experiments were 6500 and 12000, while the Re_θ of our experiments was 1760. The other experimental conditions of their measurements are presented in table 2.

D	$10.0 \cdot 10^{-2}$ m		
Re	68000	Re	135000
Re_θ	6500	Re_θ	12000
U_0	10.2 m/s	U_0	20.3 m/s
y^+	260, 650, 1040, 1820, 2600, 3000	y^+	350, 1400

Table 2. The experimental conditions of Comte–Bellot *et al.* 1986

We now turn to the results in fig.5. Broadly speaking we can distinguish two regions: $y^+ > 30$ and $y^+ < 30$. In the first region H_q for the I and III quadrant is virtually independent of height and type of surface. Similar behavior is observed for II and IV quadrants. For both surfaces $|H_4| < |H_2|$, thus indicating that the intensity of the motions represented by the IV quadrant are less than those of the II quadrant. It is also gratifying to note that the values of H_q found in this investigation are the same as the ones reported by Comte–Bellot et al., showing the reliability of the methodology.

In the second region ($y^+ < 30$) the H_q values show significant differences between the two surfaces. For the smooth surface H_1 becomes very much larger with decreasing height than $|H_3|$. This is also observed for the grooved surface, though to a much smaller extent. Thus in a boundary layer, close to the surface, the outward interactions are much stronger than the inward ones. More importantly the I quadrant result for the smooth surface is much larger than that for the grooved surface, suggesting that the latter inhibits the outward interactions. The $|H_2|$ value above the smooth surface decreases with decreasing height, until it is even less then

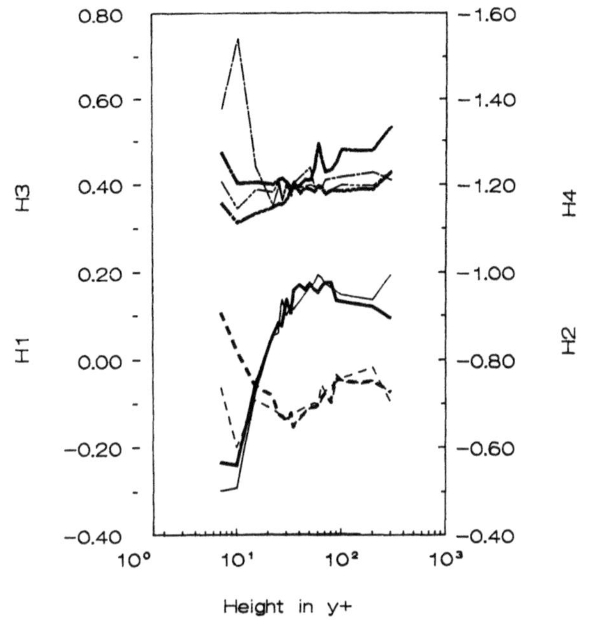

Fig. 5. The threshold values H_q, H_1 : — — , H_2 : ——— , H_3 : — — — — , H_4 : — — and bold represents the grooved and light the smooth surfaces.

$|H_4|$. This is also seen for the grooved surface, but is much less pronounced. This means that just above the viscous layer the sweeps are stronger than the ejections. The fact that H_1 increases and $|H_2|$ decreases with decreasing height indicates that there might be a correlation between the ejections and the outward interactions; after all both are outward movements of fluid. Comparing the results of the grooved surface with the smooth surface, the grooved surface somehow seems to inhibit the occurrence of outward motions.

4.2. Periods and durations of sweeps and ejections

In this section we will identify two kinds of periods. The first period will be called the single event period (SEP) and is the average period of every sweep or ejection, which can be detected using the technique of Comte–Bellot et al. The second period is what we call the group event period (GEP) which is the average period of ejections or sweeps after applying the method of Bogard and Tiederman. The groups of sweeps or ejections in general will involve one or more sweeps or ejections.

Figure 6 shows the two periods for both ejections and sweeps for a smooth surface. Two results emerge from this plot. Grouping of events drastically increases the periods by almost a factor of 5. Secondly, the dependence of the period of the single events on height for both ejections and sweeps is reminiscent to the behavior of the $|H_2|$ and $|H_4|$ respectively in figure 5. However the grouped events show a much stronger dependence on height. The GEP increases from about $80\ t^+$ *to about* $270\ t^+$ for ejections as y^+ goes from 10 to 100.

For $y^+ < 60$ the sweeps have a larger period than the ejections. This is also observed for the SEP but only up to $y^+ \simeq 20$. This coupled with the results of figure 5 shows that the sweeps are low in intensity and low in frequency. For the grooved surface the same conclusion can be drawn, with the only difference that for the GEP the sweeps become larger for $y^+ < 27$.

The fact that the period increases with height makes us believe that the spatial extent of these events increases. This is consistent with the flow visualization observations of Smith and Metzler (1983) and Blokland and Krishna Prasad (1986)

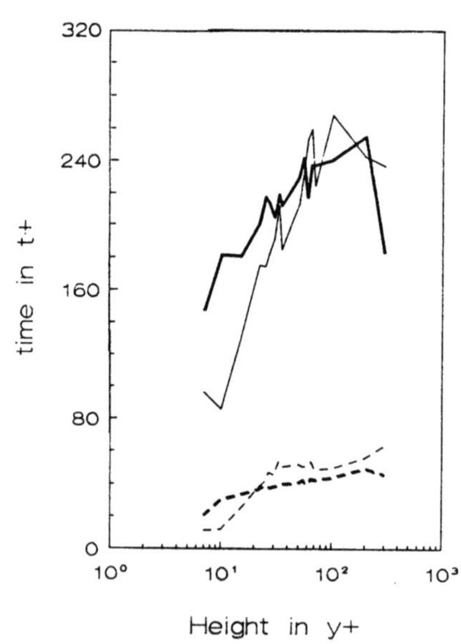

Fig.6. The period for ejections and sweeps for the smooth surface.
Single events : sweeps — —, ejections :— — — —. Grouped events: sweeps : — — —, ejections : ———.

who report an increase in the spacing and width of low speed streaks with increasing y^+.This is also recently confirmed by Choi (1989). These results conflict with the oft quoted attribute of ejections and sweeps "...*that ejected fluid* (presumably from low speed streaks) *is replaced by high momentum fluid from the edge of the wall–layer to maintain continuity...*" (Beljaars et al. 1981). The present results on the other hand suggest that very close to the wall the ejections mark the low speed streaks and the sweeps mark high speed regions. This conjecture is also supported by the work of Blokland and Krishna Prasad (1990), who describe a persistent wave–triad model for wall turbulence. The model comprises of three interactive waves; a streak wave that corresponds to low speed streaks, a symmetric wave (centered with the streak wave) and an anti – symmetric wave (whose centre is displaced half a streak wave length in the spanwise direction and has a wavelength twice as large as that of the symmetric wave). The symmetric wave is associated with the ejections and the anti – symmetric wave with the sweeps.

Figures 7 and 8 compare ejection and sweep periods respectively for smooth and grooved surfaces. The difference for single events between the two surfaces is not so easy to see on this scale. However the grouped events show some important differences for the two surfaces. The average GEP between ejections increases over the grooved surface in the region $27 < y^+ < 60$ when compared with that over the smooth surface. In other words the ejection activity is less frequent over the grooved surface. On the other hand the GEP between sweeps over the grooved surface decreases compared with that over the smooth surface, but in the region $y^+ < 25$, in contrast to ejections, there are more sweeps over the grooved surface.

Careful examination of the lower plots in the two figures which represent the time interval between single events (without the application of Bogard and Tiederman technique) shows similar trends. This confirms that the phenomena we are discerning are genuine in spite of the noisy nature of the data for grouped events. The picture that emerges from this analysis is that the grooves act as a trap for high speed fluid coming from the outer region.

A possible explanation for the previous observations lies in the association mentioned earlier of ejections with low speed streaks and sweeps with high speed regions. The riblets are presumed to inhibit the lateral movement of low speed streaks producing a stabilizing effect thus causing fewer ejections. This phenomena makes the high speed regions more pronounced. This is presumed to make the

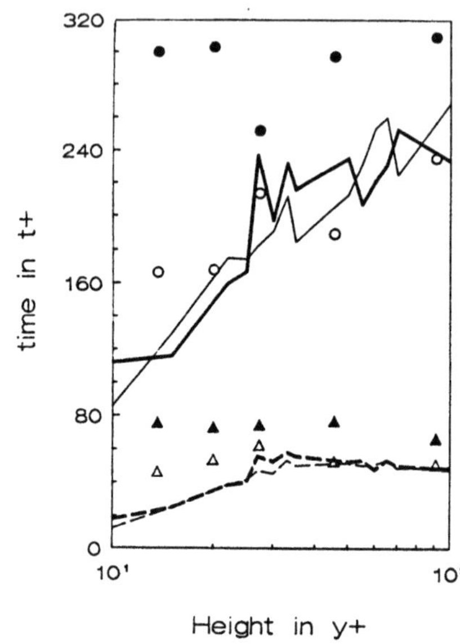

Fig. 7. The ejection period. Single events : Δ, – – –. Grouped events : O, ——. Bold and closed: grooved surface. Light and open: smooth surface. The markers are results from Schwarz–van Manen *et al.* 1989.

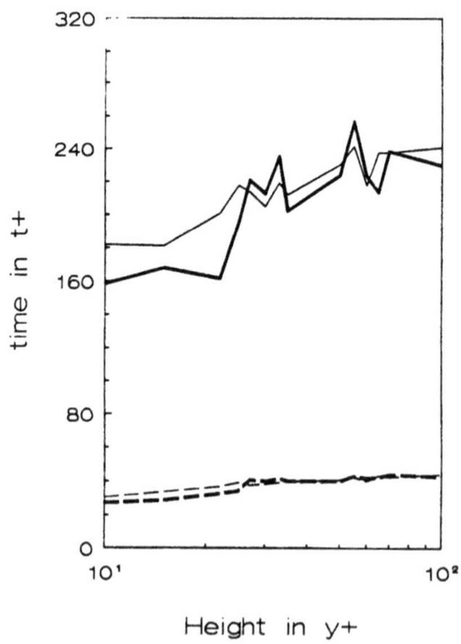

Fig. 8. The sweep periods. Single events : – – –. Grouped events : ——. Bold is the grooved and light the smooth surface.

sweep–like motion come almost down to $y^+ = 0$ and in particular is more frequent. As was pointed out earlier the present experiments were undertaken to examine in greater detail the anomalous behavior of ejection period – height curve at $y^+ \simeq 27$ (Schwarz–van Manen et al. 1989). These results are reproduced as markers in fig.7. We attribute the discrepancy between the two sets of measurements to the rather low sampling frequency (64 Hz as against 512 Hz) and fewer heights investigated in the earlier work. However it is important to note that the peaks and dips are not spurious since they are observed for both smooth and grooved surfaces, for sweeps (fig.8), for threshold levels (fig.5) and the signals before the application of Bogard and Tiederman technique.

The height $y^+ = 60$ corresponds to the point at which Sreenivasan (1987) locates a "fat vortex sheet" to mimic the behavior of the turbulent boundary layer. Using arguments of inviscid stability theory, he suggests that the two–dimensional vortex structures get kinked on two distinct scales. These are said to lead to the formation of two distinct scales of Λ shaped eddies. The smaller of the two corresponds to the so–called hairpin eddies of Head and Bandyopadhyay (1981) and the legs of the larger scale corresponds to the so–called double roller structure. The most amplified mode of the smaller of the two scales is shown to be

$$\frac{\lambda_{zs}\ U^*}{\nu} = 1.7 \left[\frac{U^*\ \delta}{\nu} \right]^{1/2} \tag{2}$$

and for the larger scale

$$\frac{\lambda_{zl}}{h} = 17 \tag{3}$$

For the present experiments these wavelengths correspond to $t_s{}^+ \simeq 1$ and $t_l{}^+ \simeq 70$. In these estimations the wave lengths have been converted into times by assuming that the wave velocities are equal to local velocities. Comparing these estimates with the results in fig.7 we come to the conclusion that the structures detected in this work correspond to the larger scale structures of the two. In fact at $y^+ = 60$ the time scales obtained in this investigation before the application of Bogard and Tiederman technique is just about 15% lower. Considering the crudeness of the estimations, this should be construed as an excellent agreement.

However Sreenivasan's approach is in conflict with the present results on two counts. Firstly it fails to say anything about the happenings at $y^+ = 27$. Secondly the Bogard and Tiederman technique of identifying grouped events has a closer relationship with the low speed streaks as confirmed by flow visualization studies. This grouping produces t^+ values that are about 4 times larger. The arguments for the "critical layer" of Sreenivasan are based on a conceptual extension of the transitional critical layer. Such a procedure has been shown to be inconsistent with the phenomena in turbulent boundary layer according to Blokland and Krishna Prasad (1990). A detailed comparison between the present work and the latter theory should be available in the near future.

Figure 9 presents the ratio of the average duration to the average period of the events (ejections and sweeps) for both the smooth and grooved surfaces. The duration is the intercept of the detected event along the zero–line. In general this ratio is higher for sweeps than for the ejections confirming the conventional wisdom that ejections are more vigorous activities. For $y^+ > 20$, the ratio is constant and

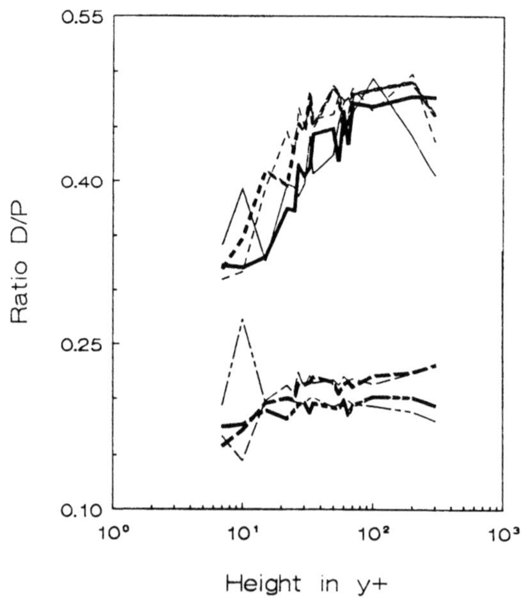

Fig. 9. The ratio of the duration to the period. Single events : sweeps — —, ejections — — — —. Grouped events : sweeps — — — , ejections ———. Bold is the grooved and light the smooth surface.

not a function of height for the detected events according to the criteria used in this investigation. The grouped events on the other hand show an increasing trend up to about $y^+ \simeq 70$. Finally the grouped ejections for the grooved surface in $27 < y^+ < 60$ show somewhat larger ratio indicating that the ejections are weakened somewhat. The opposite situation is observed for the sweeps for $y^+ < 20$.

4.3. Conditionel averages

Conditional averages were calculated to determine the possible differences for the two surfaces. The conditional averages for the u, v and uv signals are all triggered at the position of the highest $|uv|$ peak of a group of ejections or sweeps. This choice was made since the conditional averages obtained thus show similarities with those obtained by Nakagawa and Nezu (1981) with their so called Euler correlation formula, as was shown in Schwarz–van Manen et al. (1989).

At first the averages were plotted for a time period between $-60 \lesssim t^+ \lesssim 60$; hardly any difference between the two surfaces were seen. One thing was noticeable however: the peak of the u and v signal for the sweeps did not coincide with the trigger point. It is shifted slightly to the left. When the conditional averages were

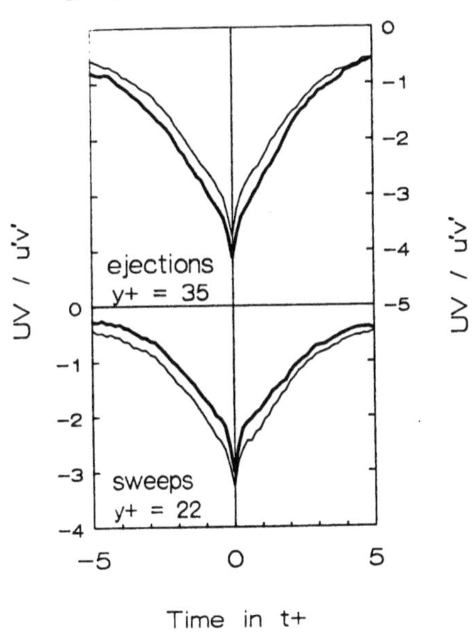

Fig. 10. The conditional averages for the uv–signal. Bold is the grooved surface and light the smooth surface.

plotted for $-5 \lesssim t^+ \lesssim 5$ (fig. 10) and the results compared with figures 7 and 8, we notice that when the period for the grooved surface is larger than that for the smooth surface, the abosolute height of the conditional average for the grooved surfaces is larger than for the smooth surface. So, the longer the period the more violent the event.

The figures 11 and 12 represent the width of the conditional averages at half the peak height versus the height (y^+) for the smooth and grooved surfaces respectively. The clearest result as far as the differences between the smooth and the grooved surface are concerned, is for $y^+ < 20$. The u signal of the conditional average for ejection above the grooved surface is about 20% narrower. The width of the v and uv signal of these averages are also narrower, but this is much less pronounced. The width of the u signals decreases between $0 < y^+ < 70$ and then increases again. Whereas the width of the v and uv signals, for both ejections and sweeps, increases between $0 < y^+ < 25$, then stabilizes up to $y^+ \simeq 100$ and then increases again. The behavior of the uv signals, for both surfaces and both ejections and sweeps, coincide with the respective behavior of the SEP's (fig. 6,7 and 8).

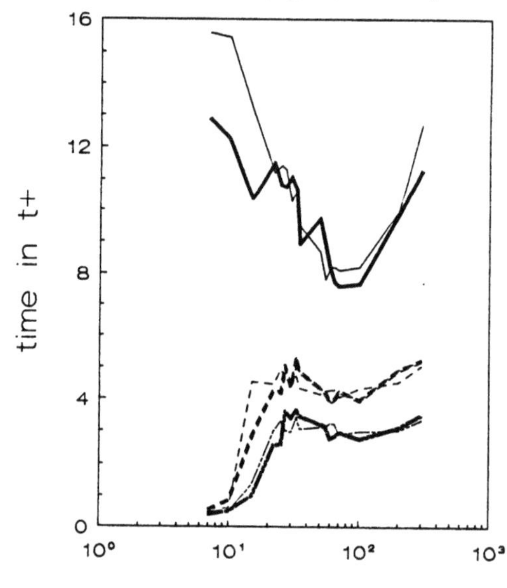

Fig. 11. The width of the conditional averages at half the height of thew peak for ejections. u ——, v – – – and uv — — — —. Bold is the grooved and light the smooth surface.

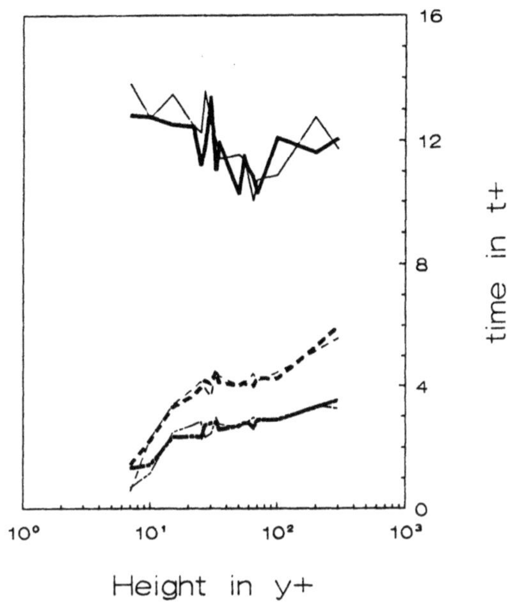

Fig. 12. The width of the conditional averages for the sweeps. u ——, v – – – and uv — – – —. Bold is the grooved and light the smooth surface.

The triggering is placed at the highest peak for a grouped event. Since the position of the highest peak in a group is arbitrary and the time between successive peaks in a group can vary randomly, the averaging procedure tends to smother out all the other peaks except the highest one. Thus while the period can be measured well, it does not guarantee a truthful representation for the signature of an ejection or a sweep.

4.4. Distribution of peak uv – values

Conditional averages are known to smother many a detail. With a view to obtaining more information about the behavior of individual events, histograms of highest uv–peaks in every group of ejections and sweeps were prepared. Two of these histograms for $y^+ = 35$ for ejections and $y^+ = 22$ for sweeps are shown in figure 13. The histogram for ejections over the smooth surface is steeper than that over the grooved surface. The reverse situation prevails for the sweeps. Another remarkable feature is that the number of ejections in the interval [–5,0] is relatively constant around 4%, while for sweeps this varies from 2% to 9%. This reconfirms the earlier observations that sweeps are generally relatively less intense motions than ejections.

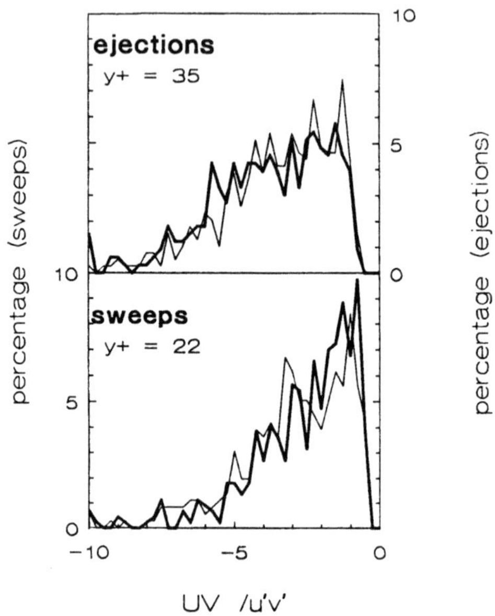

Fig. 13. The distribution of the *uv*–values of the highest peak in a grouped event. Bold is the grooved and light the smooth surface.

5. Conclusions

Two types of coherent motions over a smooth and a grooved surface have been studied in this investigation. Significant increases have been observed in the ejection periods in the region $20 < y^+ < 70$ while there are significant decreases in the sweep periods for $y^+ < 20$. In contrast the duration–period ratio for the grooved surface shows a larger value for both ejections and sweeps in these respective regions. Taken together these results imply that the grooves inhibit ejections while attracting fluid motions towards the wall region. In addition the "intensity" of these motions are significantly reduced. If we grant the assumption that ejections are firmly connected with low speed streaks, the increasing period of ejections with height implies that there are fewer streaks as we move away from the wall. This is very clear in the flow visualization studies of Blokland and Krishna Prasad (1984). This phenomenon is observed both for smooth and grooved surfaces.

A major contribution of the present study lies in the appearance of two kinks around $y^+ \simeq 27$ and $y^+ \simeq 60$ in the period–height curves. The question is whether they represent the bottom and top of a single structure. The present measurements are not sufficient to resolve this question. In this connection Sreenivasan's proposal while elegant misses some crucial points. It seems more appropriate to consider a multiple wave system such as the one proposed by Blokland and Krishna Prasad (1990). If we accept that the structures are three–dimensional, further progress seems to be constrained by the absence of any systematic measurements of spanwise velocities. More importantly if we assume these three–dimensional structures to be the hairpin type eddies, they are of such small scale that one is unable to detect them unless much higher sampling rates are employed. It is quite probable that the grooves have the strongest influence on these small scale three–dimensional structures.

Finally the conditional averaging procedure still remains a bottleneck. With the procedures employed in the present investigation, the signature of a grouped event remains shrouded in mystery. Better classification of the structures seem essential for this purpose. This will inevitably lead to much larger data series than have been employed so far.

6. Acknowledgement

The research in this article was partially supported by the Netherlands Foundation for Technical Sciences (STW) as a part of the of the Foundation for Fundamental Research on Matter (FOM).

References

Beljaars, A.C.M.: Krishna Prasad, K.: Vries, D.A. de 1981: A structural model for turbulent exchange in boundary layers. *J. Fluid Mech.* **112**, 33.

Blokland, R.: Krishna Prasad, K. 1984: Some visualization studies on turbulent boundary layers using multiwire hydrogen bubble generation. *Proc. 8th Symposium of Turbulence*, Missouri Rolla, USA.

Blokland, R.: Krishna Prasad, K. 1986: Some visualization studies on turbulent boundary layers using multiwire hydrogen bubble generation. *Agard–CPP–*413.

Blokland, R.: Krishna Prasad, K. 1990: A persistent wave triad model for wall turbulence. *To be published.*

Bogard, D.G.: Tiederman, W.G. 1986: Burst detection with single–point velocity measurements. *J. Fluid Mech.* **162**, 389.

Choi, K.S. 1989: Near–wall structure of a turbulent boundary layer with riblets. *J. Fluid Mech.* **208**, 417.

Comte–Bellot, G.: Sabot, J.: Saleh, I. 1978: Detection of intermittent events maintaining Reynolds stress. *Proc. of the Dynamic Flow Conference*, Marseille, France. 213.

Corino, E.K.: Brodkey, R.S. 1969: A visual study of turbulent shear flow. *J. Fluid Mech.* **37**, 1.

Grass, A.J. 1971: Structural features of turbulent flow over smooth and rough boundaries. *J. Fluid Mech.* **50**, 233.

Head, M.R.: Bandyopadhyay, P. 1981: New aspects of turbulent boundary–layer structure. *J. Fluid Mech.* **107**, 297.

Hoogsteen, R. 1990: Quadrant analysis in a turbulent boundary layer, *Dissertation, Eindhoven University of Technology*, The Netherlands.

Lu, S.S.: Willmarth, W.W. 1973: Measurements of the structure of the Reynolds stress in a turbulent boundary layer. *J. Fluid Mech.* **60**, 481.

Luchik, T.S.: Tiederman, W.G. 1987: Time scale and structure of ejections and bursts in turbulent channel flows. *J. Fluid Mech.* **174**, 529

Nakagawa, H.: Nezu,I. 1981: Structure of space – time correlations of bursting phenomena in an open channel flow. *J. Fluid Mech.* **104**, 1.

Pulles, C.J.A. 1988: Drag reduction of turbulent boundary layers by means of grooved surfaces. *Ph. D. Dissertation. Eindhoven University of Technology*, The Netherlands.

Schwarz–van Manen, A.D.: Thijssen, J.H.H.: Nieuwvelt, C.: Krishna Prasad, K. 1989: The bursting process over drag reducing grooved surfaces. *Second IUTAM symposium on structure of turbulence and drag reduction*, Zurich, Switzerland.

Smith, C.R.: Metzler, S.P. 1983: The characteristics of low–speed streaks in the near–wall region of a turbulent boundary layer. *J. Fluid Mech.* **129**, 27.

Sreenivasan, K.R. 1987: A unified view of the origin and morphology of the turbulent boundary layer structure. *Turbulence Management and Relaminarisation IUTAM conference*, Bangalore, India.

Some further experiments on riblet surfaces in a towing tank

F. T. M. NIEUWSTADT, W. WOLTHERS, H. LEIJDENS,
K. KRISHNA PRASAD* and A. D. SCHWARZ-VAN MANEN*

Delft University of Technology
Delft, the Netherlands

* *Eindhoven University of Technology*
Eindhoven, the Netherlands

K.-S. Choi (ed.), Recent Developments in Turbulence Management, 113–123.

Abstract

In this paper we shall discuss some experiments on riblet surfaces taken in a towing tank. One of the aims of this study is to establish that these results obtained in the towing tank are consistent with the more usual observations gathered in windtunnels. In addition we consider some effects which might play an important role when riblets are applied in practical circumstances. These are: the effect of surface roughness, the influence of the flow angle and the effect of partial surface coverage by riblets.

1 Introduction

Drag reduction by micro grooves is by now firmly established. Although the exact mechanism of this type of drag reduction is not known, experiments seem all to agree that about 5% of drag reduction can be obtained by applying micro grooves to a flat surface. For some recent reviews on different aspects of this technique we refer to Savill (1989) and Walsh (1990).

From these reviews follows that most of the experiments on the drag reducing effect of riblets were done in windtunnels. Exceptions are the studies by Choi et al. (1988) on a model sail yacht and the experiments by Nieuwstadt et al. (1989) in a towing tank. The purpose of this paper is to extend the latter study by reporting on additional measurements with a flat plate in a towing tank.

The background for our research is the possible application of riblet surfaces to ships. Such application is in principle possible and it would undoubtedly lead to economic benefit. Nevertheless, a lot of problems have to be solved before this technique can be used in practice.

Our goal in this study is twofold. First, we want to clear up some unrealistic discrepancies, which we have found in our previous investigation (Nieuwstadt et. al, 1989). Next we will consider some problems which will certainly occur when riblets are applied in practice. What is the effect of a surface roughness? How does the drag reducing capacity of riblets depend on the flow angle? What is the influence of a partial cover of the wall by riblets on the overall drag reduction?

The organization of this paper is as follows. In the next chapter we will discuss the experimental facilities and the procedures. Then we turn to the measuring program and its results.

2 Experimental facilities and procedures

The experiments were done in the small towing tank of the section shiphydromechanics of the Department of Mechanical Engineering and Marine Technology. The dimensions of this tank are: length, $85\,m$, width, $2.75\,m$ and depth $1.2\,m$. Towing speed can vary between 1 and $3\,ms^{-1}$. At the highest speed the measuring period is about $5\,s$.

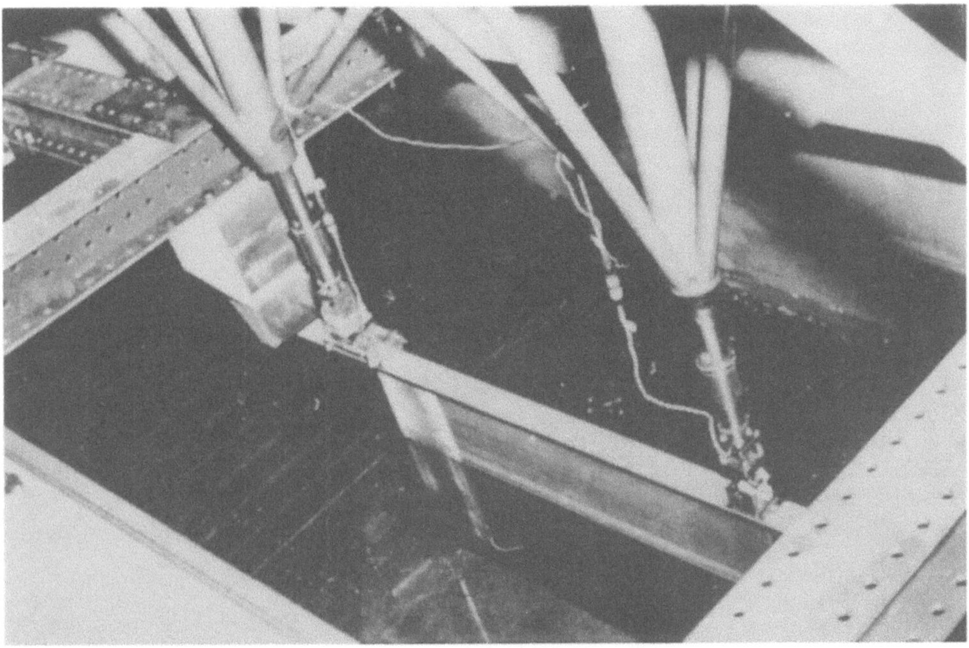

Figure 1: *The mounting of the flat plate to the carriage of the towing tank.*

The flat plate which we used in our experiments, was fastened to the towing tank carriage by means of two vertical rods. The rods are connected to the plate by hinges. This experimental set-up is shown in figure 1. In the hinges we fitted dynamometers. These instruments, which operate on the principle of a strain gauge, are used to measure the towing force on the plate.

All our experiments were carried out with a flat plate of the following dimensions: length $1\,m$, width $0.6\,m$ and thickness $0.01\,m$. The plate is slightly tapered over a distance of $0.185\,m$ toward the leading and trailing edge, which are rounded with a radius of $2\,mm$. The plate is towed in a vertical position and is immersed in the water over a depth of $0.4\,m$.

For the riblet material we use 3-M foil with a triangular riblet shape. The height of the triangle is $h = 0.11\,mm$ and the width between two tops of the triangle is $s = 0.122\,mm$.

3 Results

3.1 Flat plate

In our previous experiments described by Nieuwstadt et al. (1989) we found some results which did not agree with the usual drag measurements on flat plates. Therefore, our first goal was to clear up this inconsistency.

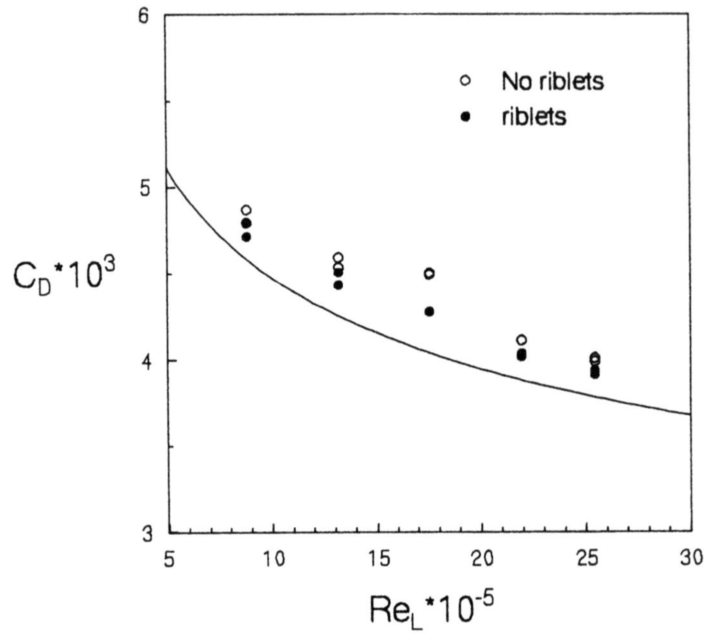

Figure 2: *The drag coefficient of a flat plate with and without riblets as a function of the Reynolds number. Solid line is the empirical relationship proposed by Prandtl and Schlichting*

To this end we carried out flat plate experiments with and without riblets at five towing speeds: 1.0, 1.5, 2.0 2.5 and 2.9 ms^{-1}. At each speed two measuring runs were done.

The observed towing force, D, on the flat plate was transformed into a drag coefficient by

$$c_D = \frac{D}{\frac{1}{2}\rho U^2 S} \tag{1}$$

where U is the towing speed. S is the wetted surface calculated as: $S = 2 \times L \times H$, where $L = 1.0\,m$ is the length of the plate and $H = 0.4\,m$ the submerged depth. This drag coefficient is plotted in figure 2 as a function of the Reynolds number $Re_L = UL/\nu$. The value for the kinematic viscosity, ν, has been corrected for temperature, which during our experiments varied between 15° C and 17° C. At 15° C the value of ν becomes $1.140\,10^{-6}\,m^2s^{-1}$ and at 17° C $1.084\,10^{-6}\,m^2s^{-1}$.

In figure 2 we also show an empirical relationship proposed by Prandtl and Schlichting (Schlichting, 1979) for the friction drag, c_f, of a flat smooth plate

$$c_f = \frac{0.455}{(\log Re)^{2.58}}. \tag{2}$$

We see that the results found from this expression are slightly lower than our mea-

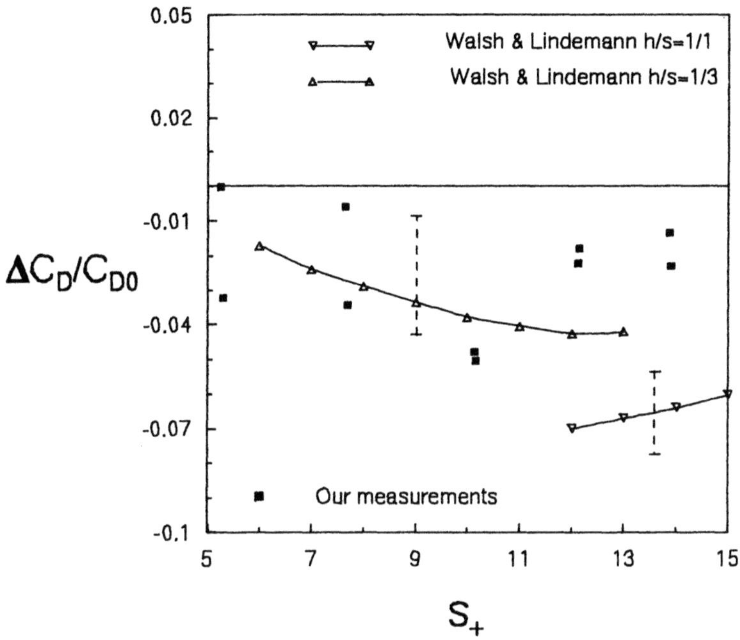

Figure 3: *The relative change in the drag coefficient of a flat plate as a result of riblets as a function of the dimensionless groove width.*

surements. An explanation of this additional drag could be a contribution due to wave drag. During the experiments we indeed observed at the leading edge of the plate a bow wave with a height, h_w, of approximately $2\,cm$. Let us estimate this additional resistance by a hydrostatic pressure difference between the leading and trailing edge. We then find

$$c_W = \frac{\frac{1}{2}\rho g h_w^2 \delta + \rho g h_w H \delta}{\frac{1}{2}\rho U^2 S} \tag{3}$$

where g is the acceleration of gravity and $\delta = 0.4\,cm$ the thickness of the leading edge. From (3) follows $c_W = 0.1 - 0.8\,10^{-3}$. The difference in figure 2 between the observations and the curve (2) falls within this range.

Furthermore, we see in figure 2 that at $Re_L = 1.75\,10^6$, i.e. at $U = 2.0\,ms^{-1}$, a somewhat higher drag coefficient is measured in relation to the observations at other speeds. We experienced during the experiments a resonance of the towing tank carriage at this speed. Therefore, we should perhaps omit the observations at this speed.

Figure 2 also exhibits the influence of the riblets on the drag. This is more clearly shown in figure 3 where Δc_D is the difference in the drag coefficient between the plate with and without riblets and c_{D0} is the drag coefficient of the plate without riblets. The results have been plotted as a function of the dimensionless groove

Figure 4: *The influence of roughness elements on the flow in the boundary layer of the flat plate.*

width $s_+ = u_* s/\nu$, where the friction velocity u_* is defined as $u_* = (D/\rho S)^{1/2}$. Each value of s_+ shown in figure 3 corresponds with a towing speed. Remember that we have performed at each speed two runs, the results of which are shown separately in figure 3. We see that at the two smallest values of s_+ (i.e. at the lowest towing speeds, $U = 1.0$ and $1.5\, ms^{-1}$) the observations made at the two measuring runs differ considerably. This must be attributed to the poor reproducibility of our experiments at low towing speeds. At higher towing speeds the reproducibility is excellent and usually within 1%. Therefore, we shall restrict ourselves in the following as much as possible to the observations made at the higher towing speeds.

Our measurements in figure 3 indicate a drag reduction between 1% and 4%, which is consistent with the results found by Nieuwstadt et al. (1989). In the figure we have also plotted some data of Walsh and Lindemann (1984) to indicate that our results fall within the scatter of these other measurements which were performed in a windtunnel. Taking into account the rather crude experimental environment of a towing tank, we conclude in our experiments seem to be consistent with other observations of drag reduction by riblets.

3.2 Roughness elements

In our previous experiment (see Nieuwstadt et al. (1989)) we used a strip of carborundum roughness elements to trip the boundary layer. We found this to be not necessary in this experiment. However, we would like to estimate the influence of these roughness elements both on the total drag and on the drag reduction.

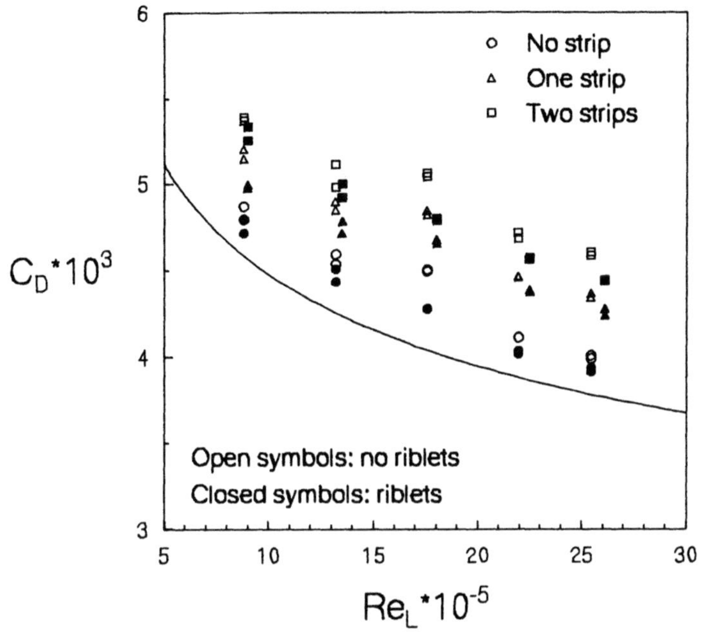

Figure 5: *The drag coefficient of a flat plate with roughness elements and with and without riblets as a function of the Reynolds number.*

To this end we have done experiments with one and two strips of carborundum grains attached near the leading edge of the plate as shown in figure 4. The figure shows also clearly that these roughness elements disturb the flow in the boundary layer considerably. Therefore, we may expect some influence on the drag.

The results of these experiments with roughness elements have been plotted in figure 5. It is quite clear that the roughness elements increase the drag substantially over the whole range of Re_L. As already well known the influence of surface roughness cannot be neglected. Furthermore, we find that the drag increase by the roughness elements is approximately constant as a function of the Reynolds number.

Another result which follows from figure 5 is that the effect of the riblets persists. In all cases the application of riblets leads to a drag reduction of about the same order of magnitude.

3.3 Flow angle

The riblets perform optimally when the flow is aligned in the direction of the grooves. However, in practical situations the flow direction is sometimes at a angle to the riblets. So it is necessary to estimate at which angle the riblets loose their effectiveness.

To investigate this we have carried out experiments with our flat plate on which the riblets were attached under an angle α (see figure 6). The results obtained with various values of α are given in figure 7.

First we note that the drag reduction at $\alpha = 12°$ seems somewhat larger than at $\alpha = 0°$. This counterintuitive result is due to the fact that the experiments at $\alpha = 0°$ and at $12°$ were done under different circumstances. The water temperature for $\alpha = 0°$ was $15° C$ and $17° C$ for the other values of α. This means that the representative Reynolds number for the experiment at $\alpha = 0°$ is slightly lower than for the other experiments.

The experiments at $U = 2.5$ and $2.9\, ms^{-1}$ show that the effect of drag reduction by the riblets changes to drag increase at an angle of $\alpha = 20\,$deg. This is consistent with experiments done in windtunnels.

The results at $U = 2.0\, ms^{-1}$ seem to differ. First of all the change in drag coefficient is much larger than at the other speeds. Secondly, the drag reduction persists for much larger values of α. We have already mentioned that we experienced a resonance of the towing tank carriage at this speed. Therefore, we believe the results at this speed to be suspect and they should be disregarded.

3.4 Partial covering by riblets

In the previous experiments, described here, we have covered the whole flat plate with riblets. However, it will be clear that in practical circumstances this will be quite impossible. In that case there will always be areas where riblets cannot be applied.

Therefore, we have done a separate investigation in which we covered only a fraction of the plate with riblets. The riblets were applied in strips of width ΔL_r with a pitch of L_r (see figure 8). The fraction of the plate covered by riblets is then $\Delta L_r / L_r \times 100\%$. The drag reduction as a function of this fraction is shown in figure 9. Remember that at the fraction 100% we should find the same results as already given in figure 3.

We find that the drag reduction changes to a drag increase even at the partial covering of 75%. This result should be interpreted with some reservation. The riblet material is stuck onto the flat plate. Therefore, a partial covering with riblet material causes edges, which have probably some influence on the flow and thus also on the drag (remember the large influence of the roughness elements discussed in section 3.2). In further experiment with partial coverage we will avoid these edges

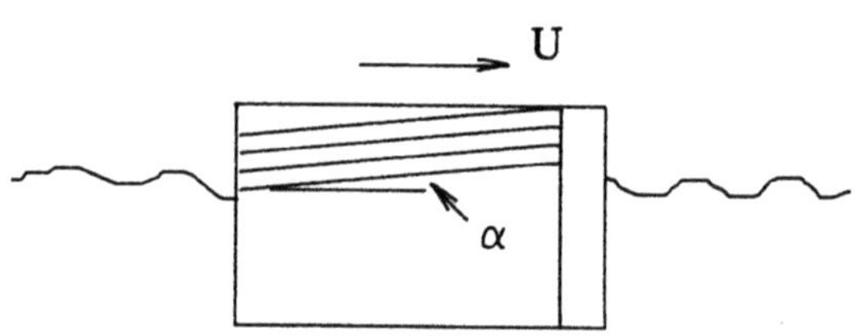

Figure 6: *Experiments with the riblets under an angle α with the towing direction.*

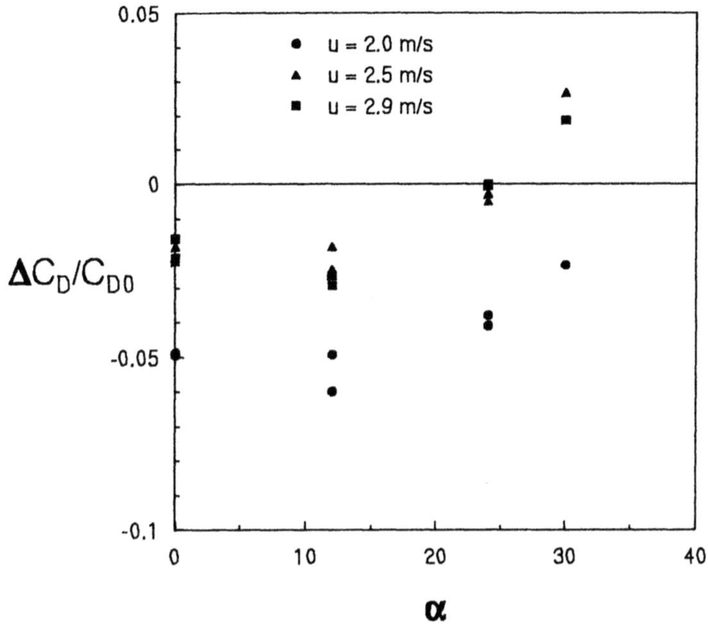

Figure 7: *The drag coefficient reduction of a flat plate as a function of the angle α to the flow direction.*

by applying smooth film in between the parts covered with riblets.

Nevertheless, one should expect that the drag reduction of a body will be sensitive to the total amount of riblet surface on this body.

4 Conclusions

Experiments done with a flat plate in a towing tank show a drag reduction by riblets of about 2% - 4%, which is consistent with our previous experiments (Nieuwstadt et al., 1989). If we allow for the rather crude experimental environment of a towing tank

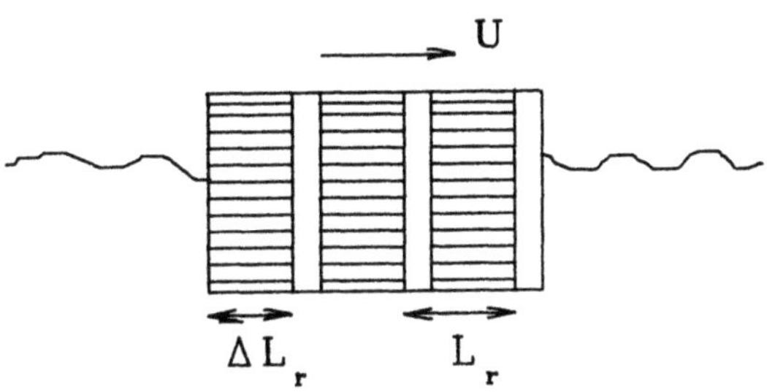

Figure 8: *Experiments with a partial coverage of the riblets on the flat plate.*

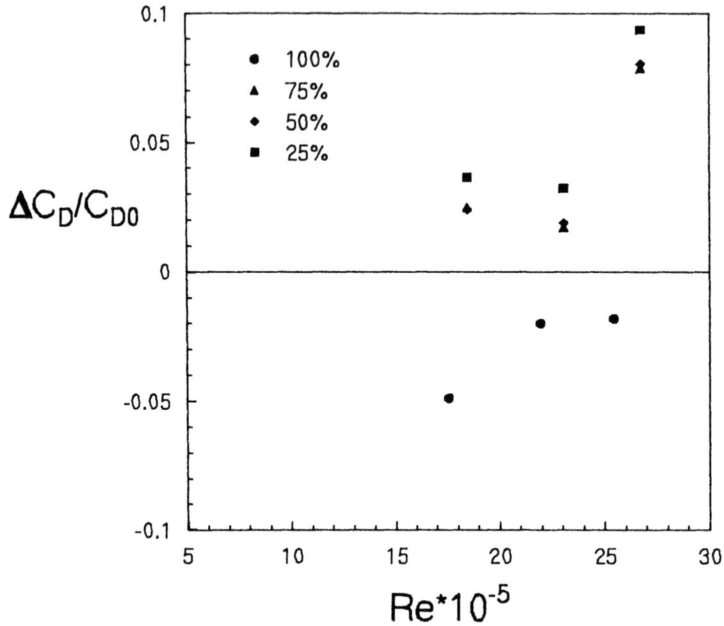

Figure 9: *The drag coefficient reduction of a flat plate as a function of the partial covering by riblet material.*

in comparison to a windtunnel, we conclude that our experiments are in reasonable agreement with other data on drag reduction by riblets.

In addition we have done some investigation into effects which might occur when we apply riblets in practical circumstances.

We find that roughness elements, such as carborundum grains used to trip the boundary layer, have a large influence on the overall drag. However, the drag reduction effect of the riblets persist at about the same magnitude as for a smooth plate.

The angle of the flow with respect to the direction of the riblets should be less than 20°. Otherwise, the drag reduction changes into drag increase. This results confirms previous experiments done in windtunnels.

A partial covering of a surface with riblets may have a large negative effect on the reduction of the total drag of a body by riblets.

Acknowledgement

We thank the Laboratory of Shiphydromechanics of the Department of Mechanical Engineering and Marine Technology for letting us use their towing tank facilities. The staff of the Laboratory of Shiphydromechanics and W. Kracht from the Laboratory of Aero and Hydrodynamics assisted with the experiments.

References

Choi, K.S., Pearcey, H.H. and Savill A.M., 1988. Test of drag reducing riblets on a third-scale racing yacht. International Conference on turbulent Conference on turbulent drag reducing by passive means. the Aeronautical Society, London, vol. 2

Nieuwstadt, F.T.M. , van der Hoeven J.G.Th., Leijdens, H., Krishna Prasad, K., 1989. Some experiments on riblet surfaces in a towing tank. In "Drag Reduction in Fluid Flows". (eds. R.H.J. Sellin and R.T. Moses), Ellis Horwood Ltd.

Savill, A.M. 1989. Drag reduction by passive devices - a review of some recent developments. In "Structure of Turbulence and Drag Reduction". (ed. A. Gyr), IUTAM Symposium Zürich, Switzerland, 1989, Springer-Verlag.

Schlichting, H. 1979. Boundary Layer Theory. Mc. Graw Hill Book Company, seventh edition.

Walsh, M.J. 1990. Riblets. In "Viscous Drag Reduction in Boundary Layers". (eds. D.M. Bushnell and J.W. Heffner), Progress in Astronautics and Aeronautics.

Walsh, M.J. and Lindemann. A.M. 1984. Optimization and application of riblets for turbulent drag reduction. A.I.A.A. paper No. 84-0347.

II. LEBUs

Analytical and experimental study of energy density spectra of the outer region of a manipulated turbulent boundary layer

E. AUGIER , H. M. ATASSI*, E. ALCARAZ, H. LADHARI and R. MOREL

Laboratoire de Mécanique des Fluides et d'Acoustique
Ecole Centrale de Lyon
Ecully, France

* *University of Notre Dame*
Notre Dame, IN, USA

K.-S. Choi (ed.), Recent Developments in Turbulence Management, 127–146.

Abstract

A model is developed for the outer region of a turbulent boundary layer manipulated by a thin device. The model assumes a statistically stationary turbulent flow convected by its mean velocity and uses the unsteady aerodynamic theory to calculate the modified velocity field downstream of the device. An expression for the modified one-dimensional energy density spectrum E_{11} is derived in terms of the unmanipulated layer spectra $E_{\infty ij}$. An experimental investigation of these spectra is also carried out to provide the upstream initial conditions for the theory and to give additional data to validate the theory. Comparison between calculated and measured spectra shows good agreement at various locations throughout the boundary layer. These results substantiate the proposed model based on the existence of wake structures coherent with upstream large structures, which induce a velocity field that may partially and selectively alter the outer layer turbulent structures.

Nomenclature

$\vec{a} = \{a_1, a_2\}$	Fourier components of $\vec{u}$
c	device chord length
δ	boundary layer thickness
γ	distance from the trailing edge, nondimensionalized with respect to δ
h	device height from the wall
$\vec{k} = \{k_1, k_2\}$	wavenumbers
$\phi_i(k_1, x_2)$	phase of the energy density spectra
t	time
θ	boundary layer momentum thickness
$\vec{u} = \{u_1, u_2\}$	fluctuating velocity
$\vec{u}^{(1)}$	irrotationnal fluctuating velocity
$\vec{u}_p$	potential part of $\vec{u}^{(1)}$
$\vec{u}_w$	wake induced part of $\vec{u}^{(1)}$
$\vec{x} = \{x_1, x_2\}$	coordinates defined Figure 1
$A_i(k_1, x_2)$	one-dimensional Fourier transform of u_i

C_0'	vortex sheet strength at the device trailing edge
H	form factor
$E_{ij}(k_1, x_2, x_2)$	energy density spectra
R_e	Reynolds number
S_{11}	$E_{11}/E_{\infty 11}$
U	mean velocity
∞	denotes the upstream unmanipulated quantities

1. INTRODUCTION

It has now been demonstrated by numerous experiments in different laboratories that by placing a thin plate (device) in the outer region of a turbulent boundary layer, the local skin friction coefficient is reduced. The experimental observations also show that such devices suppress the fluctuating vertical velocity and reduce the energy density spectra in the low frequency range. This phenomenon has been the subject of intensive research (Ref. [1]-[15]) because its practical implementation may lead to the possibility of drag reduction by passive means. Besides, it is fundamentally important to understand the basic mechanism by which such a device affects the outer region turbulent structures. The modification of these outer layer large structures certainly affects their interaction with the inner layer eddies and thus alters the turbulence production cycle (Falco, [16]).

The current paper is concerned with investigating the mechanism by which the device modifies the outer region large structures. Although there may be numerous effects to explain the device influence on the flow (Savill, [7]), theoretical models, using the concepts of unsteady thin airfoil theory, were developed to explain the dominant mechanism through which the device acts on the fluid. Dowling [17] analyzed the effect of the device on oncoming eddies and showed that the device causes a reduction in magnitude of the velocity component normal to the wall. Balakumar and Widnall [18] calculated the downstream normal component of the unsteady velocity along the mean wakeline of the device in response to a transverse sinusoidal gust and showed that it monotonically decreases as the gust reduced frequency increases. Both analyses show that the reduction in the unsteady normal velocity is induced by the wake whose effect persists over long distance as suggested by the experiments. Atassi and Gebert [19] and Gebert [20] generalized these approaches and proposed a model for lifting airfoil devices using the concepts of unsteady airfoil theory and the approximation of the rapid distortion theory of turbulence (Batchelor and Proudman, [21]). They also derived analytical expressions for the downstream modified turbulent velocity for every Fourier component of the upstream turbulent velocity. By analyzing the variation of the turbulent vertical component of the velocity throughout the boundary layer, Atassi and Gebert obtained a criterion for scaling the device chord length and its height from the wall with the turbulent structures for optimum device efficiency.

In the present paper, we extend Atassi and Gebert's model to derive the expression for the modified energy density spectrum of the streamwise fluctuating velocity $\overline{u_1^2}$. This expression naturally depends on the upstream energy density spectra. The data of Klebanoff [22] were first used to evaluate the spectral functions which then were compared with the data of Delville et al. [23] for a manipulated boundary layer. However, certain two-point correlations were needed to evaluate the modified spectrum. As a result, an experimental investigation was also carried out to measure these quantities and to provide additional data to validate the theory.

The spectral energy of $\overline{u_1^2}$, $\overline{u_2^2}$ and $-\overline{u_1 u_2}$ were measured using hot wire anemometry and signal processing throughout the turbulent boundary layer with and without a manipulator. Since the device induces an unsteady velocity field which for a certain frequency range reduces the unmanipulated turbulent velocity by cancellation, it is necessary for both velocity fields to have similar magnitudes but opposite phases. The phase and magnitude of the induced field are determined by conditions at the height h of the device. Thus, two-point correlations of the turbulent velocity were also measured with one point always located at h, while the other point's height was varied througout the thickness of the boundary layer.

In section **2** of the present paper, we briefly describe the analytical model and give the expressions for the modified unsteady velocity for every upstream Fourier component. In section **3**, we derive the expression for the modified energy density spectrum E_{11}. The experimental investigation is described in section **4** and comparison between theory and data are presented in section **5**.

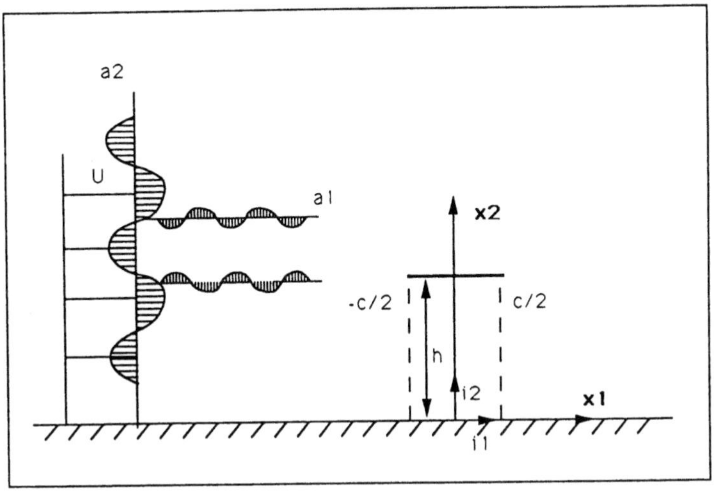

Figure 1: Schematic of a thin plate in an unsteady rotational flow in the presence of a wall.

2. THE ANALYTICAL MODEL

We consider a fully developed turbulent boundary layer of average thickness δ along a flat horizontal wall. A thin flat plate of chord length c is placed at a height h, of the wall as is shown in Figure 1. We assume that the plate is located in the outer region of the boundary layer, i.e., $0.2\delta < h < \delta$. In this region, viscous dissipation is negligible and the mean flow can be assumed to be almost uniform. As a result, the turbulent flow can be considered inviscid but rotational. Thus, the plate is subject to an inviscid incompresible turbulent flow convected with the mean velocity U. We further assume that the turbulence is statistically stationary and two-dimensional. All the lengths will be normalized with respect to $c/2$, the time with respect to $c/(2U)$, and the velocities with respect to U.

Thus the total velocity field can be written as :

$$\vec{V} = \vec{i_1} + \vec{u}(x_1 - t, x_2) \tag{1}$$

The expression for $\vec{u}$ far upstream of the thin plate can be Fourier decomposed

$$\begin{aligned} \vec{u}_\infty(x_1 - t, x_2) = 2\int_{-\infty}^{+\infty}\int_{0}^{+\infty} & [a_{\infty 1}(k_1, k_2)\cos(k_2 x_2)\vec{i_1} \\ & -j a_{\infty 2}(k_1, k_2)\sin(k_2 x_2)\vec{i_2}]\exp(jk_1[t - x_1])dk_1 dk_2 \end{aligned} \tag{2}$$

where k_1 is the usual reduced frequency, $j^2 = -1$ and $a_{\infty i}(k_1, k_2)$, $i = 1, 2$, are the Fourier transform components of $\vec{u}_\infty$. The total unsteady field can then be split into two parts [24,25] :

$$\vec{u} = \vec{u}_\infty + \vec{u}^{(1)} \tag{3}$$

where the upstream disturbance $\vec{u}_\infty$ is rotational and $\vec{u}^{(1)}$ is irrotationnal. Since the problem is linear , we can consider, without loss of generality, a single Fourier component of the upstream disturbance,

$$\begin{aligned}\vec{u}_\infty(x_1 - t, x_2) = 2[&a_{\infty 1}(k_1, k_2)\cos(k_2 x_2)\vec{i}_1 \\ &-j a_{\infty 2}(k_1, k_2)\sin(k_2 x_2)\vec{i}_2]\exp(jk_1[t - x_1])\end{aligned} \tag{4}$$

The unknown velocity field $\vec{u}^{(1)}$ is solenoidal and irrotational and therefore must satisfy Laplace's equation

$$\nabla^2 \vec{u}^{(1)} = 0 \tag{5}$$

The boundary condition for $\vec{u}^{(1)}$ along the plate are obtained by writing the impermeability condition for the total unsteady velocity $\vec{u}$, which yields

$$u_2^{(1)} = 2j a_{\infty 2}(k_1, k_2)\sin(k_2 h]\exp(jk_1[t - x_1]) \tag{6}$$

In addition, we impose the conditions that $u_2^{(1)}$ and the fluctuating pressure p' are continuous across the wakeline, $x_1 > 1, x_2 = h$. This boundary-value problem was formulated by Gebert and Atassi [26] who derived a singular integral equation which they solved by collocation.

In the present problem, we are mainly interested in the downstream expression of $\vec{u}^{(1)}$. Atassi and Gebert [19] and Gebert [20] showed that

$$\vec{u}^{(1)} = \vec{u}_p + \vec{u}_w \tag{7}$$

where $\vec{u}_p$ is a regular potential field and $\vec{u}_w$ is the fluctuating velocity induced by the wake. At large distance downstream of the plate, the potential field $\vec{u}_p \to 0$ as $1/|\vec{x}|$. Therefore, to determine the modified velocity downstream of the manipulator we only need to determine $\vec{u}_w$. To this end, we note that $\vec{u}_w$ satisfies the Laplace equation and since the pressure is continuous across the wake, $\vec{u}_w$ has a jump $\Delta u_{w1}(x_1, h, t)$ across the wake $(x_2 = h)$ satisfying the equation

$$\left(\frac{\partial}{\partial t} + \frac{\partial}{\partial x_1}\right)\Delta u_{w1}(x_1, h, t) = 0 \qquad \text{for } x_1 \geq 1 \tag{8}$$

This equation can be readily integrated and in view of (4), we get

$$\Delta u_{w1}(x_1 - t, h) = C_0' \exp(jk_1[t - x_1 + 1]) \qquad \text{for } x_1 \geq 1 \tag{9}$$

where C_0' is the jump of u_{w1} at the trailing edge of the plate.

Far downstream, the expression for u_{w1} can be readily constructed to order $1/|\vec{x}|$,

$$u_{w1} = \pm\frac{1}{2}C_0'\exp(jk_1[t - x_1 + 1])\exp(-k_1|h - x_2|) \tag{10}$$

where the sign + is taken when $x_2 > h$ and the sign − is taken when $x_2 < h$.

In order to account for the effect of the wall, we use the method of images. Thus, another plate is considered below the wall at a distance $-h$. For $x_2 < h$, the induced velocity field of the two wakes is then given by

$$u_{w1} = 2ja_{\infty 2}(k_1, k_2)C_0\sin(k_2h)\exp(jk_1 - k_1h)\cosh(k_1x_2)\exp(jk_1[t - x_1]) \tag{11}$$

where $C_0 = C_0'/X$, with $X = -2ja_{\infty 2}(k_1, k_2)\sin(k_2h)$. C_0 depends only on the reduced frequency k_1, and is determined from the solution of the integral equation by imposing the Kutta condition at the plate trailing edge.

The second component of the downstream velocity is then easily obtained by integration of the continuity equation. For $x_2 < h$, we obtain

$$u_{w2} = -2ja_{\infty 2}(k_1, k_2)C_0\sin(k_2h)\exp(jk_1 - k_1h)\sinh(k_1x_2)\exp(jk_1[t - x_1]) \tag{12}$$

Finally, combining (4,11 and 12), we arrive at the following expression for the total unsteady velocity field at large distance downstream of the device

$$\begin{aligned}\vec{u}(x_1 - t, x_2) = 2\int_{-\infty}^{+\infty}\int_{0}^{+\infty}&[a_1(k_1, k_2)\cos(k_2x_2)\vec{i}_1\\ &-ja_2(k_1, k_2)\sin(k_2x_2)\vec{i}_2]\exp(jk_1[t - x_1])dk_1dk_2\end{aligned} \tag{13}$$

with

$$\begin{aligned}a_1(k_1, k_2) &= a_{\infty 1}(k_1, k_2) + ja_{\infty 2}(k_1, k_2)C_0\exp(jk_1 - hk_1)\\ &\quad \times cosh(x_2k_1)sin(k_2h)/cos(k_2x_2)\\ a_2(k_1, k_2) &= a_{\infty 2}(k_1, k_2) - ja_{\infty 2}(k_1, k_2)C_0\exp(jk_1 - hk_1)\\ &\quad \times sinh(x_2k_1)sin(k_2h)/sin(k_2x_2)\end{aligned} \tag{14}$$

for $x_1 \geq 1$ and $x_2 < h$.

3. THE MODIFIED STREAMWISE ENERGY SPECTRUM

Equation (13) gives the expression for the modified velocity in terms of the Fourier components of the upstream velocity. The components of the latter are difficult to obtain from measurements. Experiments usually provide velocity correlations from which it is possible to directly calculate the one-dimensional streamwise energy density spectrum. Thus, in order to compare between theory and experiments, we derive the expressions for the modified one-dimensional streamwise energy spectrum. This is obtained by integrating (2) and (13) with respect to k_2. Carrying out this integration and introducing $\zeta = x_1 - t$, we get the pair of Fourier transforms,

$$u_i(\zeta, x_2) = \int_{-\infty}^{+\infty} A_i(k_1, x_2) \exp(-jk_1\zeta) dk_1 \tag{15}$$

$$A_i(k_1, x_2) = \frac{1}{2\pi} \int_{-\infty}^{+\infty} u_i(\zeta, x_2) \exp(jk_1\zeta) d\zeta \tag{16}$$

where

$$\begin{cases} A_1(k_1, x_2) = 2 \int_0^\infty a_1(k_1, k_2) \cos(k_2 x_2) dk_2 \\ A_2(k_1, x_2) = -2j \int_0^\infty a_2(k_1, k_2) \sin(k_2 x_2) dk_2 \end{cases} \tag{17}$$

The velocity correlation functions are defined as :

$$\overline{u_i(\zeta_1, x_2) u_j(\zeta_1 - \zeta, x_2')} = \lim_{T \to \infty} \frac{1}{2T} \int_{-T_1}^{+T_1} u_i(\zeta_1, x_2) u_j(\zeta_1 - \zeta, x_2') d\zeta_1 \tag{18}$$

where $T_1 > T$. If we further assume that T is large and that $u_i(\zeta_1, x_2)$ has a finite duration, and using (15) and (16), we arrive at

$$\overline{u_i(\zeta_1, x_2) u_j(\zeta_1 - \zeta, x_2')} = \frac{\pi}{T} \int_{-\infty}^{+\infty} A_j(k_1, x_2') A_i^*(k_1, x_2) \exp(jk_1\zeta) dk_1 \tag{19}$$

where the superscript * denotes the complex conjuguate. Since $\overline{u_i(\zeta_1, x_2) u_j(\zeta_1 - \zeta, x_2')}$ is real, it is convenient to write (19) in a symmetric form,

$$\overline{u_i(\zeta_1, x_2) u_j(\zeta_1 - \zeta, x_2')} = \frac{\pi}{T} \int_{-\infty}^{+\infty} Real[A_j(k_1, x_2') A_i^*(k_1, x_2) \exp(jk_1\zeta)] dk_1 \tag{20}$$

We now define the energy density spectra $E_{ij}(k_1, x_2, x_2')$ of the velocity correlation functions by

$$\overline{u_i(\zeta_1, x_2) u_j(\zeta_1, x_2')} = \frac{1}{2} \int_{-\infty}^{+\infty} E_{ij}(k_1, x_2, x_2') dk_1 \tag{21}$$

Putting $\zeta = 0$ in (20) and comparing with (21), gives the following expression between the energy density spectra and the Fourier transforms of the unsteady velocity,

$$E_{ij}(k_1, x_2, x_2') = \frac{2\pi}{T} Real[A_j(k_1, x_2')A_i^*(k_1, x_2)] \tag{22}$$

This suggests that

$$A_i(k_1, x_2) = \sqrt{\frac{T}{2\pi}} \sqrt{E_{ii}(k_1, x_2, x_2)} \exp(j\phi_i) \tag{23}$$

where $\phi_i = \phi_i(k_1, x_2)$, $i = 1, 2$ is the phase of the Fourier transforms of the unsteady velocity components. Note that $E_{\infty 12}$ can now be expressed in terms of $E_{\infty 11}, E_{\infty 22}, \phi_{\infty 1}$ and $\phi_{\infty 2}$,

$$E_{\infty 12}(k_1, x_2, x_2) = \sqrt{E_{\infty 11}(k_1, x_2, x_2) E_{\infty 22}(k_1, x_2, x_2)} \cos[\phi_{\infty 1}(k_1, x_2) - \phi_{\infty 2}(k_1, x_2)] \tag{24}$$

We now calculate the Fourier transforms of the velocity components by integrating equations (14) with respect to k_2,

$$A_1(k_1, x_2) = A_{\infty 1}(k_1, x_2) - A_{\infty 2}(k_1, h) C_0 \exp(jk_1 - hk_1) \cosh(k_1 x_2) \tag{25}$$

$$A_2(k_1, x_2) = A_{\infty 2}(k_1, x_2) - jA_{\infty 2}(k_1, h) C_0 \exp(jk_1 - hk_1) \sinh(k_1 x_2) \tag{26}$$

The expression for E_{11} can now be obtained by substituting (25) and (26) into (21) and using (22). We finally arrive at the following expression,

$$\begin{aligned} E_{11}(k_1, x_2, x_2) = E_{\infty 11}(k_1, x_2, x_2) - 2\sqrt{E_{\infty 11}(k_1, x_2, x_2) E_{\infty 22}(k_1, h, h)} \\ \times Real[C_0 \exp(j[\phi_{\infty 1}(k_1, x_2) - \phi_{\infty 2}(k_1, h) + k_1])] \exp(-hk_1) \cosh(k_1 x_2) \\ + E_{\infty 22}(k_1, h, h) C_0 C_0^* \exp(-2hk_1) \cosh^2(k_1 x_2) \end{aligned} \tag{27}$$

where

$$\begin{aligned} \phi_{\infty 1}(k_1, x_2) - \phi_{\infty 2}(k_1, h) = \cos^{-1}\left(\frac{E_{\infty 12}(k_1, x_2, x_2)}{\sqrt{E_{\infty 11}(k_1, x_2, x_2) E_{\infty 22}(k_1, x_2, x_2)}}\right) \\ + \cos^{-1}\left(\frac{E_{\infty 22}(k_1, x_2, h)}{\sqrt{E_{\infty 22}(k_1, h, h) E_{\infty 22}(k_1, x_2, x_2)}}\right) \end{aligned} \tag{28}$$

Note that, in a similar manner, the other energy spectra can also be calculated for a manipulated boundary layer. However this study and its experimental counterpart will be performed in an other article.

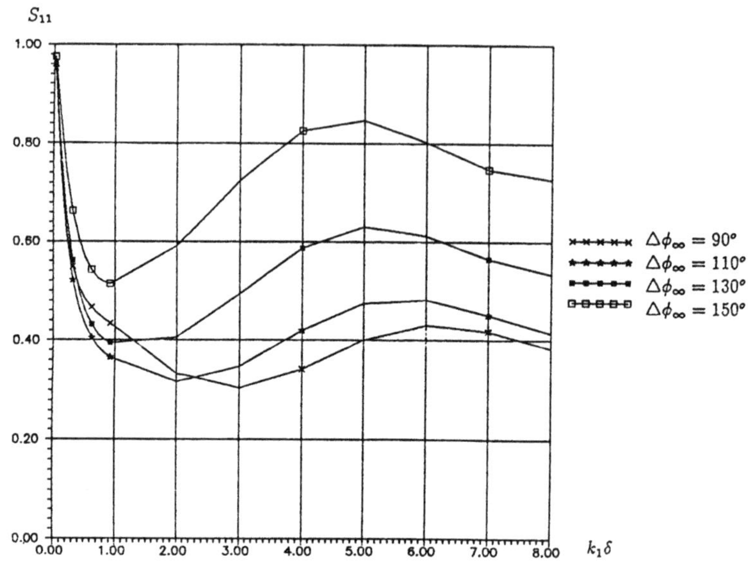

Figure 2: S_{11} versus k_1 for different $\Delta\phi_\infty$ at $x_2 = 0.8\delta$.

Equation (27) shows that modified spectrum E_{11} depends not only on $E_{\infty 11}$ but also on $E_{\infty 22}$ and the phase difference $\Delta\phi_\infty = \phi_{\infty 1} - \phi_{\infty 2}$. In order to assess the modification of E_{11}, it is convenient to consider the ratio

$$S_{11} = E_{11}/E_{\infty 11} \tag{29}$$

Since the determination of $\Delta\phi_\infty$ necessitates the evaluation of the spectra of the unmanipulated turbulent layer at two locations as shown by equation (28), we examine the values of S_{11} for different $\Delta\phi_\infty$ to assess the sensitivity of E_{11} to this parameter. For this we need to know only the ratio $E_{\infty 22}/E_{\infty 11}$ which may depend on the frequency k_1. However for simplicity, we assume that $E_{\infty 22}(k_1, x_2, x_2) \sim E_{\infty 22}(k_1, x_2, h) \sim E_{\infty 22}(k_1, h, h)$ and we take $E_{\infty 22}/E_{\infty 11} = 0.14$ which corresponds to the average value of Klebanoff's data, taken at $x_2 = 0.58\delta$, for $0.25 < k_1\delta < 3$.

For a device of chord length $c = \delta$ located at $h = 0.8\delta$, figures 2 to 4 show plots of S_{11} versus $k_1\delta$ at three different distances from the wall, $x_2 = 0.8\delta$, 0.6δ and 0.38δ, respectively. Every figure has four plots of S_{11}, corresponding to $\Delta\phi_\infty = 90^o$, 110°, 130° and 150°. These figures show that at the level of the device, S_{11} is reduced over a wide range of frequencies. As we move lower than the device height, the minimum value of S_{11} is not significantly affected, however, the reduction in the values of S_{11} covers now a smaller frequency range, indicating a more selective effect of the device throughout the boundary layer. S_{11} exhibits meaningful depedence on $\Delta\phi_\infty$ at $x_2 = h$. However, for $x_2 < h$, this

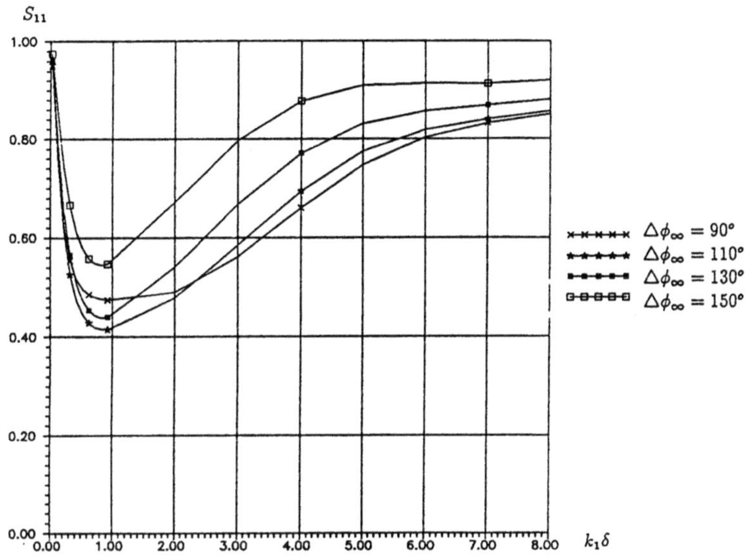

Figure 3: S_{11} versus k_1 for different $\Delta\phi_\infty$ at $x_2 = 0.6\delta$.

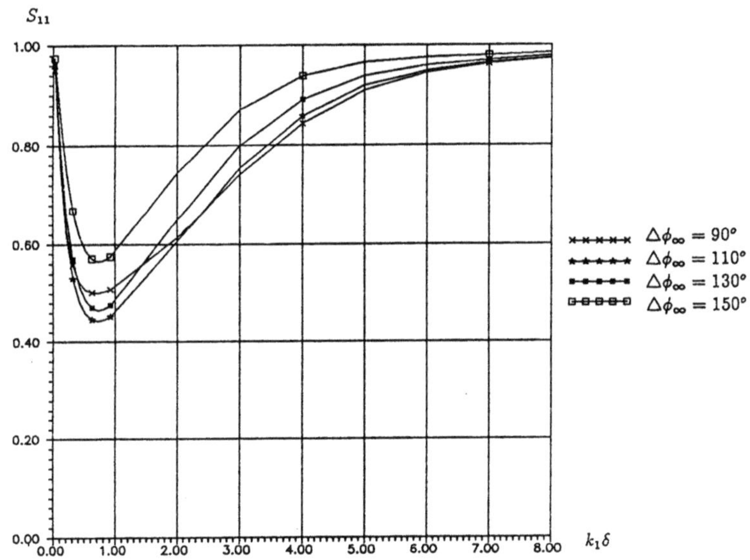

Figure 4: S_{11} versus k_1 for different $\Delta\phi_\infty$ at $x_2 = 0.38\delta$.

depedence is reduced at $x_2 = 0.38\delta$, as shown in Figure 4. We also note that the minimun value for S_{11} occurs for $\Delta\phi_\infty$ between 110° and 130°. The figure also shows that most of the reduction in E_{11} of the modified layer occurs in the range $0.2 < k_1\delta < 2$.

4. EXPERIMENTAL INVESTIGATION

4.1 Experimental Apparatus

Experiments aimed mainly at measuring the one-dimensional energy spectra of the double velocity correlations have been performed in the open wind tunnel of the Laboratoire de Mécanique des Fluides et d'Acoustique de l'E.C.L. (Figure 5). It is a low subsonic speed wind tunnel with a nearly zero-pressure gradient and a constant local temperature controlled by a thermocouple. The investigated turbulent boundary layer develops along the upper surface of a horizontal flat plate located in the middle of the tunnel. The test section is 4 m long, 50 cm wide and 25 cm high. It has transparent side walls for observation of flow visualization using tomography. The turbulence level in the free stream is less than 1% and the mean flow is two-dimensional. A 50 mm long sand paper strip is glued on the surface of the flat plate starting from its leading edge to insure a fully developped turbulent boundary layer and to increase its thickness at the location of the device. The manipulator is made from steel and has a rectangular section. In order to avoid any vibration, it is adequately tightened and controlled by a reflecting laser beam.

The main characteristics of the flow and of the device are given in the following.

Wind Tunnel

Type	Open
Test section	Area (1250 cm^2)
Contraction ratio	5.76
Pressure Gradient	0

Anemometers

Probes	Disa
Acquisition system	H.P. 35650

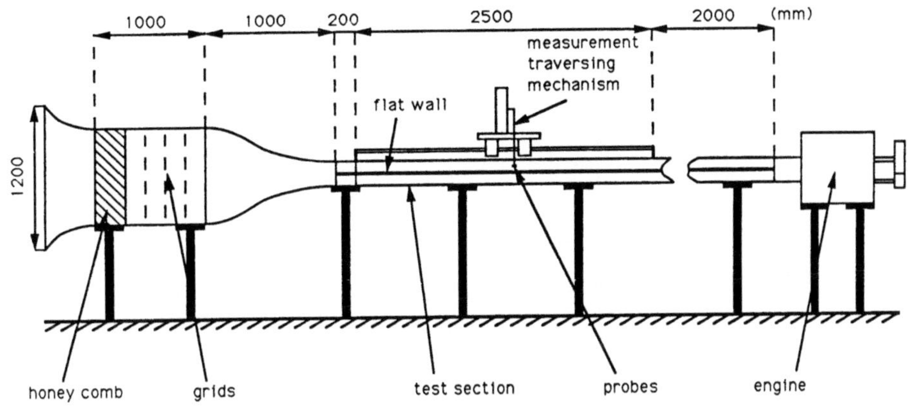

Figure 5: Schematic of the wind tunnel.

Flow

$U = 2.8$ m/s	$Re_\theta = 1030$
$\delta = 5.0510^{-2}m$	$Re_\delta = 9426$
$\theta = 5.5210^{-2}m$	$H = 1.44$

Device

Chord lenght $\frac{c}{\delta}$	1
Thickness $\frac{t}{\delta}$	0.01
Distance to the wall $\frac{h}{\delta}$	0.8
Mean velocity at the device level	2.6 m/s
$(\overline{u_1^2})^{\frac{1}{2}}$	0.12 m/s
$(\overline{u_2^2})^{\frac{1}{2}}$	0.097 m/s

4.2 Measurements

Mean flow quantities and fluctuating velocities are obtained from time averaging of the instantaneous velocity measurements with two crossed hot-wire anemometers for both the regular and manipulated boundary layers. The measurements were made at the same streamwise location but at different distances

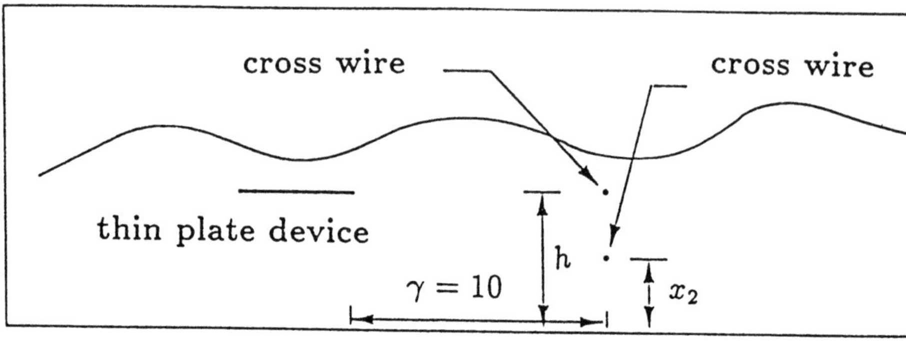

Figure 6: Schematic of the crossed hot-wire locations.

from the wall. The distance, γ, from the trailing edge of the device, nondimensionalized with respect to δ, at which the measurements were performed was chosen to be large enough so that the local effect of the manipulator does not affect the fluctuating velocity field in a significant way. This is because, for simplicity, we have neglected the near field effect in our modelling. Thus, most measurements were taken at $\gamma = 10$.

For accurate comparison with the analytical model, a crossed hot-wire was located at the device height, h, from the wall, while the other one was located at a variable height x_2, below the device location (Figure 6). Thus, velocity correlations obtained with these two wires will account for the phase difference of the fluctuating velocities at these two locations. The two components of the streamwise and spanwise unsteady velocity were obtained, for each crossed hot-wire, using the King's law for the effective cooling velocity, and by resolution of a linear system of equations. The energy density spectra were then calculated with a fast Fourier transform on a sampling of 2048 values for an acquisition frequency of 2048 Hz.

5. RESULTS AND DISCUSSION

Two sets of spectral energy density measurements were carried out for low Reynolds number turbulent boundary layer flows. The first set, denoted $E_{\infty 11}$, $E_{\infty 12}$ and $E_{\infty 22}$, was taken for an unmanipulated layer and was intended to provide upstream conditions for the theory and a comparison basis with the results

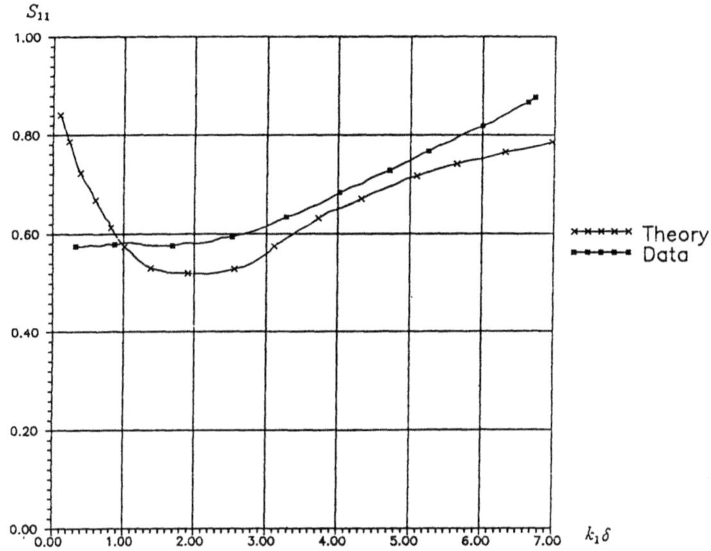

Figure 7: Comparison between calculated and measured values of S_{11} versus $k_1\delta$ at $x_2 = 0.6\delta$.

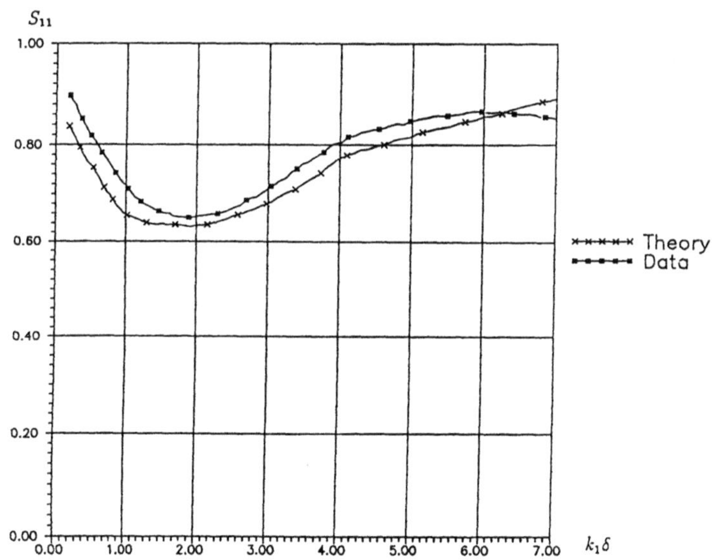

Figure 8: Comparison between calculated and measured values of S_{11} versus $k_1\delta$ at $x_2 = 0.52\delta$.

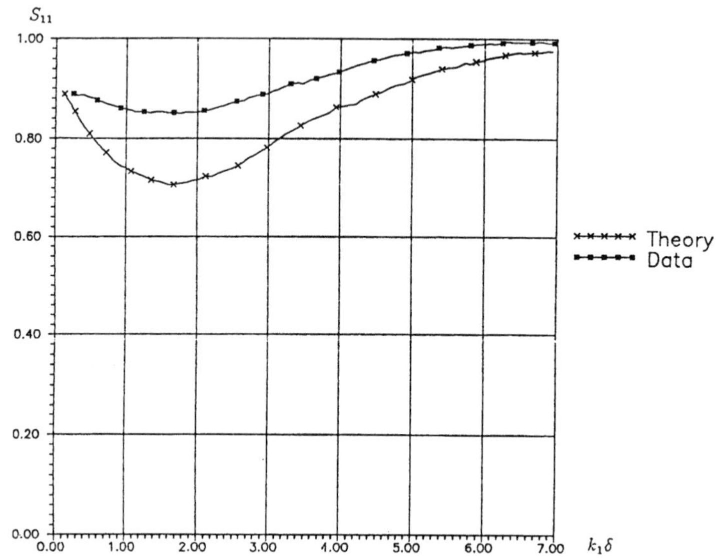

Figure 9: Comparison between calculated and measured values of S_{11} versus $k_1\delta$ at $x_2 = 0.38\delta$.

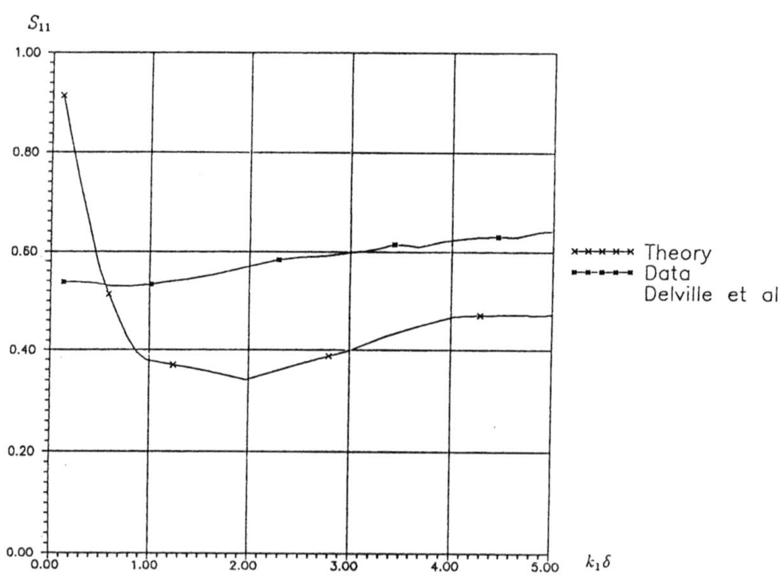

Figure 10: Comparison between theoretical and Delville et al. measured values of S_{11} versus $k_1\delta$ at $x_2 = 0.4\delta$, $h = 0.43\delta$ and $\gamma = 4$.

of manipulated layers. The second set was taken downstream of the device at $\gamma = 10$. The main characteristics of the flow and of the device were given in section 4.1. Figures 7, 8 and 9 essentially show the comparison between theory and experiments at three different heights from the wall, $x_2 = 0.6\delta$, 0.52δ, and 0.38δ, respectively. In every figure, we have plotted the experimental and the calculated values (using (27) and (28)) of S_{11} versus the reduced frequency $k_1\delta$. Note that the experimental values of the energy spectra usually exhibit large scatter particularly at low k_1 where accurate measurement is difficult. To facilitate the comparison with the theory, we have smoothed out their plots. The average scatter of the data was less than 10%. Figures 7, 8 and 9 show that the theoretical values of S_{11} are in good agreement with the experiments. If we broadly define the efficiency of a device by the amount S_{11} is reduced from unity and the broad range of k_1 over which this reduction occurs, we observe that the theory always predicts a slighty more efficient device than the data. At lower heights, the difference between theory and data increases as shown in figure 9, for $x_2 = 0.38\delta$.

Finally, we have compared our theoretical predictions for E_{11} with the data of Delville et al. [23], for which the free stream velocity $U = 26$ m/s, the boundary layer thickness $\delta = 2.35$ cm, the device chord ratio, $c/\delta = 1.1$ and $h/\delta = 0.43$. These data correspond to a flow of higher Reynolds number $Re_\delta = 40,700$ and thus it is interesting to compare them with our theoretical predictions for futher validation. The unmanipulated values of $E_{\infty 11}$ were taken from their data, while the values of $E_{\infty 12}$ and $E_{\infty 22}$ were obtained from those of Klebanoff's using the similarity relationship

$$\left(\frac{E_{\infty ij}}{\overline{u_i u_j}\ \delta}\right)_K = \left(\frac{E_{\infty ij}}{\overline{u_i u_j}\ \delta}\right)_D$$

where the subscripts D and K denote the data of Delville et al. and Klebanoff, respectively. Note that data from CEAT and CERT groups which are already in the ERCOFTAC database may be also used in a futur comparison. Figure 10 shows a comparison between the theoretical values of S_{11} and those calculated from the data of Delville et al. at $x_2 = 0.4\delta$. The theory predicts a stronger reduction in the values of S_{11}, for this high Reynolds number case. However, the agreement between theory and experiments appears to be adequate.

6. CONCLUSIONS

The theoretical model of Atassi and Gebert has been extended to calculate the streamwise energy density spectrum E_{11} for turbulent boundary layers manipulated by a streamlined device. The theory models the physical mechanism by which the device alters the turbulent boundary layer structure. It shows that the wake shed from the device's trailing edge has a structure which is coherent with the upstream turbulence. This wake induces a fluctuating velocity field which may partially cancel the convected turbulent flow and thus alter its structure. For a device whose chord length is almost equal to the turbulent boundary layer thickness, the streamwise energy density spectrum E_{11} is significantly reduced in the frequency range $0.25 < k_1\delta < 3$. This corresponds to the range of large structures with high turbulent energy.

To validate the theory and to properly calculate E_{11} downstream of a device, experiments were set up to measure the upstream values of the unmanipulated energy density spectra and the downstream values of E_{11}. The results show that the theory slightly overpredicts the reduction of E_{11}.

The current work presents a relatively simple model to explain the basic physical mechanism of outer boundary layer structure modification by the insertion of a passive streamlined device. The good agreement between the calculated streamwise energy density spectrum E_{11} and the data validates our model which can be extended to calculate other energy density spectra and the wall friction coefficient.

References

1. Hefner J.N., Weinstein L.M. and Bushnell D.M., "Large-Eddy Breakup Scheme for Turbulent Viscous Drag Reduction", Viscous Flow Drag Reduction, *Progress in Astronautics and Aeronautics*, 1972, presented at the Symposium on Viscous Drag Reduction, Dallas, Texas, (Nov. 1979).
2. Hefner J.N., Anders J.B and Bushnell D.M., "Alteration of Outer Flow Structures for Turbulent Drag Reduction", AIAA 21st Aerospace Sciences Meeting, Reno, Nevada, (Jan. 1983), Paper No. 83-0293.
3. Corke T.C., Nagib H.M. and Guezennec Y.G., "A New View on Origin, Role, and Manipulation of Large Scales in Turbulent Boundary Layers", Illinois Institute of Technology, (Feb. 1982), NASA CR-165861.
4. Bushnell D.M., "Turbulent Drag Reduction for External Flows", AIAA 21st Aerospace Sciences Meeting, Reno, Nevada, (Jan. 1983), Paper No. 83-0227.
5. Anders J.B, Hefner J.N. and Bushnell D.M., "Performance of Large-Eddy Breakup Devices at Post Transitional Reynolds Numbers", AIAA 21st Aerospace Sciences Meeting, Reno, Nevada, (Jan. 1983), Paper No. 84-0345.
6. Mumford J.C. and Savill A.M., "Parametric Studies of Flat Plate Turbulence Manipulators Including Direct Drag Results and Laser Flow Visualization", paper presented at the ASME Symposium on Laminar/Turbulent Boundary Layers

: Control Modification and Marine Applications ETCE Conference, New Orleans, Louisiana, (Feb. 1984).
7. Savill A.M. and Mumford J.C., "Manipulation of Turbulent Boundary Layers by Outer Layer Devices : Skin-Friction and Flow Visualization Results", *J.Fluid Mech.*, vol. 191, pp. 389-418, 1988.
8. Plesniak M.W. and Nagib H.M., "Net Drag Reduction in Turbulent Boundary Layers Resulting from Optimized Manipulation", AIAA Shear Flow Control Conference, Boulder, Colorado, (Mar. 1985), Paper No. 85-0518.
9. Guezennec Y.G. and Nagib H.M., "Documentation of Mechanisms Leading to Net Drag Reduction in Manipulated Turbulent Boundary Layers", AIAA Shear Flow Control Conference, Boulder, Colorado, (Mar. 1985), Paper No. 85-0159.
10. Anders J.B. and Watson R.D., "Airfoil Large-Eddy Breakup Devices for Turbulent Drag Reduction", AIAA Shear Flow Control Conference, Boulder, Colorado, (Mar. 1985), Paper No. 85-0520.
11. Coustols E. and Savill A.M., "Resume of Important Results presented at the Third Turbulent Drag Reduction Working Party", *Applied Scientific Research*, Kluwer Academic Publishers, vol. 46, pp. 183-196, 1989.
12. Coustols E. and Cousteix J., "Turbulent Boundary Layer Manipulation in Zero-Pressure Gradient", *Proc. 16th ICAS Jerusalem*, August 28-September 2, vol. 2, pp. 999-1013, 1988.
13. Lemay J., Savill A.M., Bonnet J.P. and Delville J., "Some Similarities Between Turbulent Boundary Layers Manipulated by Thin and Thick Flat Plate Manipulators", *Proc. 6th Turbulent Shear Flows Symp.*, Toulouse : Springer-Verlag, pp. 179-193, 1987.
14. Tenaud C., Lemay J., Bonnet J.P. and Delville J., "Balance of Turbulent Kinetic Energy Downstream a Single Flat Plate Manipulator : Comparisons Between Detailed Experiments and Modelling", *Turbulence Control by Passive Means*, Kluwer Academic Publishers, *Proceedings of the 4th European Drag Reduction Meeting, edited by E.Coustols*, 1990.
15. Chang S.I. and Blackwelder R.F., "Modification of Large Eddies in Turbulent Boundary Layers", *J.Fluid Mech.*, vol. 213, pp. 419-442, 1990.
16. Falco R.E., "New Results, a Review and Synthesis of the Mechanism of Turbulence Production in Boundary Layers and its Modification", AIAA 21st Aerospace Sciences Meeting, Reno, Nevada, (Jan. 1983), Paper No. 83-0377.
17. Dowling A.P., "The Effect of Large-Eddy Breakup Devices on Oncoming Vorticity", *J.Fluid Mech.*, vol. 160, pp. 447-463, 1985.
18. Balakumar P. and Widnall S.E., "Application of Unsteady Aerodynamics to Large-Eddy Breakup Devices in a Turbulent Flow", *Physics of Fluids*, vol. 29, pp. 1779-1787.
19. Atassi H.M. and Gebert G.A., "Modification of Turbulent Boundary Layer

Structure by Large-Eddy Breakup Devices", *Turbulent Drag Reduction by Passive Means*, Royal Aeronautical Society, London, England, 1987, Palo Alto, CA, pp. 432-456.

20. Gebert G.A., "Turbulent Boundary Layer Modification by Streamlined Devices", Thesis, Department of Aerospace and Mechanical Engineering, University of Notre Dame, Indiana, (Apr. 1988).

21. Batchelor G.K. and Proudman I., "The Effect of Rapid Distorsion of a Fluid in Turbulent Motion", *Q.J. Mech. Appl. Math.*, vol. 1, pp. 83-103, 1954.

22. Klebanoff P.S., "Characteristics of Turbulence in a Boundary Layer with Zero Pressure Gradient", N.A.C.A. report No. TN 3178, (Jul. 1954).

23. Delville J., Bonnet J.P. and Lemay J., "Etude Experimentale de l'Influence d'un Manipulateur de Turbulence de Type Lame Mince sur la Structure d'une Couche Limite Plane Incompressible", *Association Aéronautique et Astronautique de France*, Note Technique No. 87-18, 1987.

24. Goldstein M.E., *Aeroacoustics*, Mc Graw-Hill, 1976.

25. Atassi H.M., "Unsteady Vortical Disturbances Around Bodies", *Proceedings of the tenth U.S. National Congress of Applied Mechanics*, J.P. Lamb ed., ASME, 1986, pp. 475-484.

26. Gebert G.A. and Atassi H.M., "Unsteady Vortical Disturbances Around a Thin Airfoil in the Presence of a Wall", *AIAA Journal*, vol. 27, No. 10, (Oct. 1989), pp. 1448-1450.

Review: effect of the OLDs on near wall coherent structures; discussion and need for future work

S. TARDU and G. BINDER*

Ecole Polytechnique Fédérale
Lausanne, Switzerland

* *Institut de Mécanique de Grenoble*
Grenoble, France

K.-S. Choi (ed.), Recent Developments in Turbulence Management, 147–160.

Abstract

This paper reviews some aspects of the effect of outer layer devices on the structure of near wall turbulence. The accent is on the reaction of coherent structures inherent to a turbulent boundary layer, to such manipulations. The first part of the paper concerns the alteration of the bursting and ejection frequency. Several authors agree that the bursting frequency is not modified by outer layer manipulation, despite the differences in the detection techniques used so far, but part of the bursting process defined as frequency of the individual lift-up of low-momentum fluid near the wall varies in the same way as the drag. The second part of the paper summarises the possible interactions between the wake vortices generated by the outer layer device, and the near wall hairpin vortices. It is argued that the passage of the wake vortices in the sublayer may decrease the near wall activity by decreasing the strength of the developing hairpins. Since the outer layer devices reduce the effect of a large scale structure by shedding vorticity of opposite sign, and the break up of the near wall structures depends to some extent on the outer layer disturbances, it is possible that there is a relative stabilisation of the streaks. Some directions for future work are given in the third part, emphasising the need for studies concerning the effect of the outer layer devices in internal flows.

1. Introduction

Apart from their importance in drag reduction (see Bandyopadhyay, 1986; Gad-el Hak, 1989,1990; Savill 1990-a for recent reviews), the outer layer devices (OLDs) provide a means to determine how the outer layer affects the inner layer structures. For instance, the question of appropriate scaling of the bursting frequency with inner (Blackwelder and Haritonidis, 1983), outer (Rao et al. 1971), or mixed (Luchik and Tiederman 1987) variables remains unanswered , yet it is not the only one. The role of the outer layer in the spatio-temporal distribution of near wall coherent structures remains to be established.

This paper reviews some aspects of the response of the bursting mechanism to the manipulation (or to the large eddy break-up) of the outer layer , in a way parallel to Smith et al. (1989). The principal aim is to fix some directions for future work.

2. Alteration of the ejection and bursting frequency

There is still no clear consensus on the detection, identification and even the definition of the bursting frequency.This makes the comparison of several published results difficult because of the different detection techniques and detection parameters which have been used so far.

The first result reported by Corke et al. (1982) has shown no effect on the bursting frequency in the manipulated case f_{bm} comparing with f_{bn} in the natural boundary layer at a station where the ratio of the wall shear stresses was $\tau_m/\tau_n=0.7$ (the subscripts m and n refer respectively to manipulated and standard flow). Chang and Blackwelder (1990) indicate less than 10% variation of f_{bm} near the wall in a situation where $C_{fm(min)}/C_{fn}=0.85$. They used VITA and quadrant techniques by choosing the local values of the turbulent intensity as threshold. (Most of the ejection and sweep detection techniques are based on level crossings of u' and u'v' signals with the exception of the VITA technique which consists of detecting large variances of u' signal during a short time which has to be of the same order of the time scale of the event. For a review and definition of these schemes -and many others such as TPAV and positive slope- together with detailed comparisons with flow visualisations, the reader is referred to the paper of Bogard & Tiederman; 1986).

The definition of the bursting frequency given by Bogard & Tiederman (1986) is basically different, and less conventional in the sense that they distinguish between ejections and bursts. Ejections are the individual lift-ups of low-momentum fluid away from the wall and the bursts result from the breakup of a single streak. Generally the bursts may contain several ejections which are closely grouped together. It is then necessary to distinguish between the bursts with single (BSE) and the bursts with multiple ejections (BME). An interpretation of the BMEs will be given in the next

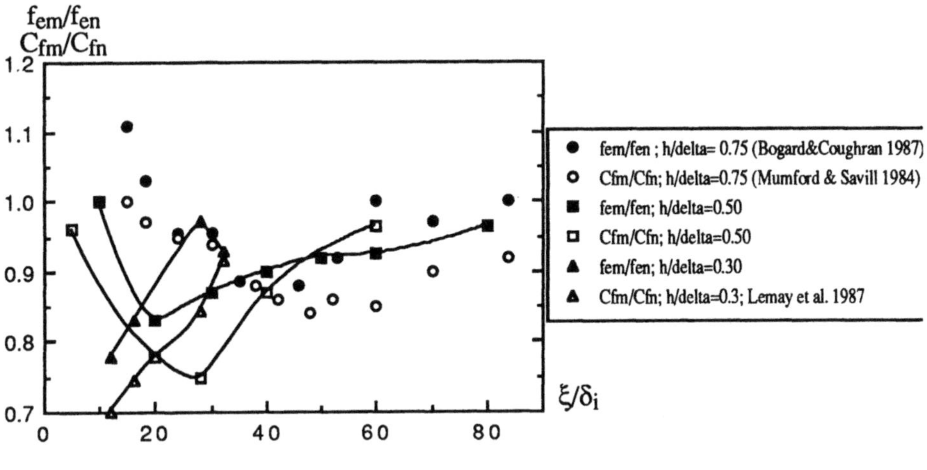

Figure 1) Profiles of the ejection frequency detected at $y^+=15$ compared with the profiles of the skin friction coefficients vs. at several streamwise distance from the LEBU and for several LEBU position h/δ in the boundary layer. Data from Bogard and Coughran (1987).

paragraph. Note that there is no direct relationship between the bursting frequency given by Bogard and Coughran (1987) and that given by Corke et al (1982) and Chang and Blackwelder (1990) . Multiple ejection bursts have not only been encountered in several experimental investigations (Kim et al., 1971; Bogard and Tiederman, 1986), but also in direct numerical simulations (Kline and Robinson,1990; Robinson et al.1988). Kline and Robinson (1990) observed in many cases two or three liftings of the low -speed-streak closely spaced in the streamwise direction. The Overal Production Module of Falco(1990) describes some possible mechanisms for the generation of multiple ejections.

Using the definition given above, Coughran and Bogard (1986) (also Bogard & Coughran, 1987) have studied in a systematic manner the behaviour of the ejection frequency f_e in a boundary layer manipulated by an OLD in tandem configuration. They have clearly shown that the profiles of f_{em}/f_{en} are closely similar to the profiles of skin friction coefficients C_{fm}/C_{fn} for different OLD's parameters (fig.1). They also compared 3 different detection schemes and showed that the results were generally independent of the technique used (fig. 2a). Note on figure 2a that the smaller variations of f_{em} have been observed with VITA. VITA contains a time scale T_v and the dependence of the

a)

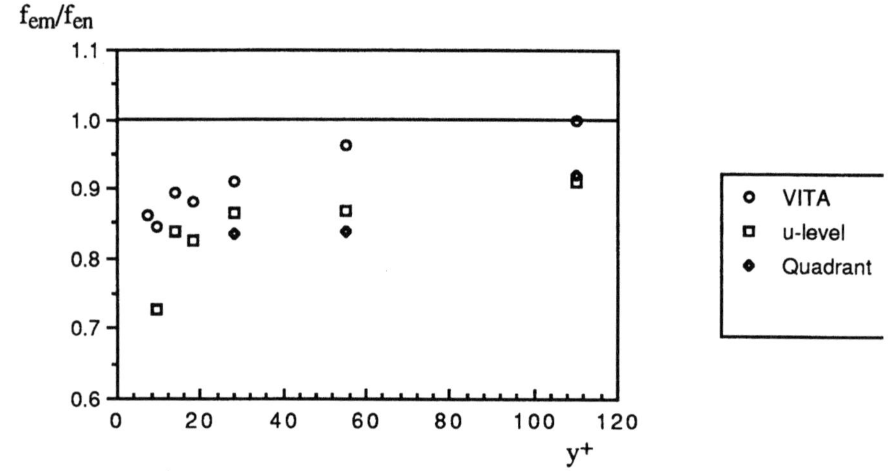

b)

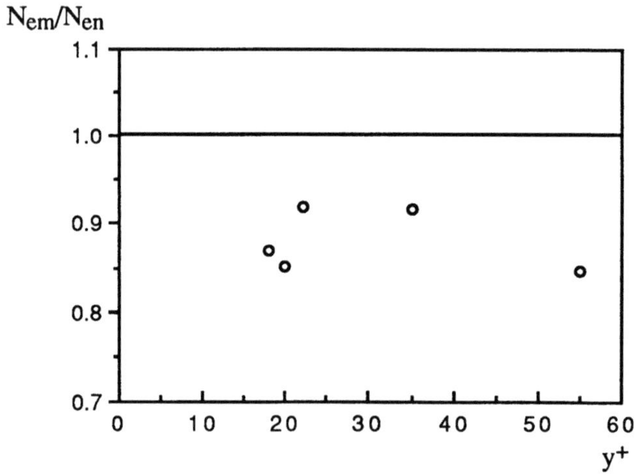

Figure 2) Comparison of ejection detection techniques (a) and profiles of the number of ejections in a burst in manipulated and unmanipulated flow (b). Data from Bogard and Coughran, 1987 and from Coughran and Bogard, 1986.

VITA integration time on other time scales of the flow is not clearly established. Instead of using a constant T_v^+ as ξ/δ, one should take T_v proportional to the duration of the event obtained with the conditional averages and iterate as done by Tardu and Binder (1991) for unsteady flow. This may have an effect on the results, since the integral time scales were found to be strongly altered.

The findings of Bogard and Coughran are important since they show the direct dependence of f_e on the wall shear stress τ. A second interesting point concerns the bursting frequency. After the grouping of the ejections into bursts has been done, f_{bm} was found to be the same as in the natural case. Since at the same time f_{em} is smaller than f_{en} the number of ejections per burst $N_{e/BME}$ must decrease (Fig. 2b). They then concluded that, although the OLDs do not affect the mechanism that stimulates a burst to occur, they affect the inner structure of the bursting mechanism.

Unfortunately no systematic studies exist on the effect of the OLDs on the magnitude, time scale and intermittence of the conditional averages. It should therefore be interesting to compare σ_{Tem}, the standard variation of the time between ejections, with the corresponding quantity in unmanipulated case. This would clarify the question of whether or not there is an ordering effect on the coherent structures, since several authors observed that the arrival of the structures happened in a more "*ordered fashion*" (Savill and Mumford, 1988; Bandyopadhyay, 1986).

3. Effect on the hairpin structures

The direct *wake effect* (Savill, 1986; Savill and Mumford, 1988) is essentially based on the interaction of the lower contrarotating wake vortices with the vortical structures existing in the near wall region studied by several research groups. The discussion here is merely based on recent findings on the dynamics of hairpin vortices reported by Smith, Walker and their co-workers . There are similarities between this model and the model of Choi (1986).

Both numerical and experimental studies concerning the hairpin vortex interaction in a shear flow have shown that, as the hairpin is advected in a shear layer , the induced pressure gradient upstream generates by viscous interaction a discrete and violent eruption of the near wall fluid. This ejection of new vorticity from the wall was interpreted as the ejection event encountered near the wall in turbulent boundary layer (Hon and Walker, 1987). Furthermore, this lift-up leads to the development of an inflectional profile in the wake of the primary hairpin, destabilising the shear layer and generating a secondary hairpin. The process is intermittent and the creation of a tertiary hairpin depends on the strength of the unsteady $\partial p/\partial x$ induced by the subsequent structures (Fig. 3). The same scheme may give an interpretation of the bursts with multiple ejections (Bogard & Tiederman 1986) i.e hairpins which are strong enough to generate secondary vortices, being close each other, since the time period required to induce an eruption (ejection) for a given vortex is inversely proportional to its strength (Walker 1989).

From this point of view, there is concordance between the results of Bogard and Coughran (1987) and the discussion of the effect of the OLDs on the near wall events reported in detail by Savill (1986) and Savill & Mumford (1988). Indeed, the decrease of the number of ejections per BME may be due to the effect of the wake vortices and their weakening effect on the primary hairpin which in turn reduces the generation of

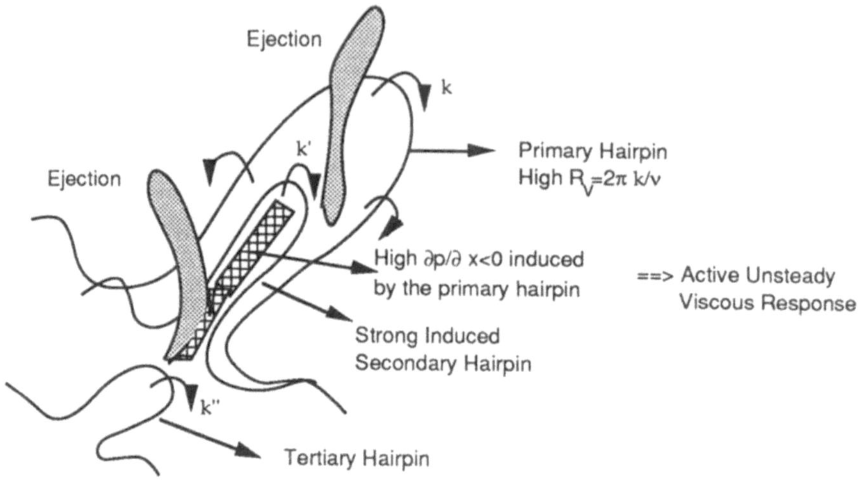

Figure 3) A model of the multiple breakup of a single streak. Adapted from Walker (1989) and Taylor and Smith (1990).

secondary vortices (Fig. 4). A critical aspect of this phenomenon is the Reynolds number $R_V=2\pi k/\nu$ (k is the strength of the vortex) since ejections only occur when R_V is greater than a threshold (Walker 1989; Walker et al.1987). The passage of the wake vortices may reduce the corresponding Reynolds number of the initiating hairpins by decreasing k, and consequently reduce the ejection rate. This interpretation is similar to that given by Anders et al.(1984) and provides also a possible explanation of the bursting modification reported by Bogard and Coughran(1987).

On the other hand, the suppression of ejection per burst indicates a tendency of the decrease of f_{BME} itself so that f_{BME} will probably scale with the local τ in the case of a more stronger interaction of wake vortices with the sublayer. It is possible that in Bogard's experiments the strength of the wake vortices at the stage when they reach the sublayer was not sufficient to give a noticeable decrease of f_{BME}. Even though the context is totally different, this is in agreement with the findings of Tardu and Binder (1989) who observed that the BMEs respond to external imposed velocity oscillations in the same manner as the wall shear stress, while a strongly different reaction of the BSEs has been observed. This suggests to investigate in more detail several aspects of the bursting mechanism in the manipulated layers.

If there exists a direct relationship between the outer manipulated layer and the near wall structures, the presence of the OLDs should also alter the streak stability and the distance λ between them. No studies exist to our knowledge on this particular point and correlation measurements with an array of wall hot film gauges should provide insight. Note that the space-time iso-correlation contours of the longitudinal fluctuating velocity, given by Lemay et al. (1987) at $y/\delta=0.5$ in the wake, have shown a reduction of the negative contours but no significant effect on the spanwise integral scale was observed, unlike the longitudinal and vertical integral scales which are considerably reduced.

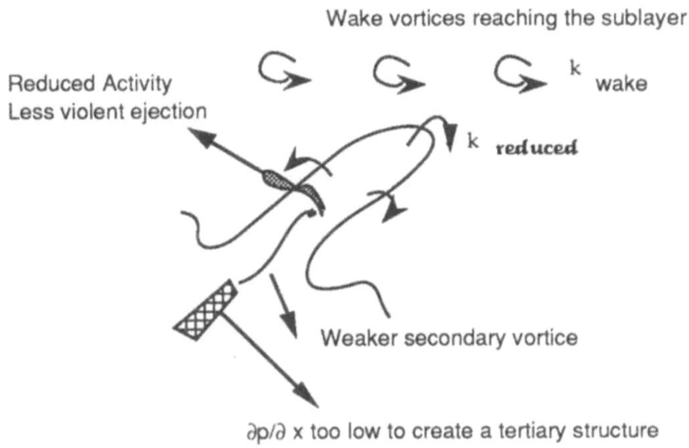

Figure 4) Possible mechanism for the interaction of the contrarotating wake vortices with the hairpins generated at the wall.

The vertical growth of a hairpin vortex is governed by the induced velocity and the shear effect (Head and Bandyopadhyay, 1981) as indicated in figure 5 . Due to the decrease of the shear and to the decrease of $\partial\delta/\partial x$, the induced velocity may dominate the shear effect and the hairpin angle may increase over 45° as noted by Savill & Mumford (1988). A similar picture has recently be given by Taylor and Smith (1990) who studied the effect of pressure gradient on the development of a single hairpin. These suggestions are yet in contradiction with the findings of Nagib et al. (1987). Their measurements indicate no alteration of the streamwise angle of the roller structures. They also report angles smaller than 45° in a regular boundary layer, contrarily to Head and Bandyopadhyay (1981) and Perry et al. (1986). One of the reasons of this discrepancy may be that, several different angles being identifiable in an hairpin during its spatial growth (Taylor and Smith,1990), the angle they measured could depend either on the part of the hairpins they detected (i.e. legs, necks or heads), or on their stage of development. The small angles that they report would probably indicate that they merely detected the legs of the hairpins which are inclined at about 10 to 20° in a standard boundary layer.

Acarlar and Smith (1987 a-b) showed that the bulges in the outer layer are the amalgamation of single hairpin vortices which interact in a complex pairing process and reinforce each other resulting in a more chaotic motion. The "pairing" process is certainly affected by the OLDs, since the hairpins were observed in the near wall region in a more organised fashion, weakened by the wake as they are convected downstream (Savill and Mumford 1988; Corke 1984), the wake imposing its own structure (Coustols and Cousteix 1989).

An interesting result of the Illinois Institute of Technology research group concerns the effect of the OLDs on the spanwise extent of the near wall roller like vortical structures (Nagib et al.,1987; Wark and Nagib, 1990).They show that the spanwise integral scale Λ_z associated with the ensemble-averaged instantaneous events scales

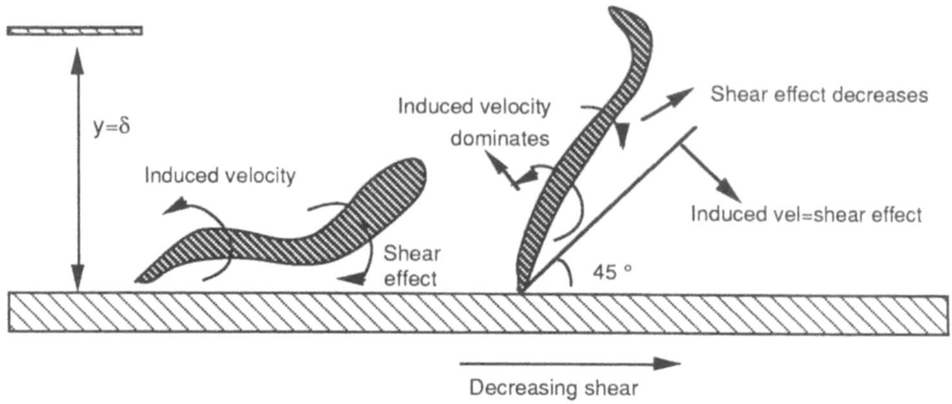

Figure 5) Possible effect on the hairpin angle; Adapted from Taylor & Smith (1990) and based on the observation of Savill and Mumford (1988)

with the outer variables over the range of Reynolds number they investigated; (Re_Θ=3400-5200). -Note, however that their closest point in the spanwise direction in their measurement grid is $\Delta z^+ = 110$, a value which is too large to detect the smaller scales- (+ designates scaling with inner variables). The computational results from NASA at $100 < Re_\Theta < 1000$ (Work and Nagib, 1990) shows that $\Lambda_z^+ = 35$, a value which is close to the vortex diameter d_w^+ of quasi-streamwise vortices computed by Robinson ($d_w^+ \approx 30$; Robinson, 1990; Fig.23). The conceptual model of Wallace (1982) indicates also the same value for d_w^+. On the other hand, there exist new evidences that the roller like quasi-streamwise vortices are indeed one-legged hairpins. Smith et al. (1991) demonstrated that, while during the initial generation process the hairpins are always symmetrical, in a high shear layer, a small degree of asymmetry of the vortical structures is highly accentuated and "the majority of vortices in a turbulent boundary layer will be one-legged hairpins" (Smith et al. 1991). These results combined with the observations of Acarlar and Smith (1987-a) give a possible explanation of the scaling of Λ_z with the streamwise distance x as shown by Nagib et al.(1987) (this scaling is of course not plausible in fully developed turbulent internal flows). The coalescence of the legs of the multiple symmetrical or asymmetrical hairpins by a complex pairing process could result in the increase of the diameter of the quasi-streamwise vortices with x. The most noteworthy effect of the OLDs (Nagib et al. 1987; Wark and Nagib ,1990) is the sudden decrease of Λ_z by about 20% immediately downstream of the device (Fig. 6). The spanwise size of the flow structures recovered their conventional boundary layer form 80 boundary layer thicknesses downstream of the manipulators. The authors conclude consequently that the main effect of the OLDs is to affect only the largest events contributing to the turbulence production. It may be questioned here whether or not the OLDs inhibit the spanwise development of the near wall structures. A second question may be arised concerning the effect of the flow acceleration between the manipulator and the wall for $\xi<5\delta$ (see for ex. Kleid and Friedrich,1990). It may play some role on the spatial growth of the hairpins, although this is not clearly established.

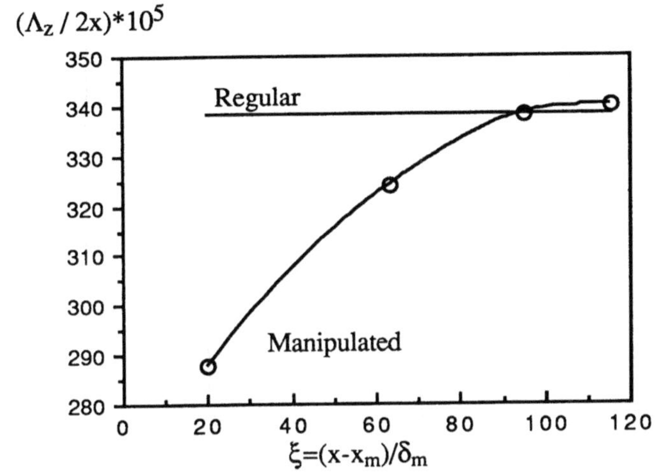

Figure 6) Wark and Nagib, 1990. Variation of the spanwise scale of the roller like vortical structures as a function of downstream distance from OLDs.

The breakup of the near wall structures is sensitive to some extent on the outer layer disturbances (Fig.7). The roll-up of a synthetic low-speed streak has been found to respond in a phase locked manner to the passage of the vortices of the outer layer (Acarlar and Smith 1987-a). Since the effect of a large scale structure passing through the OLD is reduced by the shedding of vorticity of opposite sign (Dowling 1985), the perturbations of the near wall structures produced by the outer layer are weaker i.e. there is a relative stabilisation of the streaky structure (Fig. 8). A direct conclusion of this argument would be that the amplification of the opposite sign vorticity should increase the performance of the OLDs. Some experimental results exist, indeed, showing that the manipulator profile shape has an important effect on the drag reduction. The low speed studies of Lemay and Savill indicate that the use of a single inverted cambered NACA 4415 profile is almost two times more efficient than the flat plate at the same h/δ (Savill 1990-a, Fig.13). The use of airfoils introduces other parameters as the angle of attack , the airfoil Reynolds number, etc. The alteration of the characteristics of the wake vortices have to be taken into account in addition to their capability of reducing oncoming vorticity. Bertelrud (1990) pointed out that the main future of the OLDs is due to the blockage of the phase transfer of the small scales by the production of vorticity between the inner and outer layer . Concerning this aspect, he also reported that the vortex shedding from the airfoils is significantly different compared with the flat plates.

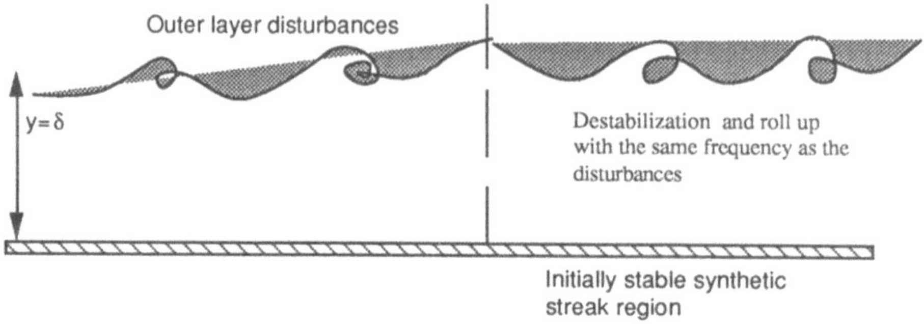

Figure 7) Effect of the outer layer disturbances on the streak stability; adapted from Acarlar &Smith (1987)

According to these arguments, the outer layer devices have a direct effect on the sublayer structures. Chang and Blackwelder (1990) have however concluded that manipulators have no direct effect on the near wall region, that the primary effect of the LEBUs is to decrease the entrainment into the boundary layer, and that the decrease of C_f is solely due to the decrease of $\partial\delta/\partial x$. They note that, the direct consequence of their point of view is that the OLDs should not decrease the drag in internal flows since there is no entrainment. Experimental evidence from channel flow in favour of or against this conjecture would undoubtedly contribute to clarify this issue. Generally speaking more efforts should be devoted to internal flows. Such efforts have been recently undertaken by several research groups (Prabhu et al., 1987; Pollard et al., 1989, 1990; Kleid and

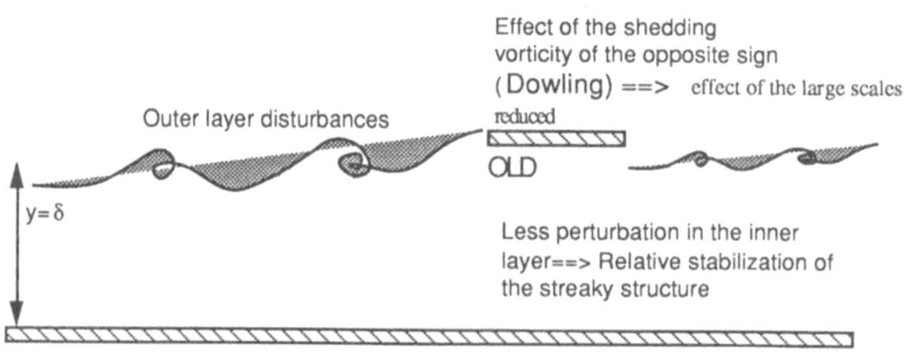

Figure 8) Possible effect of the OLDs on the stability of the near wall structures

Friedrich,1989, 1990). The large eddy simulation of Kleid and Friedrich (1989,1990) for instance, have shown that the structure of the turbulence near the wall is essentially the same in manipulated channel flow compared with manipulated flat plate boundary layer. The decrease of the local wall shear stress compares also well both qualitatively and quantitatively in internal and external flows, the maximum drag reduction taking place at larger downstream distances in channel flow (Kleid and Friedrich 1990; Figures 9 and 15). Although the smaller scales are omitted in such simulations, this first attempt is very useful at least to fix conditions for experimental setups. This also give an invaluable data base to compare the behaviour of the coherent structures in manipulated and standard boundary layer, in a way parallel to Kim and Moin (1986) for example. Pollard et al. (1989,1990) used a finite volume elliptic low-Re, k-ε model to compare computational and experimental results for both single and tandem ring devices in turbulent pipe flow . They concluded that it is unlikely that any *net* benefit could be obtained in *developing pipe* flow but that this may still be possible in *developed pipe* flow. On a fundamental basis, the net drag reduction is out of the scope of this paper, but the effect on the local value of C_f is important. The main question arised from the study of Pollard et al. (1989,1990) is that, computations gave a slight drag reduction in developed pipe flows: only 5 % of reduction of C_f is observed at $\xi=3R$ (R is the pipe radius)which is small compared with 15% at $\xi=8\delta$ in the manipulated boundary layer. This was explained by the insufficient development length both upstream and downstream of the manipulators to determine whether any net profit could be obtained (Pollard et al., 1990). To this extend, Savill (1990, this meeting) has reported an extension of higher resolution LES results concerning the effects of OLDs on channel flows. It is found that the drag is reduced, but at much larger distances than in the boundary layer case.Therefore, recent progress allow to conclude that, the main future of the OLDs is identical in both internal and external flows and that the mechanism that acts on the near wall structures is probably similar.

4. Conclusion and need for future work

The main conclusion that can be drawn from the preceding paragraphs is the consensus that the bursting frequency is not altered by outer devices. However more systematic studies are needed concerning the details of the responses of the near wall structures. This is especially so, in what is concerned with the effect on their strengths, their contribution to u'v' (in several quadrants and particularly for u'<0 v'>0 -ejections- and u'>0 v'<0 -sweeps-), the modification of their interarrival times, their spanwise extent etc. The fact that the bursting frequency is not affected does not simply mean that the coherent wall motions are not structurally altered by the manipulation of the outer layer, as was also pointed out by Bandyopadhyay (1986). The use of the wall shear stress intensity to detect the coherent structures would also provide direct information on the near wall bursts (Nagib et al.,1987; Choi 1988). Further studies, in a way parallel to Nagib et al. (1987), and concerning the strength of spanwise variations of streamwise instantaneous conditional velocity contours $\partial u'/\partial z$, would also give idea on the effect of the OLDs on the mechanisms of streak instability. Finally, more simultaneous measurements and flow visualisations are needed. Investigations of the effects of OLDs on turbulent spots which bear close resemblance with the turbulent wall structures (Perry et al., 1981) would also be welcome.

Acknowledgments

Part of this paper was prepared while the first author was at Ecole Polytechnique Fédérale de Lausanne as invited research scientist and our thanks are due to Prof. I. Rhyming and to Dr. T.V. Truong. The authors are also grateful for invaluable advices and comments of the referees.

References

1)Acarlar M.S.; Smith C.R., 1987-a "A study of hairpin vortices in a laminar boundary layer Part.1 Hairpin vortices generated by a hemisphere proturberance" J. Fluid Mech. vol. 175 p. 1.
2)Acarlar M.S.; Smith C.R. ,1987-b "A study of hairpin vortices in a laminar boundary layer Part-2 Hairpin vortices generated by fluid ejection", J. Fluid Mech. vol. 175 pp. 43-83.
3)Anders, J.B., Hefner J.N. Bushnell, D.M. ,1984 "Performance of large-eddy breakup devices at post-transitional Reynolds numbers", AIAA-84-0345.
4)Bandyopadhyay P.R., 1986 "Mean Flow in Turbulent Boundary Layers Disturbed to Alter Skin Friction", J. Fluids Eng. 106 p. 127-140
5)Bertelrud A., 1990 " The importance of LEBU device shape for turbulent drag reduction" in Turbulence Control by Passive Means; Ed. by E. Coustols; Kluwer Academic Publishers, 1990.
6)Blackwelder R. F. Haritonidis J.H., 1983 "Scaling of the bursting frequency in turbulent boundary layers", J. Fluid Mech. vol. 132 pp. 87-103
7)Bogard D.G. , Tiederman W.G., 1986 "Burst detection with single point velocity measurement" J. Fluid Mech., vol. 162 p. 389
8)Bogard D.G., Coughran M.T., 1987 "Bursts and Ejections in a LEBU modified boundary layer", 6th Symp. Turb. Shear Flows, Toulouse France 1987
9)Chang S., Blackwelder R.F., 1990 "Modification of large eddies in turbulent boundary layers", J. Fluid Mech. vol. 213 pp. 419-442
10) Choi K.S., 1986 "A new look at the near wall turbulence structure", in Advances in Turbulence Conf. Lyon,Springer-Verlag 1987
11) Corke, T.C.; Guezenec Y.; Nagib H.M., 1982 "A New View on Origine Role and Manipulation of Large Scale Structures", NASA CR165861
12) Coustols E.; Cousteix J., 1989 "Experimental manipulations of turbulent boundary layers in zero pressure gradient flows through external and internal devices", 7th symp. Turb. Shear Layers, Stanford Un. 1989 U.S.A
13) Coughran, M.T.; Bogard, D.G., 1986 "An experimental study of the burst structure in a LEBU modified layer" 10th Symp. on turbulence Rolla, Missouri.
14) Dowling, A.P., 1985 "The effect of large-eddy breakup devices on oncoming vorticity" J. Fluid Mech., vol. 160 pp. 447-464
15) Ersoy S.; Walker J.D.A, 1986 "Flow induced at a wall by a vortex pair", AIAA J. vol. 24 pp. 1597-1605
16) Falco R.E.; Klewicki J.C.; Pan; K.,1990 "Production of Turbulence in Boundary Layers and Potential for Modification of the near Wall Region" in Structure of Turbulence and Drag Reduction; Ed. by A. Gyr ; Springer-Verlag 1990.
17) Gad-el Hak M., 1989 "The Art and Science of Flow Control in Frontiers" in Exp. Fluid Mech. Springer-Verlag pp. 211-291
18) Gad-el Hak, 1990 "Flow control" App. Mech. Reviews, vol. 42 pp. 261-292
19) Hon T.L.; Walker J.D.A,1987 "An analysis of the motion and effects of hairpin vortices" Rep. FM-11 May 1987 Lehigh Un. Bettlehem USA

20) Head, M.R.;Bandyopadhyay P.R., 1981 "New aspects of turbulent boundary-layer structure" J. Fluid Mech. vol. 107 pp. 297-338

21) Kim, H.T.; Kline S.J.; Reynolds, W.C., 1971 " The Production of Turbulence Near a Smooth Wall in a Turbulent Boundary Layer" J. Fluid Mech. , vol. 50, part 1, p.133,1971.

22) Kim J.; Moin, P., 1986 " The Structure of the Vorticity Field in Turbulent Channel Flow. Part 2: Study of Ensemble Averaged Fields" J. Fluid Mech. , vol. 162, p. 339, 1986.

23) Kline S.J.; Robinson S.K.,1990 " Turbulent Boundary Layer Structure: Progress, Status, and Challenges" in Structure of Turbulence and Drag Reduction; Ed. by A. Gyr ; Springer-Verlag 1990.

24) Klein H.; Friedrich R.,1989 "Manipulating large scale turbulence in a channel and a boundary layer" 7th Symp. Turb. Shear Layers, Stanford Un. 1989 U.S.

25) Klein H.; Friedrich R,1990 "Large-Eddy Simulation of Manipulated Boundary Layer and Channel Flows" in Turbulence Control by Passive Means; Ed. by E. Coustols; Kluwer Academic Publishers, 1990.

26) Lemay J.; Savill A.M; Bonnet J.P.; Delvile J., 1987 "Some similarities between turbulent boundary layers manipulated by thin and thick flat plate manipulators" 6th Symp. on Turb. Shear Flows, 1987 Toulouse France

27) Lu L.J., Smith C.R. , 1988 "Image processing of hydrogen buble flow visualization for quantitative evaluation of hairpin-type vortices as a flow structure of turbulent boundary layers" Rep. FM-14 August 1988 Lehigh Un. Bettlehem USA

28) Luchik T.S. Tiederman W.G., 1987 "Timescale and structure of ejections and bursts in turbulent channel flows" J. Fluid Mech. vol. 174 p. 529.

29) Nagib, H.M.; Wark, C.E.; Guezennec Y.G., 1987; "Documentation of Turbulence Producing Structures in Regular and Manipulated Turbulent Boundary Layers", in Turbulence Management and Relaminarisation; Edited by Liepmann, H.W. and Narasimha R. Springer Verlag, 1987.

30) Perry A.E., Lim T.T. , Teh E.W. ,1981 "A visual study on turbulent spots" J. Fluid Mech. vol. 104 pp. 285-403.

31) Perry, A.E., Henbest, S. and Chong, M.S., 1986 "A Theoretical and Experimental Study of Wall Turbulence",. J. Fluid Mech., vol. 165, p. 163.

32) Pollard A., Savill A.M.; Thomann H.,1989 " Turbulent pipe flow manipulation: some experimental and computational results for single manipulator rings" Appl. Sc. Research, vol. 46. No:3 p.281

33) Pollard A., Thomann H., Savill A.M., 1990 "Manipulation and modelling of turbulent pipe flow: some parametric studies of single and tandem devices" in Turbulence Control by Passive Means; Ed. by E. Coustols; Kluwer Academic Publishers, 1990.

34) Prabhu, A.; Kailas Nath, P.; Kulkarni R.S.; Narasimha R., 1987 " Blade Manipulators in Channel Flow" in Turbulence Management and Relaminarisation; Edited by Liepmann, H.W. and Narasimha R. Springer Verlag, 1987

35) Rao K.N.; Narasimha R.; Badri Narayanan M.A., 1971 "The bursting phenomenon in a turbulent boundary layer" J. Fluid Mech. vol. 48 p. 339

36) Robinson, S.K.; Kline S.J.; Spalart, P.R., 1988 "Quasi-Coherent Structures in the Turbulent Boundary Layer: Part II. Verification and New Information from a Numerically Simulated Flat Plate Layer," in Near Wall Turbulence: 1988 Zaric Memorial Conference, Hemisphere, 1988.

37) Robinson, S.K.,1990 "A Review of Vortex Structures and Associated Coherent Motions in Turbulent Boundary Layers" in Structure of Turbulence and Drag Reduction; Ed. by A. Gyr ; Springer-Verlag 1990.

38) Savill A.M., 1986 "On the Manner in which outer layer disturbances affect turbulent boundary layer skin friction" In Advances in Turbulence Conf. Lyon, pp.533-545, Springer Verlag.

39) Savill A.M. , Mumford J.C. ,1988 "Manipulation of Turbulent Boundary Layers by Outer-Layer Devices: Skin Friction and Flow Visualization results" J. Fluid Mech. vol. 191, pp. 389-418

40) Savill A.M.,1990-a " Drag Reduction by Passive Devices- A Review of Some Recent Developments"; in Structure of Turbulence and Drag Reduction; Ed. by A. Gyr ; Springer-Verlag 1990.

41) Savill A.M., 1990-b "Further Computational Results for Manipulated Pipe Flow" 5th European Drag Reduction Meeting 15-16 Nov. 1990, London, U.K.

42) Smith C.R. Walker J.D.A, Haidari A.H., Taylor B.K. ,1989 "Hairpin Vortices in Turbulent Boundary Layers: The Implication for reducing Surface Drag" 2nd IUTAM Symp. on Structure of Turb. and Drag Reduction Springer-Verlag 1989

43) Smith C.R. Walker J.D.A, Haidari A.H., Sobrun U., 1991 "On the Dynamics of Near-Wall Turbulence" to appear in Philos. Trans. Roy. Soc. A August 1991.

44) Tardu S., Binder G., 1989 "Ejections and Bursts in pulsatile turbulent wall flow Measurements and visualization" in 7th symp. Turb. Shear Layers, Stanford Un. 1989 U.S.A

45) Tardu S., Binder G., 1991 "Response of Bursting to imposed velocity oscillations" Submitted to the J. Fluid Mech.

46) Taylor B.K.; Smith C.R., 1990 "Pressure Gradient Effects on the developpment of hairpin vortices in a initially laminar boundary layer" Rep. FM-15 April 1990 Lehigh Un. Bettlehem USA

47) Walker, J.D.A., 1989 "Wall layer eruptions in turbulent flows" 2nd IUTAM Symp. on Structure of Turb. and Drag Reduction Springer-Verlag 1989

48) Walker J.D.A., Smith C.R.; Cerra A.W.; Doligalski T.L., 1987 "The impact of a vortex ring on a wall" J. Fluid Mech. vol. 181, pp. 99-140

49) Wallace J.M.,1982 "On the Structure of Bounded Turbulent Shear Flow. A Personal View" in Theoretical and Applied Mech. XI, Un. of Alabama, Huntsville, p.509, 1982

50) Wark C.E.; Nagib, H.M.,1990 "Relation between outer structures and wall-layer events with and without manipulation", in Structure of Turbulence and Drag Reduction; Ed. by A. Gyr ; Springer-Verlag 1990.

III. Surface Roughness

Turbulent drag reduction of a d-type rough wall boundary layer with longitudinal thin ribs placed within the traverse grooves

H. OSAKA and S. MOCHIZUKI

Yamaguchi University
Ube, Japan

K.-S. Choi (ed.), Recent Developments in Turbulence Management, 163–180.

ABSTRACT

In order to control the turbulent energy production and transport processes due to the coherent vortices associated with the bursting phenomenon in a d–type rough wall turbulent boundary layer, longitudinal thin ribs were placed within the transverse grooves with a suitable spanwise spacing. Direct measurement of the local skin friction coefficient evidently shows the effectiveness of drag reduction using the longitudinal ribs. Maximum drag reduction rate to the d–type rough wall flow and to the smooth wall flow are 10% and 3%, respectively. The drag reduction rate can be reasonably expressed in terms of the rib Reynolds number. Comparisons of some mean flow properties between the d–type rough wall flow with and without the longitudinal ribs provide evidence that the present passive control device also reduces the turbulent energy production rate.

1.INTRODUCTION

Turbulent drag reduction due to passive control devices, such as "riblets", has been investigated energetically in many countries for the past decade (Walsh, 1982, Walsh & Lindermann, 1984). Experimental facts and knowledge obtained from available studies are summarized in several reviews (Bandyopadhyay, 1986, Wilkinson et al., 1988, Coustols & Savill, 1989). Nowadays, drag reduction phenomenon over the riblet surface is essentially established on endeavor by many research groups (Sawyer & Winter, 1987). However, direct drag measurement has been rarely done because of the difficulty. Thus, fundamental data are still lacking to sufficiently describe the most reasonable geometry for the riblets or the effect of the other external conditions such as pressure gradient or surface curvature.

Possible explanations concerned with the mechanism of the drag reduction phenomenon have been proposed by a few researchers. For example, Bacher & Smith(1985) reported a weakening effect on the streamwise vortices in the near–wall. Choi(1988) suggested that restriction of spanwise movement of the streamwise vortices associated with the near–wall burst had a primary importance. Bechert et at. (1985) noticed the reduction of spanwise momentum transfer due to the riblets. Although most of investigators are interested in correlation between the bursting phenomenon and the drag reduction (Choi, 1988), evidence to support this expectation has not been obtained yet. Concerning the turbulent drag reduction mechanism over the riblet surface, studies of the interaction process between steady streamwise vortices and a turbulent boundary layer (Osaka & Fukushima, 1990) may give a valuable information concerned with alteration of the turbulent structure.

Several kinds of control device for wall bounded shear flows have been mostly developed over the smooth surface (Bushnell & McGinley, 1989). Few applications to the flow over a rough surface can be found (Bandyopadhyay, 1985, Pineau et al., 1987). Nevertheless, the skin friction control over a rough surface has more importance for practical application to engineering problems and also for modelling of the drag generating mechanism. The passive drag reduction device like rilbets gives not only substantial benefit without extra energy supply but also more advanced understanding of the turbulent producing motion.

Let us propose a drag reducing method for a d–type rough wall boundary layer.

The present authors examined the bursting phenomenon in the near–wall region over a d–type rough surface using conditional sampling techniques and suggested the existence of similar streamwise vortex motion to that observed over smooth surface (Osaka & Mochizuki, 1987, 1990). The vortex structure is assumed to have hairpin shape and plays a dominant role on the turbulent production and transport processes. We will describe the following method based on that observation and recognition of momentum exchange between the boundary layer flow and the low momentum fluid in the transverse groove (Townes & Sabersky, 1966, Osaka et al., 1986, Osaka & Mochizuki, 1989). Figure 1 illustrates a momentum transfer mechanism due to an unsteady streamwise vortex pair and a control method with the longitudinal thin ribs placed within the transverse grooves. Spanwise movement and shedding of fluid in the transverse grooves necessarily affect motion of the hairpin vortices which directly influences the bursting activity over the d–type rough surface. When the longitudinal thin ribs restrict the fluid movement in the grooves, it is expected that turbulence producing motion associated with the hairpin vortices is weakened and then reduction of skin friction and turbulent intensities takes place at the same time.

Paticular attention in the present study is paid to two objects. The first is the possibility of turbulent drag reduction by the control method mentioned above, and the second is verification of the assumed turbulent production model. Direct measurement of the skin friction, which is the most reliable method to determine the skin friction was carried out. Measurement of the mean flow properties was also made at two representative Reynolds numbers concerning with the effectiveness of the longitudinal ribs for the drag reduction.

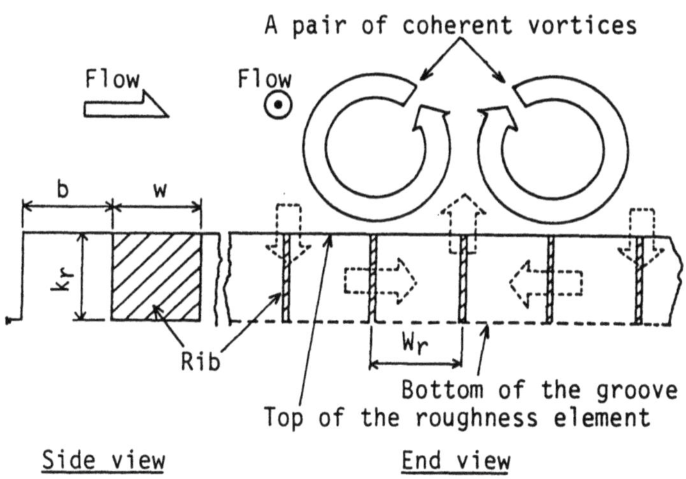

Fig.1 Momentum transport mechanism due to a pair of unsteady coherent vortices and control with the longitudinal thin ribs.

Table 1 Experimental conditions.

		800	1800
Measurement location	x_o(mm)	800	1800
Boundary layer thickness	δ_o(mm)	20	40
Rib Reynolds number	$W_r u_\tau/\nu$	35–150	35–150
Momentum thickness Reynolds number	$R_\theta(=U_1\theta/\nu)$	700–3000	1200–5000

Table 2 Experimental uncertainty.

	U	$\overline{u^2}$	$\overline{v^2}$	$\overline{w^2}$	$-\overline{uv}$	C_f
Uncertainty (%)	1.5	1.7	11	9.3	5.0	5.0

2.EXPERIMENTAL FACILITY AND TECHNIQUE

2.1 Experimental Facility and Condition

The experiment was made in a low turbulence wind tunnel of fluids engineering laboratory of Yamaguchi university. The test section has 300mm width, 500mm height and 4m long. The free stream turbulence level is maintained within 0.2%. The present d–type rough surface consists of two–dimensional repeated square ribs which are carefully machined out and its material is bakelite. The roughness height kr, element width b and groove width w are all 3mm. The longitudinal thin ribs were made of 0.1mm thickness copper sheet. The ribs were placed in the transverse grooves with a spanwise spacing Wr=3mm in 305mm streamwise and 150mm spanwise area. The measurement was made at two streamwise distances from the leading edge so that results in wider range of the momentum thickness Reynolds number were obtained. The experimental conditions are specified in Table 1. The fully developed turbulent boundary layer was formed over the d–type rough surface under the zero pressure gradient in this experimental Reynolds number range.

2.2 Skin Friction and Velocity Measurements

The direct measurement device which had 60mm diameter (10 times of the roughness pitch) floating element was prepared for skin friction measurement (Osaka & Mochizuki, 1988). The mean value was obtained as an average of output signal from a diffrential transducer for 20 sec. Constant temperature anemometers with both I–type and X–type hot–wire probes were used for velocity measurement. Tungsten wire which had 5μm diameter and 1mm sensor length was welded to the tip of the probes. Because the sensor length was less than 40 viscous wall units in any case, the hot–wire sensor had reasonable spatial resolution (Ligrani & Bradshaw, 1987). Output signal from the anemometers was digitized and then analyzed with a personal computer system. The measurements were made mostly at a streamwise distance Δx=225mm measured from the beginning of the longitudinal ribs. This distance corresponds to 5.6 and 11 times of initial boundary layer thickness at x_O =800mm and 1800mm respectively, and also 2000 viscous wall units. It is reasonable to expect that turbulent structure in the wall layer at least adjusts to the local boundary condition at the wall after traveling the streamwise distance. The experimental uncertainty for skin friction and velocity measurements were estimated according to the literature (Yavuzkurt, 1984) and are written in Table 2.

3.RESULTS AND DISCUSSION

3.1 Local Skin Friction Coefficient

The local skin friction coefficient is plotted versus the momentum thickness Reynolds number in Fig.2. The solid line in the figure represents the Kármán–Schoenherr's flat plate boundary layer result. When the Reynolds number is larger than 2000, no effect of the longitudinal ribs is found and the skin friction coefficient of both the d–type rough surfaces is larger than that of the smooth surface. When the Reynolds number is less than 2000, it can be seen that the longitudinal ribs reduce the skin friction of the d–type rough wall flow. Remarkably, at the certain range of Reynolds number R_θ=750–1200 the skin friction coefficient of the d–type rough surface with the longitudinal ribs is smaller than that of the smooth surface. The smaller skin friction flow over a rough surface compared to the smooth surface has been found in a few papers. Tani(1988) re–evaluated Nikuradse's rough pipe flow data and predicted that at small roughness Reynolds number drag reduction occured even over the sand grain roughness. Matsumoto et al.(1986) disturbed a d–type rough wall boundary layer with a circular cylinder and found a condition of the disturbance for the drag reduction in the farthest downstream region. The three kinds of flow under consideration take almost the same skin friction coefficient at about R_θ=800. The reduction rate of the skin friction coefficient to the present d–type rough

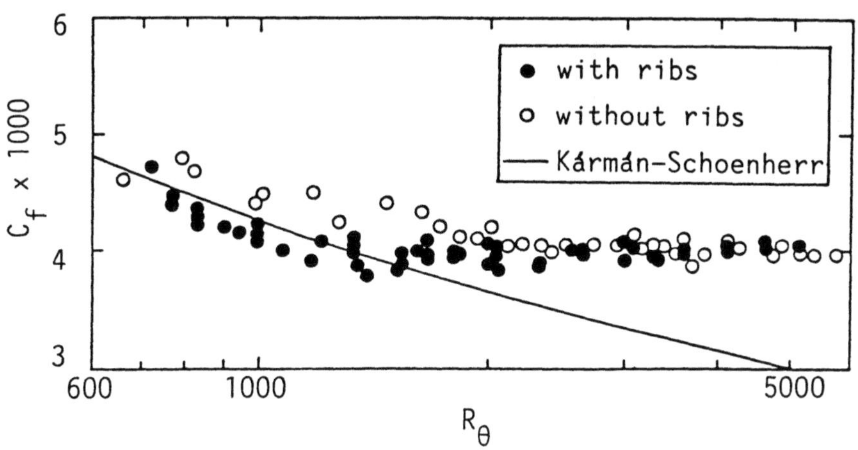

Fig.2 Local skin friction coefficient.

wall flow defined as equation(1) is shown in Fig.3.

$$\Delta C_{fd} = (C_{fR} - C_{fd})/C_{fd} \times 100 \quad (\%) \qquad (1)$$

Here, C_{fR} and C_{fd} are the local skin friction coefficient of the d–type rough surface with and without the longitudinal ribs respectively. The dotted area shows the data scattering. The maximum scattering of ΔC_{fd} is ±6%. The effectiveness of the longitudinal ribs can be definitely seen in the Reynolds number range of Rθ=750–2000. The maximum reduction rate is about 10% at Rθ=1200.

Now, we will examine the validity of the assumption of turbulent production model. Rib Reynolds number $Wr^+(=Wru_\tau/\nu)$ is introduced as a non–dimensional parameter to describe the effect of longitudinal ribs. Wr^+ is considered as a ratio of representative length scale of the ribs Wr to the fundamental length scale for the streamwise vortex pair ν/u_τ. Figure 4 shows the reduction rate versus the rib Reynolds number. Both results obtained x_0=800mm and 1800mm are given in the same figure. The drag reduction over the riblet surface occurs up to the non–dimensional spanwise spacing $S^+ \fallingdotseq 20$ (Walsh & Anders, 1989). In the present result, when the rib Reynolds number is smaller than about 100 at which spanwise spacing of the ribs and spanwise scale of the streamwise vortex pair are of the same order (Osaka & Mochizuki, 1987), larger drag reduction occurs at smaller rib Reynolds number in both results. This directly means the verification of the assumed turbulent production model as a basis of the present control method. However, both results at x_0=800mm and 1800mm are not of the same in tendency and magnitude of the reduction rate profile. As shown in Table 1, the momentum thickness Reynolds number at x_0 =800mm is smaller than

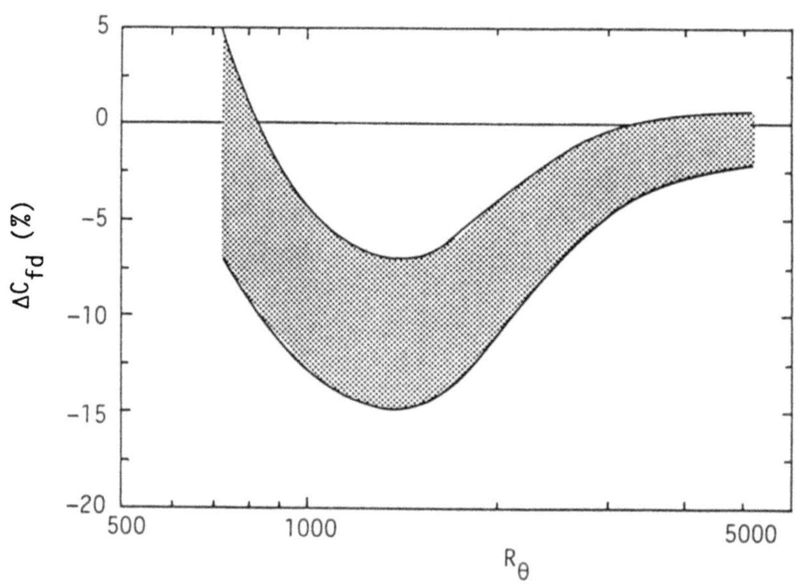

Fig.3 Reduction rate of the local skin friction coefficient to the d–type rough wall flow.

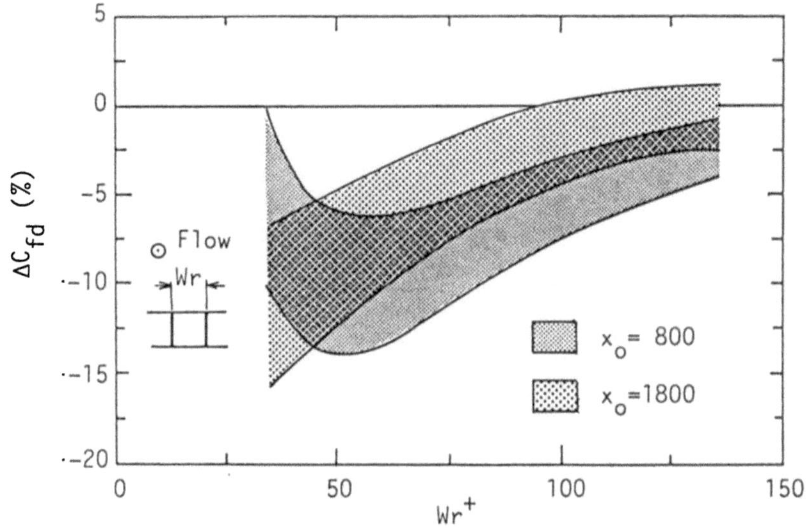

Fig.4 Reduction rate of the local skin friction coefficient to the d–type rough wall flow.
(The reduction rate is expressed in terms of the rib Reynolds number)

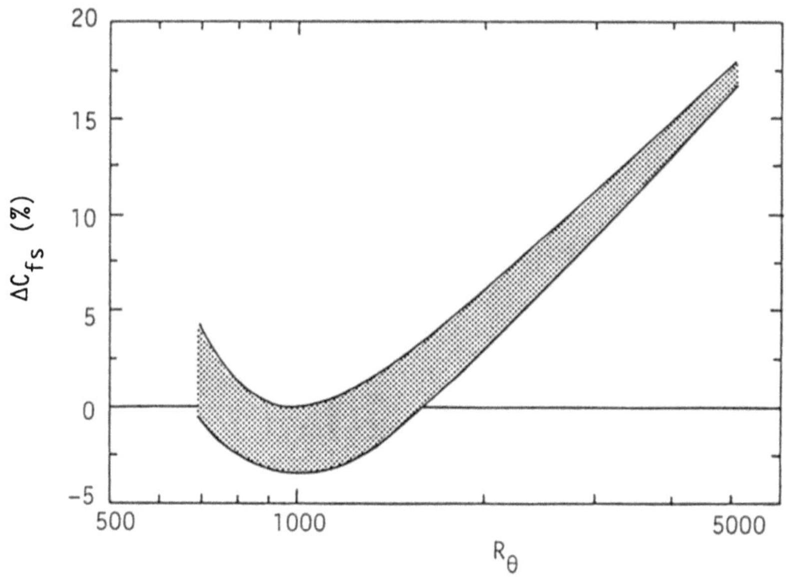

Fig.5 Reduction rate of the local skin friction coefficient to the smooth wall flow.

that at x_o =1800mm. We have already found in the previous paper (Osaka & Mochizuki, 1990) that the ejection event of the present d–type rough wall flow becomes more intense at lower Reynolds number. Because the longitudinal ribs are assumed to restrict the motion associated with the ejection event, we can expect larger effectiveness at a lower Reynolds number. As the rib Reynolds number decreases in the region of $Wr^+<50$, absolute value of the reduction rate at x_o =800mm becomes smaller. The effectiveness of the longitudinal ribs is relatively low in such very low Reynolds number about $R\theta<800$. We may consider unimportance of the streamwise vortex pair or the momentum exchange between the boundary layer flow and the vortex in the grooves as the reason for the reduced effectiveness. This reason is in a discrepancy with our conditional sampling result which indicates a dominant contribution of the streamwise vortices to the turbulent production process at Rθ=800 (Osaka & Mochizuki, 1987). The latter finding can be attributed to the fact that the skin friction coefficient of the original d–type rough wall flow is almost the same as that of the smooth wall flow at Rθ=800. According to the result at x_o =1800mm, which hardly includes the low Reynolds number effect a larger reduction rate is seen at the lower rib Reynolds number. We can not suggest an optimum value of the rib Reynolds number for the drag reduction, because no results were obtained in the region of $Wr^+<35$ in the present study.

The drag reduction rate to the smooth wall flow defined as equation(2) is shown versus Rθ in Fig.5.

$$\Delta C_{fs} = (C_{fR}-C_{fs})/C_{fs} \times 100 \quad (\%) \tag{2}$$

Here, C_{fs} is the skin friction coefficient of the smooth wall flow, which is determined from the Kármán–Schoenherr's experimental formula. It is clearly seen that the skin friction coefficient of the d–type rough wall flow with the longitudinal ribs is smaller than that of the smooth wall flow in the region of Rθ=750–1200. The maximum reduction rate is about 3% at Rθ=1000. This result directly indicates that the present passive control device utilizing the longitudinal thin ribs is effective for the drag reduction of the d–type rough wall boundary layer.

3.2 Mean Velocity Profile

Effect of the longitudinal ribs on the mean velocity profile at two representative Reynolds numbers, Rθ=1300 and 5000, which correspond to $Wr^+=35$ and 120 respectively is examined in Fig.6. It is supposed that an alteratin of the turbulent structure due to the passive control appears in the mean properties through the turbulent production term. The Reynolds shear stress production can be explained by means of the streamwise vortex pair in our turbulent production model. Taking account this consideration, the outer layer variables, free stream velocity and boundary layer thickness, are employed to normalize the profiles. The profiles are plotted in a semi–logarithmic diagram to examine the alteration which may be limited in the wall layer. The symbol y_t expresses the distance from the top of roughness element normal to the wall. At Rθ=1300 the mean velocity close to the wall $y_t/\delta<0.02$ slightly increases over the d–type rough surface with the longitudinal ribs. At Rθ=5000 no difference between the d–type rough wall with and without the longitudinal ribs is found. The step change of surface roughness (Antonia & Luxton, 1972) is not recognized

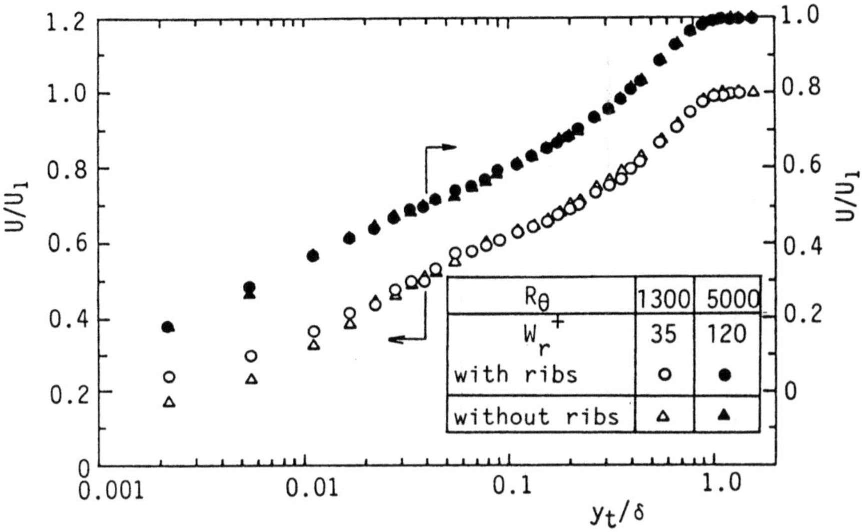

Fig.6 Mean velocity profiles normalized with the outer variables.

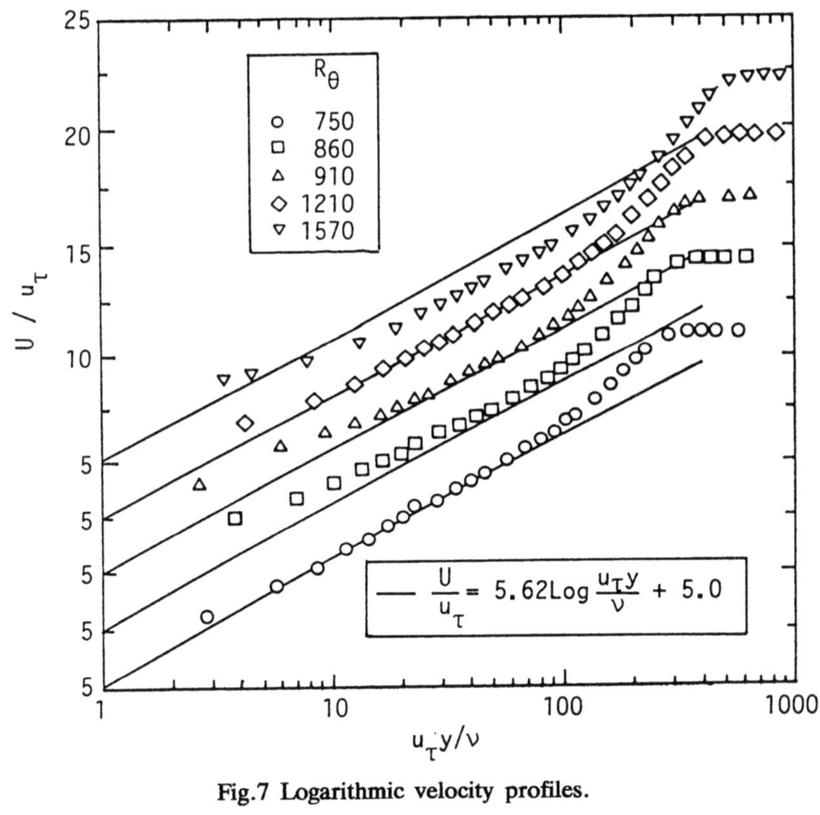

Fig.7 Logarithmic velocity profiles.

in the profiles of both Reynolds numbers. The results discussed here were measured at the center of neighboring longitudinal ribs and over a transverse groove. The results depend on the relative longitudinal location to the roughness element and also the relative spanwise location to the longitudinal ribs. The dependence of the profiles on various locations should be discussed.

The logarithmic velocity profile in the Reynolds number range where the skin friction coefficient over the d–type rough surface with the longitudinal ribs is smaller than that of the smooth surface is examined in Fig.7. Before the rough surface data are plotted in the figure, we need to determine a virtual origin. If the boundary condition at the wall acts on turbulent eddies in the outer layer only through magnitude of the friction velocity, the velocity defect law could be used (Furuya & Fujita, 1966) for determining the friction velocity. However, we employed the method proposed by Monin & Yaglom(1973), because the Reynolds number range of the present study is rather low. There is no guarantee that turbulence keeps a similar structure even in the low Reynolds number flow. Actually, we will find a small amount of change in the outer layer structure in the following discussion. The error in origin was determined from the friction velocity obtained by the direct measurement so that the logarithmic velocity region was as long as possible. The solid line represents the profile of flat plate boundary layer. The logarithmic velocity region can be recognized in all the profiles in this figure. The slope and intercept of the logarithmic profile over the d–type rough surface with longitudinal ribs is almost the same as that over smooth surface at Rθ=750 and 1210. The skin friction coefficient of both the flows have almost the same value at these Reynolds numbers. For Rθ=860 and 910 where the skin friction coefficient of the

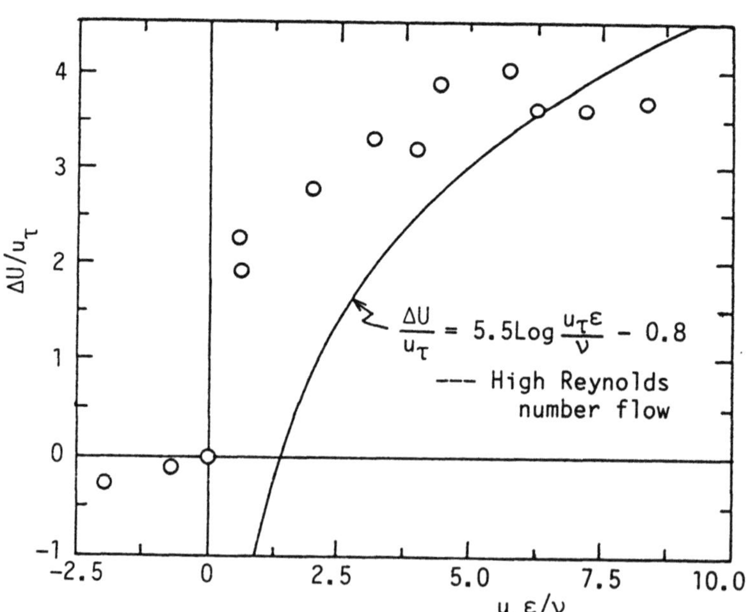

Fig.8 Roughness function.

d–type rough surface with the longitudinal ribs is smaller than that of smooth surface, it is clearly seen that the logarithmic profiles are slightly shifted upward to the smooth wall flow profile. The roughness function closely correlated to an increased drag over a rough surface has usually a negative sign (Rotta, 1962). Upward shift of the logarithmic profile was also found in the drag reduction phenomenon by the large eddy break–up device (LEBU's) (Bandyopadhyay, 1985). At Rθ=1570 the roughness function has a negative sign and indicates the drag increase by the surface roughness. The mean velocity close to the wall seems to increase over the d–type rough surface with the longitudinal ribs. Such tendency had never been observed over the riblets surface (Choi, 1989). It may be possible to interpret that the smaller skin friction causes the velocity increase close to the wall. We need to examine the turbulent structure vicinity to the wall in order to understand in more detail in that phenomenon. Figure 8 shows the roughness function versus the Reynolds number based on the error in origin which may be considered as the representative scale of a d–type rough wall. The Kármán constant is assumed to be constant value, κ=0.41. The logarithmic relation indicated by the solid line was obtained in a fully rough regime (Osaka & Mochizuki, 1988). The data in a transitionally rough regime, $\varepsilon u_\tau/\nu < 5$, deviates from the solid line. The Reynolds number $\varepsilon u_\tau/\nu$ also takes a negative sign when the roughness function has a negative sign. It means that the virtual origin is located above the top of roughness element. Note that the virtual origin is determined associated with logarithmic layer and the mean velocity does not take zero at y=0. In addition, the logarithmic function is unacceptable to represent the correlation between the roughness function and the roughness Reynolds number in the wide range from the drag increase to the drag reduction.

The outer layer velocity profiles are examined in detail according to the law of the wake proposed by Coles(1956). The law of the wake for a rough wall

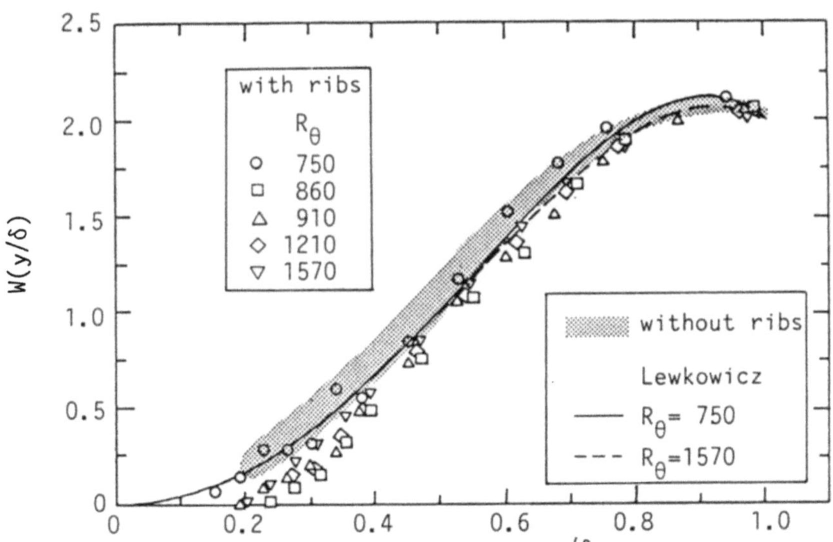

Fig.9 Wake function.

turbulent boundary layer can be written as the following equation(3).

$$\frac{U}{u_\tau} = 5.62 \operatorname{Log} \frac{u_\tau y}{\nu} + 5.0 - \frac{\Delta U}{u_\tau} + \frac{\Pi}{\kappa} W\left(\frac{y}{\delta}\right) \tag{3}$$

Here, $W(y/\delta)$ and Π are the universal wake function and the wake parameter respectively. The wake functions calculated at five Reynolds numbers shown in Fig.8 are plotted in Fig.9. The profiles of the d–type rough wall flow in low Reynolds number range fall into the dotted area (Osaka & Mochizuki, 1988). The solid and broken lines are drawn according to Lewkowicz's empirical equation (Lewkowicz, 1982) for the smooth wall given by the following equation(4).

$$W\left(\frac{y}{\delta}\right) = 2\left(\frac{y}{\delta}\right)^2\left(3-2\frac{y}{\delta}\right) - \frac{1}{\Pi}\left(\frac{y}{\delta}\right)^2\left(1-\frac{y}{\delta}\right)\left(1-2\frac{y}{\delta}\right) \tag{4}$$

We reported that the profiles of the d–type rough wall flow have a small upward deviation from that of smooth wall flow, especially in the region of $y/\delta<0.5$. While, the profiles over the d–type rough surface modified by the longitudinal ribs have a small downward deviation from that of the smooth wall flow. It is seen that the outer layer turbulent structure is slightly altered by changing the boundary condition at the wall. This result raises the question of the degree of reliability of the method to determine the friction velocity based on the outer layer similarity law. The wake parameter is given in Fig.10. The solid line represents Coles's empirical formula (Coles, 1956) for the flat plate boundary layer given by equation(5).

$$\Pi = 0.55[1-\exp(-0.243\xi^{\frac{1}{2}}-0.298\xi)]$$
$$\text{where } \xi = R_\theta/425 - 1 \tag{5}$$

The wake parameters of both the d–type rough surfaces in the region of $R_\theta>4000$

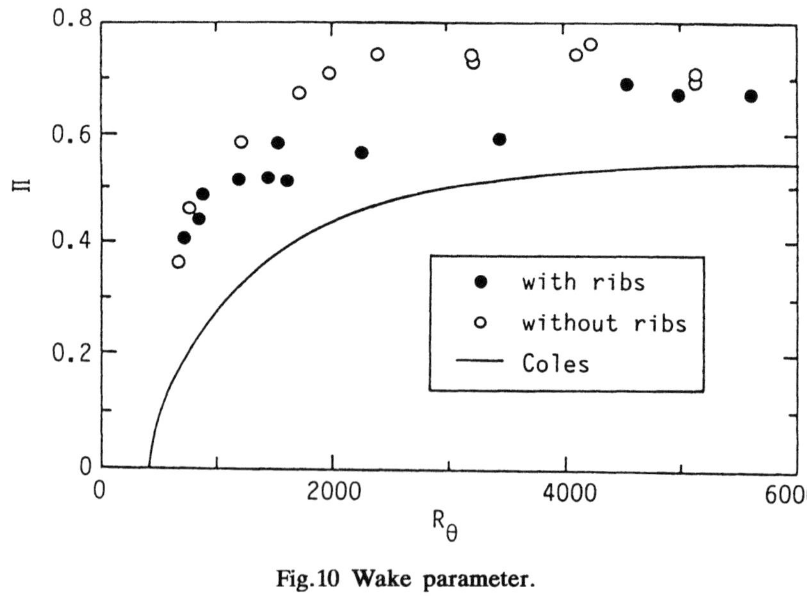

Fig.10 Wake parameter.

where the longitudinal ribs hardly reduces the drag, are in good agreement. Three kinds of flow take different values of the wake parameter regardless of the same skin friction coefficient at Rθ=800. The wake parameter becomes smaller value, when the skin friction coefficient reduces in the Reynolds number range of Rθ=1000–3000. This may be interpreted as an effect of the parameter derived from the dimensional argument for the outer layer velocity profile, u_τ/U_1 (Rotta, 1962).

3.3 Turbulence Characteristics

It is fairly expected that the present control device reduces the skin friction, turbulent intensities and Reynolds shear stress at the same time. We have already seen that the rib Reynolds number is an appropriate parameter to evaluate the effectiveness of the longitudinal ribs in Fig.3. The turbulent characteristics at two representative Reynolds numbers, Wr^+=35 and 120, are examined. The effect of the longitudinal ribs will be discussed on the comparison between the d–type rough wall flow with and without the longitudinal ribs at the same momentum thickness Reynolds number. The flow conditions of Wr^+= 35 and 120 correspond to Rθ=1300 and 5000 respectively. The streamwise velocity component profiles of turbulent intensities normalized with the outer layer parameters are given in Fig.11. The turbulent intensity is reduced over the d–type rough surface due to the longitudinal ribs in the near–wall region. The maximum reduction rate of the turbulent intensity is about 45%. On the other hand, the turbulent intensity slightly seems to increase at the outer egde of the inner layer. This result can be related to the alteration of the outer layer velocity profile as seen in section 3.2 and suggests that the longitudinal ribs act on the large–scale turbulent structure in the outer

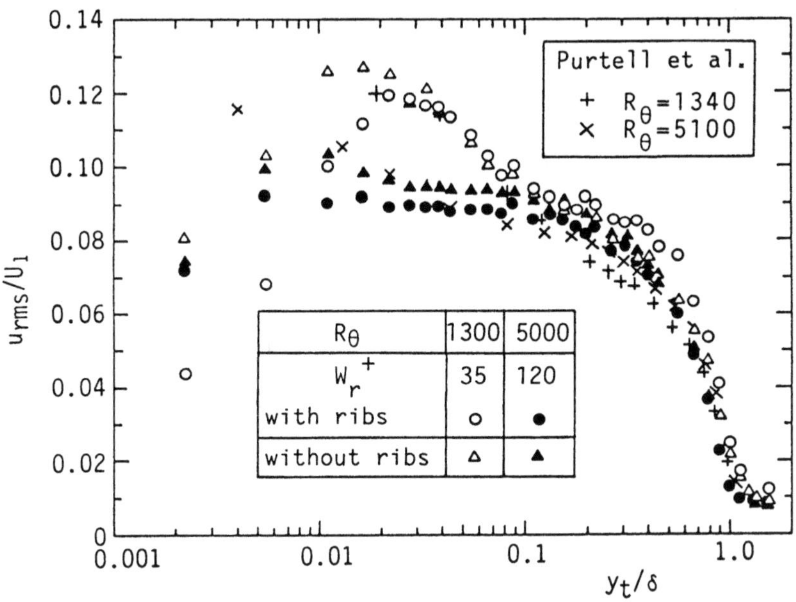

Fig.11 Turbulent intensity profiles of the streamwise velocity component.

layer. It is possible to explain that the effect of the longitudinal ribs appers in the outer layer structure through the motion of turbulent production. A study of the drag reduction phenomenon by LEBU's concluded that the reduction of entrainment rate was the primary effect (Chang & Blackwelder, 1990). The turbulent intensity is slightly reduced at Rθ=5000. The profiles over a smooth surface measured by Purtell et al.(1981) are plotted in the figure for the

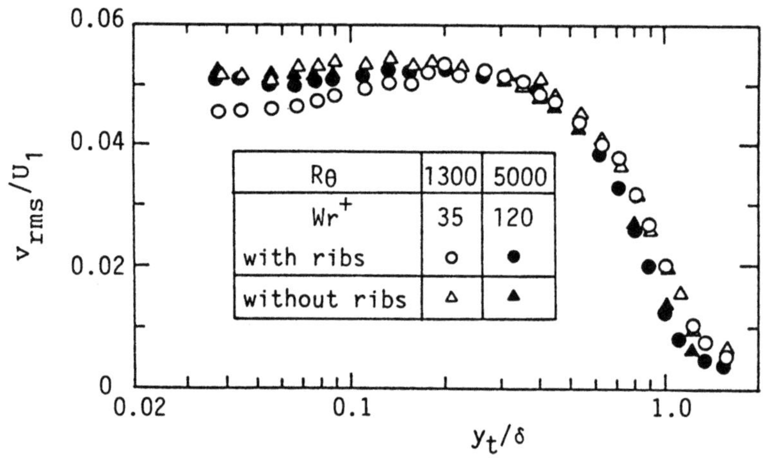

Fig.12 Turbulent intensity profiles of the vertical velocity component.

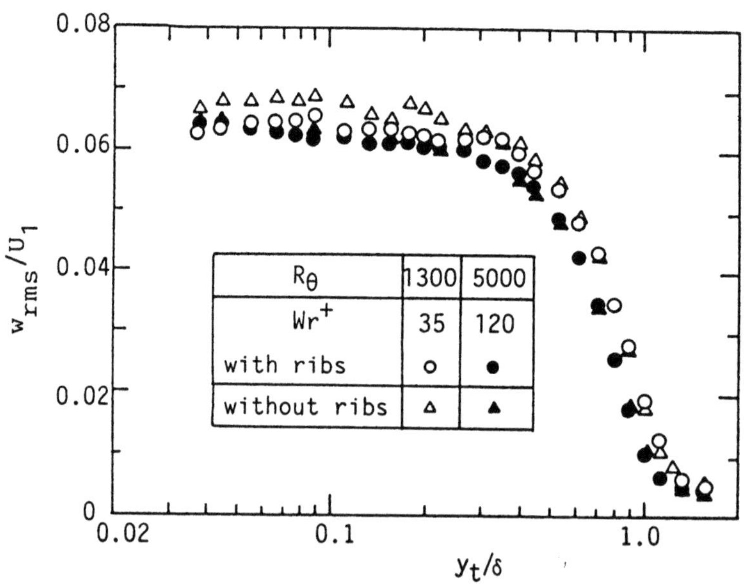

Fig.13 Turbulent intensity profiles of the spanwise velocity component.

comparison. The magnitude of turbulent intensity of the d–type rough wall flow with the longitudinal ribs is larger than that of the smooth wall flow in the outer layer at Rθ=1300. The maximum turbulent intensity occurs in almost the same magnitude and at the same non–dimensional distance from the wall in the d–type rough wall flow modified by the longitudinal ribs and the smooth wall flow, which have almost the same skin friction coefficient. Both the d–type rough wall flows have smaller turbulent intensity close to the wall at Rθ=5000 compared to that of the smooth wall flow.

The vertical velocity component profiles of turbulent intensity are given in Fig.12. The reduction of the turbulent intensity can be seen in the inner layer at Rθ=1300. The maximum reduction rate of v_{rms}/U_1 is about 12% and smaller than that of u_{rms}/U_1. The larger reduction rate in the vertical component was expected on the assumption that the longitudinal ribs restrict the motion of the streamwise vortex pair close to the wall. A larger reduction of the vertical component has been found in the wall layer of polymer additive flow. This difference of reduction rate in the streamwise and vertical fluctuating velocity components may give the information about relation between the coherent motion and the production mechanism of the Reynolds shear stress. We expect that the present control method weakens the ejection–like motion. While, the large–scale sweep–like motion is restricted in the polymer additive turbulent flow (Usui & Sano, 1988). The relative large reduction in the streamwise component suggests that the turbulence production process can be

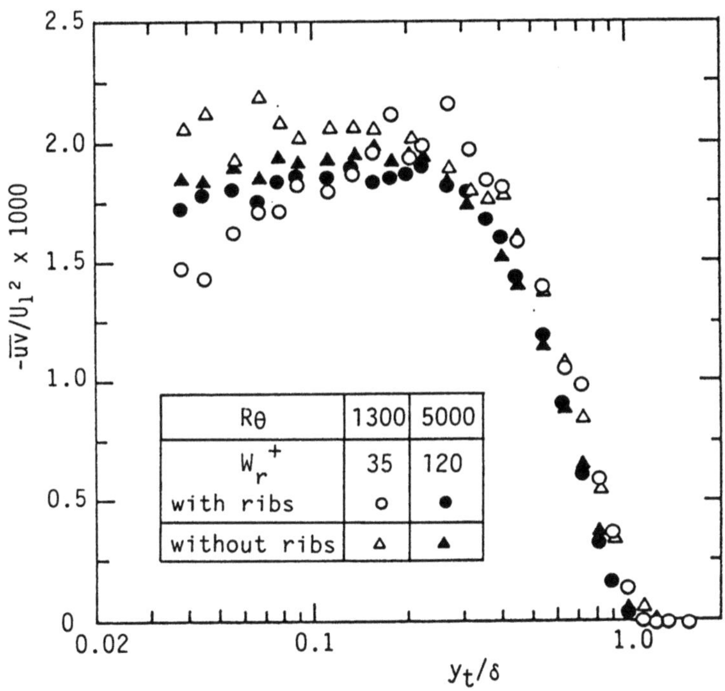

Fig.14 Reynolds shear stress profiles.

still understood by means of the dominant term $-\overline{uv}dU/dy$. At Rθ=5000 no alteration due to the longitudinal ribs is found. The spanwise velocity component profiles of turbulent intensity are given in Fig.13. The turbulent intensity w_{rms}/U_1 is also reduced in the inner layer at Rθ=1300. While, no reduction is found at Rθ=5000. It is seen from the above results that reduction of the turbulent intensity occurs in all three velocity components, when the skin friction coefficient is reduced. This experimental fact means that the present passive device controls the turbulent production mechanism and that the assumed turbulent production model is reasonable one.

The Reynolds shear stress profiles are plotted in Fig.14. The Reynolds shear stress is reduced by a large amount of 25% in the inner layer over the d–type rough surface with the longitudinal ribs at Rθ=1300. The smaller Reynolds shear stress causes less momentum flux toward the wall. The decay of turbulent transport activity suggests an alteration of the triple velocity correlations which include the Reynolds shear stress and turbulent kinetic energy fluxes. According to the results over several kinds of rough surface (Bandyopadhyay & Watson, 1988, Osaka & Mochizuki, 1990), it is expected that the Reynolds shear stress flux $-\overline{uv^2}$ is reduced in the inner layer. The magnitude of the Reynolds shear stress at the outer edge of the inner layer is slightly larger as seen in the streamwise turbulent intensity profile. At Rθ=5000 no alteration is found. The Reynolds shear stress normalized with the friction velocity is of order one in the wall region. This means that a discussion based on the constant stress layer is possible even in the present d–type rough wall flow modified by the longitudinal ribs.

4.CONCLUDING REMARKS

The passive control method with the longitudinal thin ribs was proposed for the drag reduction of a d–type rough wall boundary layer and tested experimentally. The main results are summarized as follows.

(1)The present control device is effective in reducing the skin friction of the d–type rough wall boundary layer. The skin friction coefficient of the d–type rough surface with longitudinal ribs is definitely small compared to that of the smooth surface in the Reynolds number range of Rθ=750–1200. The effectiveness of the longitudinal ribs can be reasonably expressed in terms of the rib Reynolds number, Wr^+.

(2)The mean velocity profiles of the d–type rough wall flow modified by the longitudinal ribs are examined with the friction velocity determined by the direct skin friction measurement in detail. A logarithmic velocity profile still exists over the d–type rough surface with the longitudinal ribs. When the skin friction coefficient of the d–type rough wall flow modified by the longitudinal ribs is smaller than that of the smooth wall flow, the logarithmic profile is shifted upward to the smooth wall flow profile. Study of the outer layer velocity profile suggests that the effect of changing the boundary condition at the wall appears in the outer layer turbulent structure.

(3)The turbulent intensities and Reynolds shear stress are also reduced in the wall region, when the drag reduction occurs. Comparing the results of the d–type rough wall flow, the maximum reduction rate of the streamwise turbulent intensity and Reynolds shear stress profiles are about 45% and 25%

respectively. It is seen that the longitudinal ribs placed within the grooves sufficiently controls the turbulent production rate.

REFERENCES

Antonia,R.A. & Luxton,R.E. 1982 The response of a turbulent boundary layer to a step change in surface roughness. part 2. rough–to–smooth. *J.Fluid Mech.* 53, 737–757.

Bacher,E.V. & Smith,C.R. 1985 A combined visualization–anemometry study of turbulent drag reducing mechanisms of triangular micro–groove surface modification. *AIAA Paper* No.85–0548.

Bachert,D.W., Hoppe,G. & Reif,W.–E. 1985 On the drag reduction of the shark skin. *AIAA Paper* No.85–1546.

Bandyopadhyay,P.R. 1985 The performance of smooth–wall drag reducing outer–layer devices in rough–wall boundary layers. *AIAA Paper* No.85–0558.

Bandyopadhyay,P.R. 1986 Review–Mean flow in turbulent boundary layers disturbed to alter skin friction. *Trans.ASME J.Fluids Engng* 108, 127–140.

Bandyopadhyay,P.R. & Watson,R.D. 1988 Structure of rough–wall turbulent boundary layer. *Phys.Fluids* 31, 1877–1883.

Bushnell,D.M. & McGinley,C.B 1989 Turbulence control in wall flow. *Ann.Rev.Fluid Mech.* 21, 1–20.

Chang,S.–I. & Blackwelder,R.F. 1990 Modification of large eddies in turbulent boundary layer. *J.Fluid Mech.* 213, 419–442.

Choi,K.–S. 1988 The wall–pressure fluctuations of modified turbulent boundary layer with riblets. In *Turbulence management and relaminarization* (ed. H.W.Liepmann & R.Narashimha). 149–160, Springer.

Choi,K.–S. 1989 Near–wall structure of a turbulent boundary layer with riblets. *J.Fluid Mech.* 208, 417–458.

Coles,D.E. 1956 The law of the wake in the turbulent boundary layer. *J.Fluid Mech.* 1, 191–226.

Coustols,E. & Savill,A.M. 1989 Résumé of important results presented at the third turbulent drag reduction working party. *App.Sci.Res.* 46, 183–196.

Furuya,Y. & Fujita,H. 1966 Turbulent boundary layer over rough gauze surface. *Trans.JSME* 32, 724–733 (in japanese).

Lewkowicz,A.K.Z. 1982 An improved universal wake function for turbulent boundary layers and some of its consequences. *Z.Flugwiss, Weltraumforesch* 6, 261–266.

Ligrani,P.M. & Bradshaw,P. 1987 Spatial resolution and measurement of turbulence in the viscous sublayer using subminiature hot–wire probes. *Exp.Fluids* 5, 407–417.

Matsumoto,A., Munakata,H. & Abe,K. 1986 Relaxation behavior of smooth wall and rough–wall turbulent boundary layers perturbed by a circular rod. *Proc.the 18th turbulent symposium*, 159–166 (in japanese).

Monin,A.S. & Yaglom,A.M. 1973 Statistical fluid mechanics of turbulence. 284, the MIT Press.

Osaka,H. Nakamura,I. & Kageyama,Y. 1984 Mean flow properties of the turbulent boundary layer over a d–type rough surface. *Trans.JSME* B50, 2299–2306 (in japanese).

Osaka,H. & Mochizuki,S. 1987 Streamwise vortical structure associated with the bursting phenomenon in a d–type rough wall boundary layer at a low Reynolds number. *Proc.the 6th Turbulent Shear Flows in Toulouse*, 16.7.1–16.7.6.

Osaka,H. & Mochizuki,S. 1988 Turbulent structure of a d–type rough wall boundary layer in a transitionally rough regime. *Proc.the 1st KSME–JSME Thermal and Fluids Engineering Conference in Seoul,* 2.88–2.93.

Osaka,H. & Mochizuki,S. 1989 Turbulent structures in the near wall region for a boundary layer over a d–type rough surface at a low Reynolds number. *Proc.the International Conference on Fluid Dynamic Measurement and its Application in Beijing,* 260–265.

Osaka,H. & Mochizuki,S. 1990 Statistical quantities of a d–type rough wall boundary layer in a transitionally rough regime. *Proc.the 2nd KSME–JSME Fluids Engineering Conference in Seoul,* 1.202–1.207.

Osaka,H. & Fukushima,C. 1990 Effect of Controlled longitudinal vortex arrays on the development of turbulent boundary layer. *Proc.the Engineering Turbulence Modelling and Experiments in Dubrovnik,* 593–602.

Pineau,F., Nguyen,V.D., Dickinson,J. & Belanger,J. 1984 Study of flow over a rough surface with passive boundary layer manipulators and direct wall drag measurements. *AIAA Paper* No.**87–0357**.

Purtell,L.P., Klebanoff,P.S. & Buckley,F.T. 1981 Turbulent boundary layer at low Reynolds number. *Phys.Fluids* **24**, 802–811.

Rotta,J.C. 1962 Turbulent boundary layer in incompressible flow. *Progress in Aeron. Sci.* 2, 14, Pergamon Press.

Sawyer,W.G. & Winter,K.G. 1987 An investigation of the effect on turbulent skin friction of surfaces with streamwise grooves. *Proc.the Turbulent Reduction by Passive Means.* The Royal Aeronautical Society, 330–362.

Tani,I. 1988 Drag reduction by riblet viewed as roughness problem. *Proc. Japan Acad.* **B64**, 21–24.

Townes,H.W. & Sabersky,R.H. 1966 Experiments on the flow over a rough surface. *Int.J.Heat & Mass Transfer* 9, 729–738.

Usui,H. & Sano,Y. 1988 Effect of injected polymer thread on turbulence in a pipe flow. In *Transport Phenomena in Turbulent Flow* (ed. M.Hirata & N.Kasagi). 299–309, Hemisphere.

Walsh,M.J. 1982 Turbulent boundary layer drag reduction using riblets. *AIAA Paper* No.**82–0169**.

Walsh,M.J. & Lindermann,A.M. 1984 Optimization and application of riblets for turbulent drag reduction. *AIAA Paper* No.**85–0548**.

Walsh,M.J. & Anders,J.B.Jr 1989 Riblet/LEBU research at NASA Langley. *App.Sci. Res.* **46**, 255–262.

Wilkinson,S.P., Anders,J.B., Lazos,B.S. & Bushnell,D.M. 1988 Turbulent drag reduction research at NASA langley: progress and plans. *J.Heat and Fluid flow* **9**, 266–277.

Yavuzkurt,S. 1984 A guide to uncertainty analysis of hot wire data. *Trans.ASME, J.Fluids Engng* **106**, 181–186.

The correlation of added drag with surface roughness parameters

R. L. TOWNSIN

University of Newcastle
Newcastle upon Tyne, U.K.

K.-S. Choi (ed.), Recent Developments in Turbulence Management, 181–191.

Abstract

Drag reduction in the marine environment means inhibiting fouling and reducing paint surface roughness. The conventional method of measuring ship hull surface roughness is discussed followed by an account of the available experimental evidence where the roughness functions and corresponding painted surface roughness characteristics have been measured in the laboratory. Twenty eight such surfaces are identified. The correlation of the roughness functions with surface geometry is then discussed and the most recent results described. Possible future developments are outlined to eliminate fouling, reduce roughness to the 'hydraulically smooth' datum and then to further reduce drag noting that most fish do not foul and there is some evidence of drag reducing mechanisms.

1. Introduction

In what follows the emphasis is on underwater surfaces, typically the wetted surfaces of a ship. Within a few hours of a clean surface being immersed in coastal waters colonization by micro-organisms begins. These organisms exude a mucilage and a slime coating is formed which may grow to a few millimetres in thickness. The slime film is the bed from which plants grow to form fouling on the sides of the hull especially near the waterline. In the darker bottom regions, shell will also develop. The drag penalty from developed fouling is economically intolerable. Hence a ships underwater surfaces are usually antifouled and usually by means of a toxic paint film. It should be noted that even an antifouled surface may develop a green slippery slime coating which could incur some drag penalty although this is not properly understood; however the toxin is there to prevent further development of grasses or barnacles.

It follows that marine surfaces are painted and their drag will be related to the roughness of the paint film, leaving aside any question of fouling. The first step in the drag reduction of ship's hulls and propeller surfaces is therefore to ensure that they are adequately smooth.

The roughness of hulls is a wide-banded, random distribution of crest-to-trough heights and wavelengths. The sample spectral energy itself varies considerably over the hull - some parts of the hull may be regarded as being as smooth as possible using airless spray techniques, whereas other parts of the same hull may be as rough as is found on any hull. This broad spectrum of hull condition is due to the nature of the painting process and the service conditions under which a ship operates. The paint surface may therefore suffer from drips, runs, sags, orange peel and overspray and be damaged by the deployment of cables, fenders tug nosings and by groundings and docking blocks.

2. The Conventional Approach

In attempting numerical solutions to the integral boundary layer equations for a rough hull, some roughness function must be devised which relates to the topography of the surface. Unfortunately, as described earlier, the surface texture varies considerably over a hull, presenting many difficulties such as the effect of say an extreme upstream roughness on a modest downstream roughness.

Up to the present an average roughness measure for the whole hull has had to be taken and this is at its most realistic for a new, freshly painted hull.

For quality control purposes at least, the conventional one-dimensional measure of hull roughness has been the highest peak to lowest trough measure in a distance along the surface of 50 mm. This is known as $R_t(50)$ fig.1. The local average of $R_t(50)$ is the mean hull roughness, MHR, and usually some dozen values are measured by a stylus instrument in one pass at one location. Such measures are made at some 100 stations equally disposed over the whole hull and averaged to give the average hull roughness, AHR.

Leaving aside the questions arising from averaging, the first question is whether there is any correlation possible between $R_t(50)$ and its consequential added drag? Can $R_t(50)$ be used as the determining parameter in a roughness function? All authorities agree that for the generality of rough surfaces, $R_t(\lambda)$ or any other height measure like R_a or R_q, are inadequate to correlate with added drag and that at least one other statistical measure is required, to account for texture.

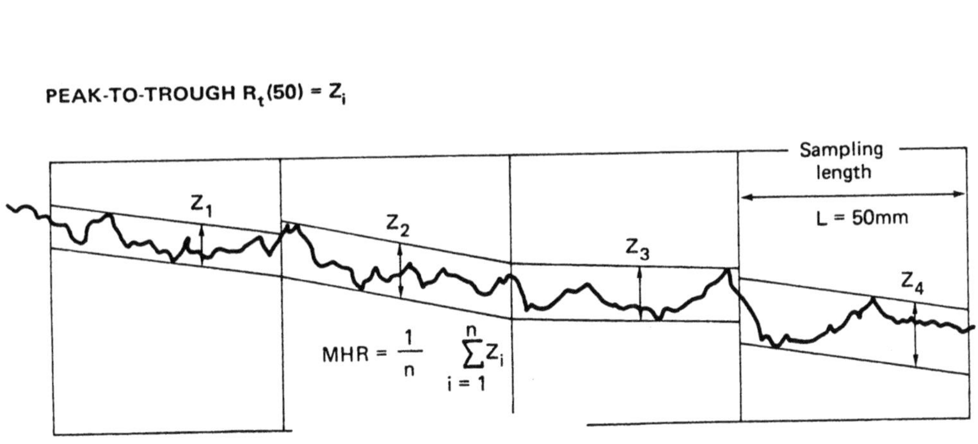

Figure 1 : $R_t(50)$ and the Mean Hull Roughness.

3. The Experimental Evidence

In order to study the correlation between added drag and rough surface characteristics, there needs to be not only a trustworthy evaluation of the roughness function $\frac{\Delta u}{u_\tau}$ but also a complete statistical description of the rough surface: unfortunately some data sets are lacking sufficient surface descriptions. Adequate data may be found from Todd's flat plane towing tank tests ref.(1), Musker's original pipe flow work ref.(2), Walderhaug's floating element measurements on 6 surfaces in a cavitation tunnel ref. (3), Johansson's floating element measurements ref.(4), Dey's measurements in a rotor apparatus, ref.(5), the six surfaces studied by Okuno, Lewkowicz and Nicholson ref. (6) and Sarabchi's painted sand surfaces ref. (7), all studied in a pipe flow rig. A summary and critique of this work may be found in Townsin and Dey ref. (8). In total 28 painted surfaces have been tested for which the roughness functions and surface statistics are known.
Any rough surface statistic is influenced by the long and short wavelength cut- offs used in the measurement or analysis. The short wavelength cut off is of vanishing importance as the higher frequencies are reached but the long wavelength cut off is of crucial importance.

4. The Correlation

Conventionally the correlation may be attempted by seeking, through trial and error, a relationship between the roughness function $\frac{\Delta u}{u_\tau}$ and the roughness Reynolds number $\frac{hu_\tau}{\nu}$ with h adequately determined from the surface statistics.
An early attempt was by Musker ref. (2) in which he defined h in terms of root mean square height R_q modified by the average slope S, the skewness S_k and kurtosis K_u of the roughness height distribution:

$$h' = R_q(1 + aS)(1 + bS_kK_u) \tag{1}$$

but the correlation was based only on his own 5 surfaces.
A different approach was made by Townsin & Dey, ref.(8), in which the characteristic length h was determined from the spectral moments of the height distribution but also the long wavelength cut off for calculating the moments varied with the scale of the roughness.

Figs.2A and 2B provide useful relevant relationships.

Many of the rough surface statistics for geometrically similar surfaces, especially those concerned with texture e.g. slope, will only remain constant if the long wavelength cut-off is also changed by the geometrical scale factor. This recognition led to the thought that surfaces which have a greater roughness in terms of height ought to be defined by statistics with a greater long wavelength cut-off.

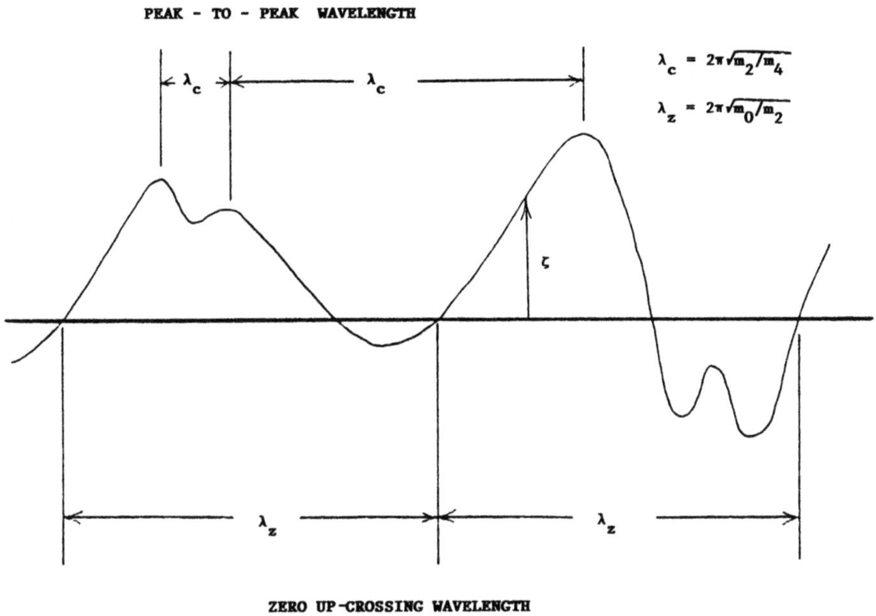

Figure 2A : A wide-banded, stationary, random process.

$$\text{BAND WIDTH RATIO} = \frac{\text{EXPECTED NUMBER OF ZERO UPCROSSINGS PER UNIT LENGTH}}{\text{TOTAL NUMBER OF PEAKS PER UNIT LENGTH (OF ANY SIZE)}}$$

Clearly, for a narrow-banded process, the band width ratio = 1

AVERAGE WAVE LENGTH BETWEEN SUCCESSIVE PEAKS OF ANY SIZE $= 2\pi\sqrt{m_2/m_4}$

AVERAGE WAVE LENGTH BETWEEN SUCCESSIVE ZERO UP-CROSSINGS $= 2\pi\sqrt{m_0/m_2}$

$$\text{BAND WIDTH RATIO} = \sqrt{1-\epsilon^2} = 1/\sqrt{\alpha} = \frac{\sqrt{m_2/m_4}}{\sqrt{m_0/m_2}} = \frac{m_2}{\sqrt{m_0/m_4}}$$

The mean square amplitude or variance of $\varsigma = m_o$ and hence

$$h = m_o\sqrt{m_4/m_2} = \frac{\text{MEAN SQUARE AMPLITUDE}}{\text{AVERAGE WAVE LENGTH BETWEEN PEAKS}} = \text{VARIANCE X PEAK FREQUENCY}$$

Figure 2B : Some statistical parameters related to the spectral moments.

After many trials, the most satisfactory correlation between $\Delta u/u_\tau$, and log hu_τ/ν was achieved when

$$h^2 = \alpha m_o m_2 \tag{2}$$

so that

$$h = m_o\sqrt{m_4/m_2} \tag{3}$$

Thus $\sqrt{m_o}$ becomes the height measure modulated by $\sqrt{\alpha m_2}$ for texture, where α is the band width parameter.

However the correlation was only satisfactory when λ was taken as 50mm for the rougher surfaces reducing to 2.5mm for the smoother surfaces. The best correlation was achieved by trial and error when the effective low frequency cut-off $\omega = 2\pi/\lambda$ was defined as

$$\omega^{0.25} = 1.3 - 0.0065\sqrt{(m_o m_2)_{50}} \tag{4}$$

where $(m_o m_2)_{50}$ is the value of the spectral moments for a 50mm cut off.

The analysis routine is, therefore, first to calculate the spectral moments for the surface using a long wavelength cut-off = 50mm and then to estimate the effective long wavelength cut-off for the surface using eqn.4. The spectral moments are then recalculated at the effective long wavelength cut-off and the roughness Reynolds numbers determined using eqn.2.

The resulting correlation of $\Delta u/u_\tau$ is shown in fig.3. The trend of the data can be represented in Colebrook-White form where

$$\frac{\Delta u}{u_\tau} = \frac{1}{\kappa} ln(0.18\frac{hu_\tau}{\nu} + 1) \tag{5}$$

This curve is drawn in fig.3 noting that the abscissa is base 10 logarithmic. As Grigson reminds us (ref.9), a power series within the logarithm may improve representation. It should again be noted that since these relationships are derived from 28 surfaces which were painted, they may not apply to other forms of roughness. The range of $R_t(50)$ among the 28 surfaces is from 45 μm to 863 μm i.e. from blasted and primed new plate to a very bad hull surface, and therefore comprehensively covers the total ship range.

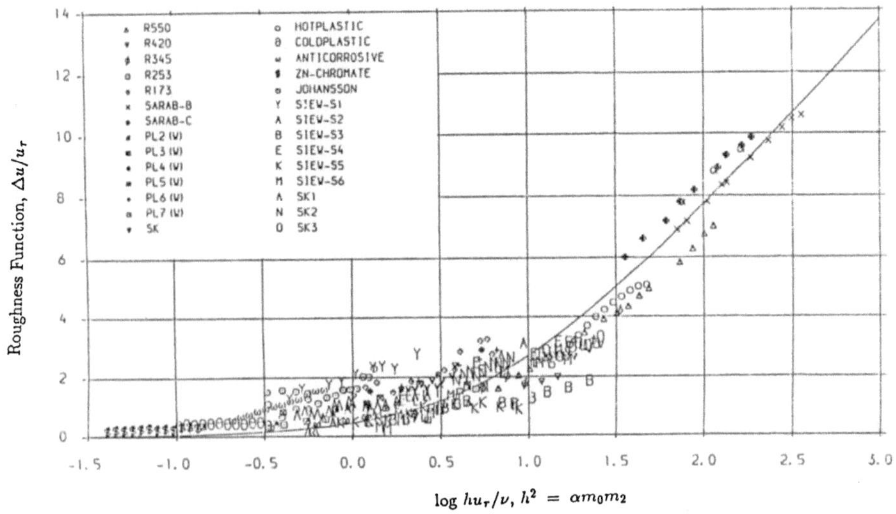

Figure 3 : The roughness function correlated with spectral moments when a longer wavelength cut-off is used for rougher surfaces.

Figures 4 and 5 show the data plotted, first when a fixed cut-off of 2.5mm is used and then when the fixed value is 50mm. In neither case is there a correlation and comparison with fig.3 shows the importance of using a varying long wavelength cut-off.

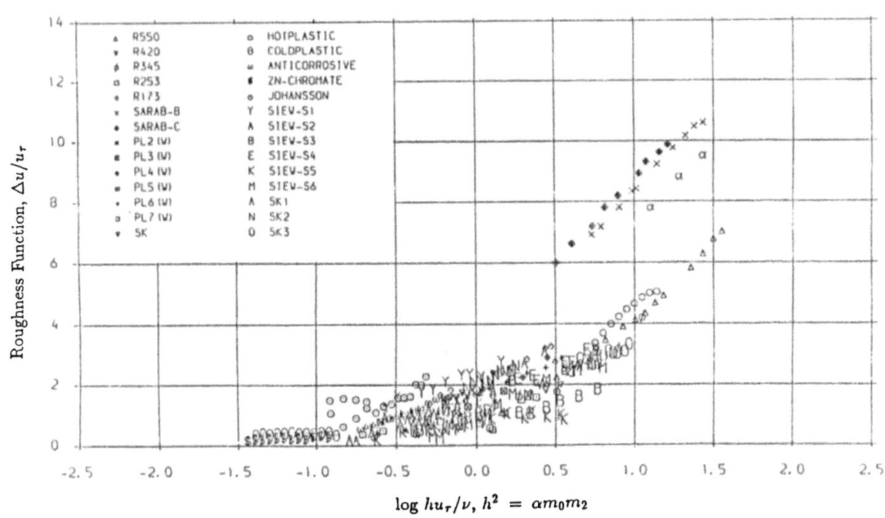

Figure 4 : The roughness function correlated with spectral moments at 2.5mm cut-off

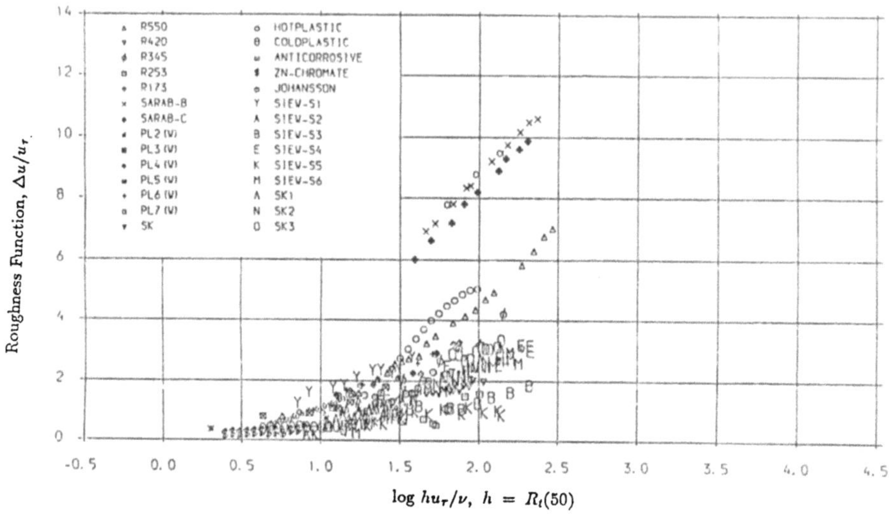

Figure 5 : The roughness function correlated with $R_t(50)$.

For relatively smooth ship painted surfaces however, say surfaces with $R_t(50)$ less than about 200 μm, an adequate correlation between $\frac{\Delta u}{u_\tau}$ and $R_t(50)$ can be detected, fig.6, so that for the new ship condition some simplification of calculation is possible and this is important in connection with meeting contractual speed and power conditions. Techno-economic assessments of the importance of ship bottom condition also require some simple reasonable method of estimating rough surface drag.

5. Future Developments

As described earlier the first step is to ensure that fouling organisms cannot survive on the hull. Whilst present organo-metallic biocides are most effective in preventing fouling the growing environmental lobby is resulting in the banning of the most effective toxins like tributyl tin. The overriding concern at present is therefore to develop more acceptable anti foulings.

The second step is to ensure that any antifouling coating is applied with supervised care and robotic application is one possibility. Another relevant consideration is the further development of ablative antifoulings so that some of the applied roughness, like overspray, can polish away in service. Self polishing copolymer paints perform in this manner to some extent.

The third step is the protection and maintenance of the smooth surface. Nowadays it is possible to keep ships out of dry dock for periods up to 5 years without fouling but due to surface damage, occurring principally when berthing, hull surfaces tend to deteriorate at 20μm to 40 μm AHR per year.

Since new shot blasted and primed steel plate starts life in the shipyard with a roughness of about 45 μm AHR and the smoothest new ship ever measured had an outturn of 78 μm AHR and that average new building has an outturn of about 110 μm AHR, it will be seen that there is room for improvement in the management of present coating practice.

The employment of robotic application could improve the final finish to the presently best possible but any further smoothing would require the development of a new breed of lower build coatings for multi coat application by robots, as in the automobile industry, but this is likely to be costly despite the robotics.

Cupro-nickel clad steel is a solution which has been tried with some success on small vessels e.g. the Copper Mariner, ref.10, but presently the cost of the material, which requires no coating and can present a smooth surface, is prohibitive.

Even if a maintainable, 'hydraulically smooth', fouling repellent surface could be developed would it be possible to achieve further drag reductions?

Riblets offer ideal conditions for slime accretion and fouling and therefore would only be suitable for short periods of immersion, hours rather than days in coastal waters, before losing their effectiveness. Riblets in a copper containing surface would inhibit fouling growth but diamataceous slimes are still likely to fill the valleys.

Full scale drag reduction has been achieved on a frigate by injecting Polyox WSR 301 into the boundary layer but the mass to be carried for a long lasting effect outweighed the gains in performance.

Whilst these and other results, at least initially, have been negative in the marine environment, there are lessons to be learnt from the natural inhabitants of the sea. It can be observed that the majority of fish do not foul. Fish antifouling is most effective because fish often inhabit just those coastal regions where contamination is most likely: the antifouling mechanism is unknown. Fish are coated with a slime film : the effect of the slime coating on the drag and propulsion characteristics is unknown. There is evidence that fish can exhibit low drag properties possibly in short bursts : the mechanisms are unknown but there are some theories and research results, see for examples pages 58/61 of ref.11. There is even evidence that fish in the same predator/prey group exhibit the same level of drag reduction and that there is a substantial variation in effectiveness between groups. Flat fish that protect themselves by concealment in the sandy bottom exhibit no drag reduction characteristics. The student of fish locomotion should however start by reading about the modes and efficiencies of propulsion e.g. ref.12, for it may be that high performance, like 2m tunnyfish achieving 40 knots, is less to do with drag reduction than with high propulsive efficiency. In any event the first requirement is no fouling.

Notation

$R_t(\lambda)$	Highest peak to lowest trough in a length λ mm (λ is thus also the long wavelength cut-off)
R_a	Arithmetic mean of the departures of the roughness profile from the mean line
R_q	The root mean square parameter corresponding to R_a
S	Average slope
S_k	Skewness
K_u	Kurtosis
h	Characteristic measure of roughness having the dimensions of length
h'	Musker's roughness parameter
a, b	Constants
m_n	Moments of the spectral distribution of the roughness profile amplitude
$\propto$	Band width of the frequency distribution of the roughness profile $m_o m_4/m_2^2$
ω	Low frequency cut off
$\frac{\Delta u}{u_\tau}$	Velocity shift due to roughness in the non-dimensional logarithmic plot of the velocity distribution through the boundary layer, $\frac{u}{u_\tau}$ versus log $\frac{hu_\tau}{\nu}$ (One form of the Roughness Function).
u_τ	Friction velocity $\sqrt{\tau/\rho}$
ρ	Density
ν	Kinematic viscosity
τ	Shear stress
κ	Von Karman constant

References

1. Todd, F.H. 1951 Skin friction resistance and the effects of surface roughness. *Trans.SNAME* **59**.
2. Musker, A.J. 1977 Turbulent shear-flows near irregularly rough surfaces with particular reference to ships' hulls. *PhD Thesis, University of Liverpool.*
3. Walderhaug, H. 1986 Paint roughness effects on skin friction. *Int.Shipbuilding Progress* **33** 382.
4. Johansson, L-E. 1985 The local effect of hull roughness on skin friction. Calculations based on floating element data and three-dimensional boundary layer theory. *Trans.RINA* **127**.
5. Dey, S.K. 1989 Parametric representation of hull painted surfaces and the correlation with fluid drag. *PhD Thesis, University of Newcastle upon Tyne.*
6. Okuno, T., Lewkowicz, A.K. & Nicholson, K. 1985 Roughness effects on a slender ship hull. *Proc.2nd Int.Symp. on Ship Viscous Resistance, Goteborg.*
7. Musker, A.J. & Sarabchi, K. 1980 Wall-friction and profilometry aspects of coating an irregularly rough surface. *Int.Shipbuilding Progress*, 27.
8. Townsin, R.L. & Dey, S.K. 1990 The correlation of roughness drag with surface characteristics. *Proc.RINA Int.Workshop on Marine Roughness & Drag, London.*
9. Grigson, C.W.B. 1987 The full-scale viscous drag of actual ship surfaces and the effect of quality of roughness on predicted power. *Jour. of Ship Research SNAME* 31.
10. Manzolillo, J.L., Thiele, E.W. & Tuthill A.H. 1976 CA 706 Copper-nickel alloy hulls: The Copper Mariner's experience and economics. *Trans.SNAME 84*.
11. Bone, Q. & Marshall, N.B. 1985 Biology of fishes, *Blackie.*
12. Lighthill, J. 1977 Aquatic animal locomotion. 11th Blackadder Lecture *Trans. NECIES*, 93.

IV. Compliant Surfaces

The optimisation of compliant walls for drag reduction

P. W. CARPENTER, A. D. LUCEY and A. E. DIXON

University of Warwick
Coventry, U.K.

K.-S. Choi (ed.), Recent Developments in Turbulence Management, 195–221.

Abstract

Previous theoretical studies have shown that sufficiently compliant walls can suppress the Tollmien-Schlichting waves (TSW) completely. However, full advantage cannot be taken of this in practice owing to the occurrence of other instability modes. These are of two main types; namely travelling-wave flutter (TWF), which is a convective wall-based instability, and divergence, which is an absolute instability. It is argued that the optimum wall properties with respect to transition delay must be such that there is marginal stability with respect to both these wall-based instabilities. In this paper we develop highly efficient methods for the prediction of TWF and divergence on single- and double-layer compliant walls made of viscoelastic materials. For the present application, the double-layer walls are restricted to those comprising a thin stiff upper layer supported by a thick soft substrate since it these types of wall that hold most promise in practical transition delay. Together with an efficient Orr-Sommerfeld solver for TSW eigenvalues, the above methods are used to establish a systematic means of optimization for double-layer walls. It is found that the inclusion of wall damping is desirable since it permits the use of a softer lower layer which is beneficial for the suppression of the TSW. A sample optimization is carried out for a compliant wall of prescribed thickness. It is found that, based on a conservative value of $n = 7$, the e^n method predicts that the greatest transitional Reynolds number achievable is about 4.5 times the rigid wall value. This factor could be slightly increased by optimising over wall thickness. However, the present results strongly suggest that the way forward is to use optimised multi-panel compliant walls designed using the procedures established here.

1. Introduction

There is little doubt that for boundary layers in water the use of appropriately designed compliant walls can lead to substantial postponement of laminar-turbulent transition. This has been established both theoretically and experimentally. The current status of the subject has been reviewed recently by Riley *et al.* (1988) and Carpenter (1990). It is quite likely that doubts remain for some readers as to whether it is really possible to manufacture compliant walls with the properties required to achieve substantial transition delay at the flow speeds typical of marine applications of practical interest. It is hoped that the present paper will help to dispel some of these doubts. However, it is accepted that nothing short of actual application of a compliant wall at the appropriate speed and in the appropriate marine environment will be fully convincing. Meanwhile the emphasis of the research in this area is now passing from the question of whether or not compliant walls are a viable method of achieving transition delay to the development of design methods with the object of determining the optimal wall properties for achieving the greatest possible transition delay and reduction in skin-friction drag. The present paper describes a methodology for determining such optimal properties for a particular class and arrangement of compliant walls.

Some aspects of the optimization procedure are quite general. However, the process is most easily understood with reference to a particular theoretical model of the compliant wall. A choice of model has to be made, in any case, at some stage in order to obtain results. For the present work a two-layer compliant-wall model is used – see Fig. 1. This figure illustrates the use of this type of compliant wall in the experimental study of Gaster (1987) and Daniel *et al.* (1987). The wall comprises a relatively thick and soft lower layer covered with a relatively thin and stiff top layer. Boundary-layer stability and transition over this type of compliant wall has been previously studied theoretically by Fraser and Carpenter (1985), Willis (1986) and Yeo (1988). Further details of the theoretical model are given in §2.

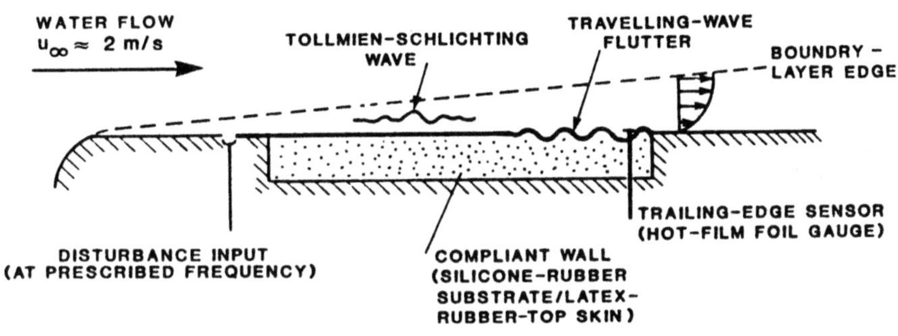

Fig.1. Schematic of the experimental set-up of Gaster et al..

It is well known that, in theory, Tollmien-Schlichting waves (TSW) can be completely suppressed by using a sufficiently compliant wall. This is plain from Fig. 11 of Carpenter and Garrad (1985). It is also implicit in the earlier theoretical work of Benjamin (1960) and Landahl (1962). What prevents this course of action succeeding in practice is the appearance of other modes of instability. Two additional modes of instability are commonly found in boundary layers over compliant walls. These are both essentially instabilities in the wall itself. One, termed travelling-wave flutter (TWF), is convective in nature. The other, termed divergence, is absolute. It is the appearance of these two instabilities that limits the transition delay attainable by use of compliant walls. In fact, it is now known (Lucey *et al.* 1991) that the TWF instability, rather than the TSW, was the route to transition for many of the compliant panels studied experimentally and theoretically by Gaster (1987). The two main wall instabilities will now be described in more detail.

The TWF is a convective instability like the TSW. In other respects, however, it is quite different from the TSW. It is destabilized when there is irreversible energy transfer to the wall from the main stream. For the type of wall in question this occurs through the work done by the fluctuating pressure forces. If the boundary layer were removed leaving an unsteady, but purely potential, flow, the perpendicular wall velocity and pressure would be exactly 90 degrees out of phase, implying no net energy transfer due to pressure work when averaged over a cycle. The presence of a shear layer leads to a critical point, where the phase speed of the disturbance equals the local undisturbed boundary-layer velocity, provided, that is, that the phase speed does not exceed the free-stream speed. In the vicinity of the critical point a jump in phase occurs in the pressure which, as first shown by Benjamin (1963), leads to non-zero irreversible energy transfer when averaged over a cycle. For low flow speeds the phase speed of the TWF is well above the free-stream value. As the flow speed increases the phase speed falls. When it reaches the free-stream speed a critical point appears and the TWF is destabilised. This fact allows relatively simple estimates to be made for the onset speed of TWF, as shown by Carpenter and Garrad (1986).

The presence of a critical point is essential for the destabilization of both the TSW and TWF. In other respects the two destabilization mechanisms are quite different. Viscous effects are essential for the TSW, whereas the mechanism for the TWF is essentially inviscid. This difference accounts for the contrast in form exhibited by their respective neutral curves – see Fig. 2. The TWF is, in fact, a much more dangerous instability than the TSW. Once a constant-frequency disturbance crosses the neutral boundary of the TWF in its passage downstream it will remain unstable and, moreover, will grow at an almost constant rate. Furthermore, as shown by Lucey *et al.* (1991) the growth rate increases extremely rapidly once the critical flow speed for the onset of the TWF is exceeded. The TSW, on the other hand, will ultimately cross the neutral curve again and decay, provided it remains within the linear regime of transition. When in the unstable region its growth rate rises steadily reaching a maximum and then falls. These comparisons are illustrated schematically in Fig. 2.

The amplitude profiles of the disturbance velocity amplitude perpendicular to the wall (the eigenfunctions) for the TSW and TWF are compared schematically in Fig. 3. As can be readily seen the maximum is at the wall for the TWF, indicating that it is essentially a wall-based instability, whereas the maximum for the TSW is located well above the critical point, indicating that it is essentially a flow-based instability. This provides a simple explanation for the fact that the TWF responds in the opposite manner from the TSW to irreversible energy-transfer to/from the compliant wall. For example, as indicated above and as shown by Carpenter and Morris (1990) and Carpenter and Gajjar (1990), work done by the fluctuating wall pressure is destabilizing to the TWF, but stabilizing to the TSW. Wall damping provides a further example. It is destabilizing to the TSW, as first shown by Benjamin (1960), but stabilizing to the TWF (Carpenter and Garrad

1985). The effects of wall damping on the neutral-stability boundaries are shown schematically in Fig. 2. Damping can increase the critical Reynolds number for the TWF, although the behaviour at infinite Reynolds number remains unchanged. The asymptotic theory of Carpenter and Gajjar (1990) may be used to predict the effects of damping on the TWF. It should be noted that the response of the various instabilities over a compliant wall to energy transfer has been studied in a much more fundamental way by Landahl (1962) and Benjamin (1963).

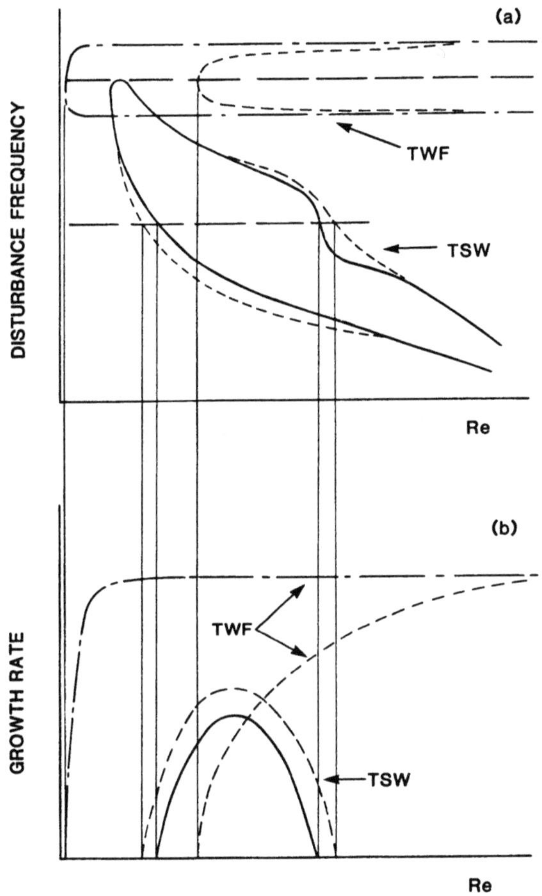

Fig.2. Comparison of neutral stability loops in the form of dimensional frequency versus Reynolds number based on boundary-layer thickness and variation of growth rates with same for TSW and TWF. Spring-backed flexible-plate compliant wall; – – –, damping present; otherwise no damping.

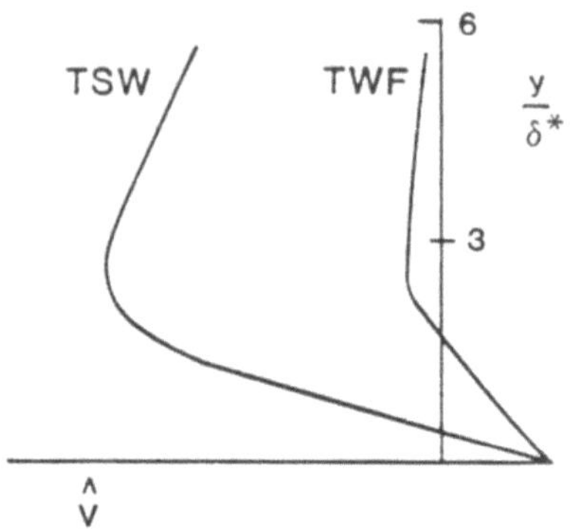

Fig.3. Comparison of disturbance-velocity profiles.

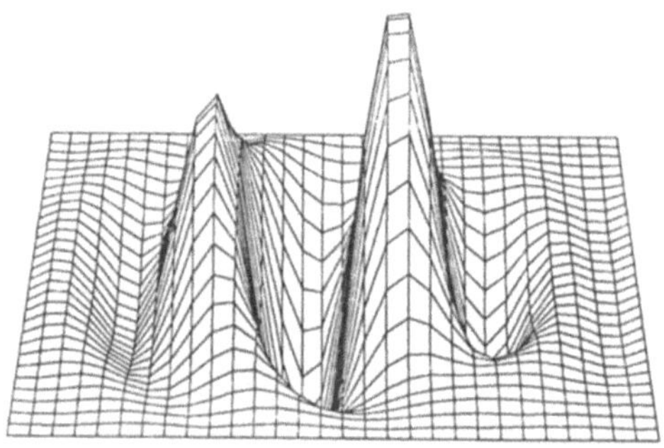

Fig.4. Numerical simulation of divergence waves on a spring-backed flexible-plate compliant panel. The free stream flows from left to right. The figure shows the surface displacement (the vertical axis has been scaled by a factor of 3000) some time after a disturbance in the form of a small bump located at the centre of the panel has been initiated. The simulation indicates that the waves travel very slowly in the downstream direction.

The instability mechanism for the divergence instability is comparatively simple. It occurs when the destabilizing pressure force created in the flow by a disturbance exceeds the restorative structural force within the compliant wall. It usually takes the form of a wave travelling very slowly downstream at a speed of between one and five percent of the free stream speed. This is evident in the experimental investigation of Gad-el-Hak *et al.* (1984) and in the numerical simulations of Lucey and Carpenter (1991a,b). An example of the numerical simulations is shown in Fig. 4. Results like this confirm that the instability is absolute in nature, as shown by Carpenter and Garrad (1986), and that it is almost unaffected by damping. This latter feature is consistent with an instability mechanism which involves conservative forces. The conservative nature of the instability mechanism also makes it possible to derive simple formulae, based on assuming unsteady potential flow, for the critical wavelength and onset speed. Such results were shown by Garrad and

Carpenter (1982) to give reasonable agreement with experimental observations.

The question of optimizing compliant walls for transition delay was first considered in a comprehensive parametric study undertaken by Gyorgyfalvy (1967). He used a membrane/spring theoretical model and based his calculations on the approximate theory of Landahl. His results were very promising, but possibly somewhat optimistic. Nevertheless the more recent optimization studies have been based, in part at least, on Gyorgyfalvy's approach. Carpenter (1985,1987) devised a scheme for optimizing the plate/spring compliant-wall model with respect to reduction of TSW growth rate and transition delay. The essence of this scheme is to require the optimal compliant wall to be marginally stable with respect to both TWF and divergence. This gives a one-parameter family of wall properties in the case when wall damping is absent. This remaining parameter can then be chosen to obtain the greatest possible transition delay. This scheme was extended to two-panel compliant walls by Carpenter (1991) and adapted for anisotropic wall compliance by Carpenter and Morris (1990). Joslin and Morris (1989) have developed numerical procedures for determining the optimal wall properties for anisotropic compliant walls. Their results are close to those obtained by following the procedure used by Carpenter and Morris. The present methods extend these optimization procedures to the more practical, but theoretically and computationally more demanding, case of the two-layer compliant walls used in the experiments of Gaster (1987) and others.

In the present work the well-known e^n method is used to predict the transitional Reynolds numbers. This method is based on the premises that for most practical applications the initial disturbance has the approximately the same very small amplitude; and that the length of the linear regime is a more or less fixed proportion of the total length of the transition region. Thus the transition process is assumed to be complete when the amplitude of the growing disturbance, as calculated by linear theory, has grown to e^n times its initial level. For a wide range of practical applications the location of the end of the transition process is satisfactorily predicted using a value of n between 9 and 10. Plainly the method makes no allowance for higher levels of initial disturbance, different receptivity mechanisms, or for the strong possibility that the nonlinear processes vary from case to case . The robustness and success of the method mainly stem from the fact that the linear regime constitutes 70 to 80 percent of the total length of the transition region for the sort of low-disturbance environment found in free flight. In the present case it is considered to be highly likely that wall compliance will significantly affect the nonlinear transition processes as well as the linear regime. For this reason $n = 7$ (corresponding approximately to the length of the linear regime alone) has been used rather than the usual value of 9 to 10. However, recent numerical simulations carried out by Metcalfe *et al.* (1990,1991) strongly suggest that the sort of compliant wall which is effective in the linear regime continues to have a favourable effect well into the nonlinear regime of transition. Consequently the use of $n = 7$ is probably very conservative.

2. Theoretical compliant-wall model

The central feature of the general linear asymptotic theory of Carpenter and Gajjar (1990) is that perturbations to the fluid normal and shear stresses can be written as linear functions of the horizontal and vertical displacements at the wall/flow interface. This allows the interacting wall/flow system to be modelled solely by the wall equations subject to modified boundary conditions. These interfacial boundary conditions require continuity of normal, (σ), and shear, (τ), stresses:

$$\sigma_{wall} = \sigma_{fluid} \quad , \quad \tau_{wall} = \tau_{fluid} \tag{1a,b}$$

at $y = 0$.

Expressions for the fluid stresses are discussed in §3; here, attention is concentrated upon finding appropriate wall solutions for the two-layer wall illustrated in Fig. 1. Consider, first, a single-layer wall represented by the relatively soft substrate in the absence of the thin upper layer. Vertical and horizontal wall displacements, ξ and η respectively, are determined by the Navier equations. In the absence of body forces, the equations of motion (Navier equations) and the normal and shear stresses are given by:

$$\frac{\partial^2 \xi}{\partial t^2} = c_T^2 \left(\frac{\partial^2 \xi}{\partial x^2} + \frac{\partial^2 \xi}{\partial y^2} \right) + (c_T^2 - c_L^2) \frac{\partial \Delta}{\partial x} \tag{2a}$$

$$\frac{\partial^2 \eta}{\partial t^2} = c_T^2 \left(\frac{\partial^2 \eta}{\partial x^2} + \frac{\partial^2 \eta}{\partial y^2} \right) + (c_T^2 - c_L^2) \frac{\partial \Delta}{\partial y} \tag{2b}$$

$$\sigma = 2\rho_s c_T^2 \frac{\partial \eta}{\partial y} + \rho_s (c_L^2 - c_T^2) \Delta \tag{2c}$$

$$\tau = \rho_s c_T^2 \left(\frac{\partial \xi}{\partial y} + \frac{\partial \eta}{\partial x} \right) \tag{2d}$$

in which the shear and dilational free-wave speeds and dilatation in the solid are given by:

$$c_T^2 = \frac{E_s}{2\rho_s(1+\nu_s)} \quad , \quad c_L^2 = 2c_T^2 \left(1 + \frac{\nu_s}{1 - 2\nu_s} \right) \quad , \quad \Delta = \frac{\partial \xi}{\partial x} + \frac{\partial \eta}{\partial y} \tag{3a,b,c}$$

and where E_s, ν_s and ρ_s are, respectively, the modulus of elasticity, Poisson ratio and density of the substrate. Visco-elastic effects can be modelled by using a complex modulus of elasticity in the above, as in Carpenter and Garrad (1985), so that:

$$E_s = (E_s)_R (1 - i\gamma_s) \tag{4}$$

where the real part $(E_s)_R$ is the so-called *storage modulus*, which is equivalent to the conventional elastic modulus; and the imaginary part $-\gamma_s(E_s)_R$ is the so-called *loss modulus*, which is a measure of the level of viscoelastic losses. By choosing the appropriate functions of frequency for $(E_s)_R$ and the *loss factor*, γ_s, it would be easy to adapt Eqn. (4) so that it corresponded to the Standard Linear Solid or the Voigt models of viscoelasticity. Nevertheless, we have chosen to make both the storage modulus and loss factor invariant with frequency because, in reality, both quantities are probably rather complicated functions of vibrational frequency and no data are available for determining how they vary with frequency. However, they would not be expected to vary greatly over the range of frequencies of interest.

Equations 2a,b can be more succinctly written in vector form; following the method described in Duncan et al. (1985), the Helmholtz decomposition is used to write the displacement vector as a sum of its irrotational and rotational parts. This procedure leads to the Navier equations being reduced to an independent pair of second-order differential equations. These can easily be solved by assuming that the displacement perturbations take a travelling-wave form. For consistency, the wall- and fluid-stress perturbations will also take this form. Thus we have that:

$$\{\xi,\eta\} = \delta^*\{\bar{\xi}(\bar{y}),\bar{\eta}(\bar{y})\}\exp\left(i\alpha(x-ct)\right) \tag{5a}$$

$$\{\sigma,\tau\} = \rho_f U_\infty^2\{\bar{\sigma}(\bar{y}),\bar{\tau}(\bar{y})\}\exp\left(i\alpha(x-ct)\right) \tag{5b}$$

where α is the wholly real wave number, c is the complex wave speed and ρ_f is the fluid density. For non-dimensionalisation, it is convenient to use the local boundary-layer displacement thickness, δ^*, as a length scale and U_∞ as a velocity scale. Overbars denote non-dimensional quantities. It is assumed that $\bar{\xi}$ and $\bar{\eta}$ are sufficiently small for linear theory to hold. Using these travelling-wave forms, displacement perturbations in the solid are found to be given by:

$$\begin{aligned}\bar{\xi} =& i\bar{\alpha}\sinh(\bar{\alpha}K_L\bar{y})A_1 + i\bar{\alpha}\cosh(\bar{\alpha}K_L\bar{y})A_2\\ &\bar{\alpha}K_T\cosh(\bar{\alpha}K_T\bar{y})A_3 + \bar{\alpha}K_T\sinh(\bar{\alpha}K_T\bar{y})A_4\end{aligned} \tag{6a}$$

$$\begin{aligned}\bar{\eta} =& \bar{\alpha}K_L\cosh(\bar{\alpha}K_L\bar{y})A_1 + \bar{\alpha}K_L\sinh(\bar{\alpha}K_L\bar{y})A_2\\ &- i\bar{\alpha}\sinh(\bar{\alpha}K_T\bar{y})A_3 - i\bar{\alpha}\cosh(\bar{\alpha}K_T\bar{y})A_4\end{aligned} \tag{6b}$$

where $K_T^2 = 1-(\bar{c}/\bar{c}_T)^2$ and $K_L^2 = 1-(\bar{c}/\bar{c}_L)^2$. and $\bar{\alpha} = \alpha\delta^*$, $\bar{y} = y/\delta^*$, $\bar{c} = c/U_\infty$, $\bar{c}_T = c_T/U_\infty$ and $\bar{c}_L = c_L/U_\infty$. A_i, for $i = 1,2,3,4$, are the constants of integration arising from the pair of second-order differential equations defining the irrotational and rotational parts of the displacement vector.

In order to evaluate the constants A_i, four boundary conditions are required. The interfacial boundary conditions, Eqns. 1, (with $\sigma_{wall} = \sigma_s$ and $\tau_{wall} = \tau_s$ for this single-layer wall) provide two of these. Using Eqns. 2c,d to write the wall stresses explicitly, we have that:

$$\bar{\sigma}_s = (\bar{c}_L^2 - 2\bar{c}_T^2)i\bar{\alpha}\bar{\xi} + \bar{c}_L^2\frac{d\bar{\eta}}{d\bar{y}} = \bar{\sigma}_{fluid} \quad , \qquad \bar{\tau}_s = \bar{c}_T^2\left(\frac{d\bar{\xi}}{d\bar{y}} + i\bar{\alpha}\bar{\eta}\right) = \bar{\tau}_{fluid} \tag{7a, b}$$

in which the expressions on the left-hand sides are evaluated at $\bar{y} = 0$.

The remaining two boundary conditions are found by imposing the condition of zero displacement at the bottom of the solid where $y = -d$, d being the thickness of the solid. This assumes perfect bonding between the compliant wall and the rigid supporting base.

$$\bar{\xi}(-\bar{d}) = 0 \quad , \quad \bar{\eta}(-\bar{d}) = 0 \tag{8a, b}$$

Introducing the displacement forms given by Eqns. 6a,b into the boundary conditions above and rearranging yields a linear system of equations for the constants A_i. It is noted at this point that, for the present solution method to proceed, it is crucial that each of the expressions for σ_{fluid} and τ_{fluid} can be written as a linear combination of $\bar{\xi}$ and $\bar{\eta}$ (and, hence, A_i). The linear system is written in matrix form as:

$$S_{ij}A_j = 0 \tag{9}$$

where repeated suffices imply summation. The elements of S_{ij} are listed below. For this single-layer wall, the terms R_{Pv} and R_{Ph} are equivalently zero and, for the wall *in vacuo*, $P_v = P_h = T_v = T_h = 0$.

$$S_{11} = K_L(R_{Pv} + P_v) \qquad S_{12} = \bar{c}_T^2(1 + K_T^2) + iP_h$$
$$S_{13} = K_T(-2i\bar{c}_T^2 + P_h) \qquad S_{14} = i(-R_{Pv} - P_v)$$
$$S_{21} = K_L(2i\bar{c}_T^2 - T_v) \qquad S_{22} = i(R_{Ph} - T_h)$$
$$S_{23} = K_T(R_{Ph} - T_h) \qquad S_{24} = \bar{c}_T^2(1 + K_T^2) + iT_v$$
$$S_{31} = -i\sinh(\bar{\alpha}K_L\bar{d}) \qquad S_{32} = i\cosh(\bar{\alpha}K_L\bar{d})$$
$$S_{33} = K_T\cosh(\bar{\alpha}K_T\bar{d}) \qquad S_{34} = -K_T\sinh(\bar{\alpha}K_T\bar{d})$$
$$S_{41} = K_L\cosh(\bar{\alpha}K_L\bar{d}) \qquad S_{42} = -K_L\sinh(\bar{\alpha}K_L\bar{d})$$
$$S_{43} = -i\sinh(\bar{\alpha}K_T\bar{d}) \qquad S_{44} = -i\cosh(\bar{\alpha}K_T\bar{d})$$

Non-trivial solutions to Eqn. 9 are determined by:

$$\det\{S_{ij}\} = 0 \tag{10}$$

For the single-layer wall *in vacuo*, the above equation yields the well-known Rayleigh modes. A typical solution for the fundamental upstream and downstream waves is included in the dispersion diagram of Fig. 5.

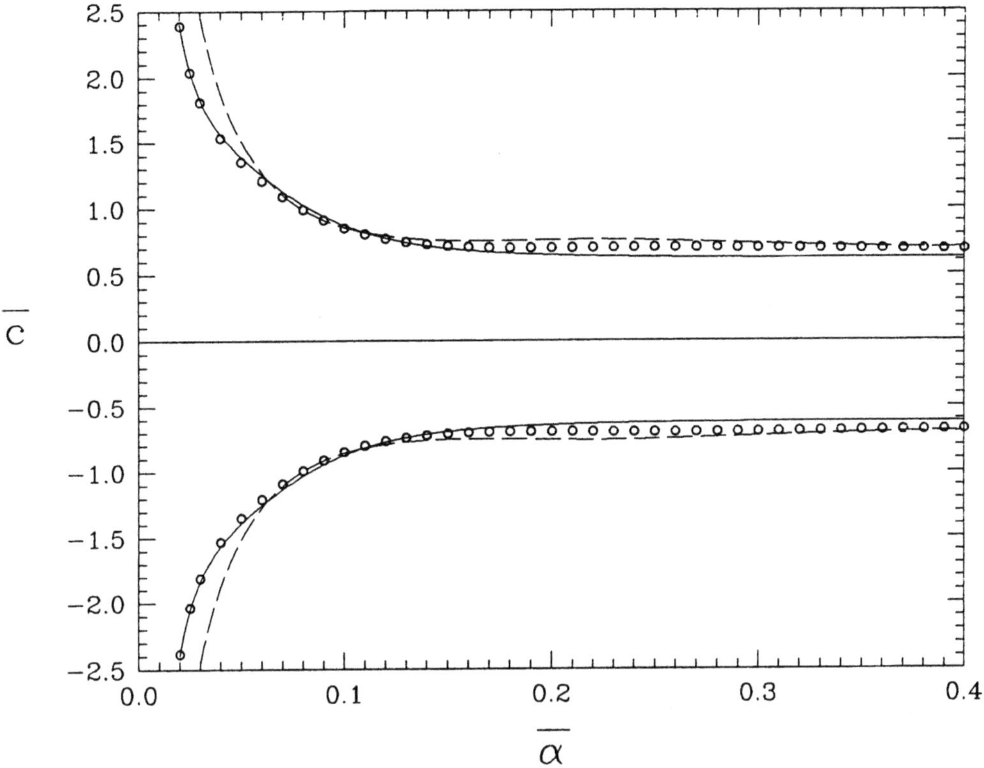

Fig.5. Free waves for single- and double-layer compliant walls. Single-layer: $E_s = 45000\ N/m^2$, $\nu_s = 0.49$, $d = 1.5\ cm$ and $\gamma_s = 0$. The double-layer wall comprises a substrate with the wall properties above plus an upper layer with $E_p = 4 \times 10^6\ N/m^2$, $\nu_p = 0.5$, $b = 0.1\ cm$ and $\gamma_p = 0$. $\rho_p = \rho_f = 1000\ kg/m^3$ and $U_\infty = 6\ m/s$ and $\delta^* = 0.0006\ m$ have been used for non-dimensionalisation. ——, single-layer wall; — — double-layer wall; o, double-layer wall with upper-layer horizontal stiffness suppressed.

Attention is now turned to the type of double-layer wall seen in Fig. 1; this can modelled as a simple modification of the single-layer wall discussed above. Since the upper layer is both thin and relatively stiff – for example, in the experiments of Gaster (1987), typical upper-to-lower layer ratios of thickness and elastic modulus are respectively 1:60 and 300:1 – its deformation can be closely modelled using classical thin-plate theory. The normal and shear stresses associated with the deflecting plate are thus considered to act as discontinuities between the substrate and fluid stresses at the wall/flow interface, $y = 0$. The modified form of Eqns. 1

are, therefore:

$$\sigma_s + \sigma_p = \sigma_{fluid} \quad , \quad \tau_s + \tau_p = \tau_{fluid} \tag{11a, b}$$

the suffices s and p referring, respectively, to the substrate and plate contributions. The substrate terms are identical to those given above for the single-layer model; it remains to develop suitable forms for the plate terms.

Considering vertical displacements of the plate, the equation of motion gives that:

$$\rho_p b \frac{\partial^2 \eta}{\partial t^2} + B_p \frac{\partial^4 \eta}{\partial x^4} = \sigma_p \tag{12}$$

where the flexural rigidity is defined by:

$$B_p = \frac{E_p b^3}{12(1 - \nu_p^2)}$$

and where E_p, ν_p and b are, respectively, the modulus of elasticity, Poisson ratio and thickness of the plate. By considering the dynamic compression/extension of the plate in the horizontal direction, the required shear-stress term can be found:

$$\rho_p b \frac{\partial^2 \xi}{\partial t^2} - E_p b \frac{\partial^2 \xi}{\partial x^2} = \tau_p \tag{13}$$

The travelling-wave displacement and stress forms given in Eqns. 5 are used in Eqns. 12 and 13 to yield the non-dimensional normal- and shear-stress amplitudes due to the plate deformation. Thus:

$$\bar{\sigma}_p = \bar{\alpha} R_{Pv} \bar{\eta} \quad , \quad \bar{\tau}_p = \bar{\alpha} R_{Ph} \bar{\xi} \tag{14}$$

where:

$$R_{Pv} = \frac{\bar{\alpha}}{\delta^*} \left\{ -C_{m1} b \bar{c}^2 + \frac{1}{F_S} \frac{B_p \bar{\alpha}^2}{\rho_s U_\infty^2 \delta^{*2}} \right\} \tag{15a}$$

$$R_{Ph} = \frac{\bar{\alpha}}{\delta^*} \left\{ -C_{m1} b \bar{c}^2 + \frac{E_p b}{\rho_s U_\infty^2} \right\} \tag{15b}$$

in which $C_{m1} = \rho_p / \rho_s$. Damping in the plate can be included by using a complex elastic modulus, E_p, in a manner analogous to that employed for the substrate in Eqn. 4. The damping factor for plate damping is written as γ_p. It should be noted that although the upper-layer contribution to the wall forces is being included as a stress discontinuity at $y = 0$, the inertial effects of the plate are included; these are evident in the first terms of the right-hand sides of Eqns. 15.

In the above derivation of normal stress due to the plate deflection, the primary mode of deformation has been taken to be plate bending. Of course, the plate may also experience vertical shear: this contribution to the deformation has been taken into account by including the factor F_S in Eqn. 15a. This term is given by:

$$F_S = 1 + \frac{(b\alpha)^2}{4(1-\nu_p)} \tag{16}$$

and serves to introduce an 'effective' flexural rigidity allowing for vertical shear. For the thin upper layer considered here, together with the low wave-number approximation implicit in the flow solution to be used, the value of F_S remains very close to unity.

The solution procedure follows that described for the single-layer wall with the boundary conditions of Eqns. 11 replacing those of Eqns. 7. In the elements of the matrix S_{ij}, given below Eqn. 9, R_{Pv} and R_{Ph} now take non-zero values. Figure 5 shows *in vacuo* dispersion curves for both a single-layer wall and the double-layer wall developed by adding a thin upper layer using the flexible-plate/substrate model described above. The wall properties are as given in the figure caption. They have been chosen so as to be relevant to the types of wall found in the optimization discussed in §4. As one would anticipate, the double-layer wall possesses increased restorative stiffness, so yielding higher values of $\bar{c}$ than the single-layer wall. What is interesting to note is that when the contributions of the plate horizontal stresses are suppressed, the plate/substrate result becomes almost identical to that of the substrate alone. Thus, for this type of wall, the most important effect of including the thin stiff upper layer is that it restricts horizontal motion of the substrate and thereby increases the overall restorative stiffness of the compliant wall.

Lastly, it is remarked that a general solution for the double-layer compliant wall could have been found by solving the Navier equations in each of the two media and imposing continuity of both displacement and stress at the solid/solid interface (eg. see Willis 1986 and Yeo 1988). Adopting this strategy introduces a further four constants of integration, A_i and, thus, a matrix, S_{ij}, of order 8. The present plate/substrate model is appropriate to the type of compliant walls used in the experiments and has the merit that the repeated solution of the characteristic equation is extremely efficient when the order of S_{ij} is kept down to 4 for the double-layer wall. Since a large number of computations are required in order to carry out the optimization procedures, it is essential to make the computer codes as efficient as possible.

3. Theoretical and Numerical Methods

The foregoing discussion has focussed upon solving the equations of motion for the compliant wall. The fluid forcing terms are now introduced as the normal- and shear-stress boundary conditions for the wall solution at the wall/flow interface. Expressions for $\bar{\sigma}_{fluid}$ and $\bar{\tau}_{fluid}$ at $y = 0$ are obtained from the general asymptotic theory of Carpenter and Gajjar (1990) which exploits the multi-deck structure of the boundary layer at high Reynolds numbers. This flow solution models the bulk of the boundary layer as an essentially inviscid shear layer but viscous effects in the wall layer are rigorously included. Two-dimensional forms of the fluid-stress perturbations at the compliant wall are given by:

$$\bar{\sigma}_{fluid} = \bar{\alpha}\left[P_v\bar{\eta} + P_h\bar{\xi}\,\right] \quad , \qquad \bar{\tau}_{fluid} = \bar{\alpha}\left[T_v\bar{\eta} + T_h\bar{\xi}\,\right] \tag{17a, b}$$

where:

$$\begin{aligned} P_v =& C_{m2}(1-\bar{c})^2 + \bar{\alpha}C_{m2}\left\{2\bar{c} - \frac{1}{H} - (1-\bar{c})^2\Phi_\infty(\bar{c})\right\} \\ & - \epsilon\bar{\alpha}^{\frac{1}{2}}C_{m2}\frac{(1-\bar{c})^4}{\bar{c}^{\frac{5}{2}}}\frac{(1+i)}{\sqrt{2}} \end{aligned} \tag{18a}$$

$$P_h = -\,\epsilon\bar{\alpha}^{\frac{1}{2}}C_{m2}\frac{(1-\bar{c})^2}{\bar{c}^{\frac{1}{2}}}\frac{(1-i)}{\sqrt{2}} \tag{18b}$$

$$T_v = \epsilon\bar{\alpha}^{\frac{1}{2}}C_{m2}\frac{(1-\bar{c})^2}{\bar{c}^{\frac{1}{2}}}\frac{(-1+i)}{\sqrt{2}} \tag{18c}$$

$$T_h = \epsilon\bar{\alpha}^{\frac{1}{2}}C_{m2}\bar{c}^{\frac{3}{2}}\frac{(1+i)}{\sqrt{2}} \tag{18d}$$

C_{m2} is the ratio of fluid-to-substrate densities, H is the shape parameter ($H = 2.591$ for the Blasius profile) and $\Phi_\infty(\bar{c})$ is a known complex function which is assumed (in this application) to be dependent only on the real part of the complex phase speed. Interpolation functions for $\Phi_\infty(\bar{c})$ are given in Carpenter and Gajjar. The small parameter, $\epsilon = 1/\sqrt{Re_{\delta^*}}$, where Re_{δ^*} is the Reynolds number based on the boundary-layer displacement thickness, obtains from the scalings in the multi-deck structure.

The first term on the right-hand side of Eqn. 18a represents the contribution from the purely potential flow outside the boundary layer which yields pressures exactly in phase with vertical displacements of the wall. The second term accounts for the inviscid shear layer; this applies a phase shift to the otherwise conservative forces through the complex function, Φ_∞. It is this phase shift to the pressure which allows the irreversible energy transfer from the flow into the wall that can lead to travelling-wave flutter. The third term in Eqn. 18a together with all of Eqns. 18b,c,d account for the effect of the viscous wall layer. These terms also provide a mechanism for irreversible energy transfer between the flow and wall. Whether TWF occurs depends on the balance of all of these rates of wall/flow irreversible energy transfer and the rate of dissipation of wall energy through damping.

The forms of fluid normal and shear stresses in Eqns. 17 are seen to be suitable for inclusion in the matrix method of solution described in §2 contributing the terms P_v, P_h, T_v, T_h to the matrix S_{ij}. The wall/flow system eigenvalue problem is, therefore, given by Eqn. 10. This is more conventionally written as:

$$\mathcal{E}(\bar{\alpha}, \bar{c}, Re_{\delta^*}) = 0 \tag{19}$$

the eigenvalue being the complex wave speed, $\bar{c}$. For a given flow speed, Reynolds number and wave number, Eqn. 10 is numerically solved using the method of false position. Initial guesses are found by first finding the free-wave solutions. The fluid forces are initially included at wavenumbers for which the wall forces are dominant and the solution then tracked to reach those obtaining at higher values of the wavenumber.

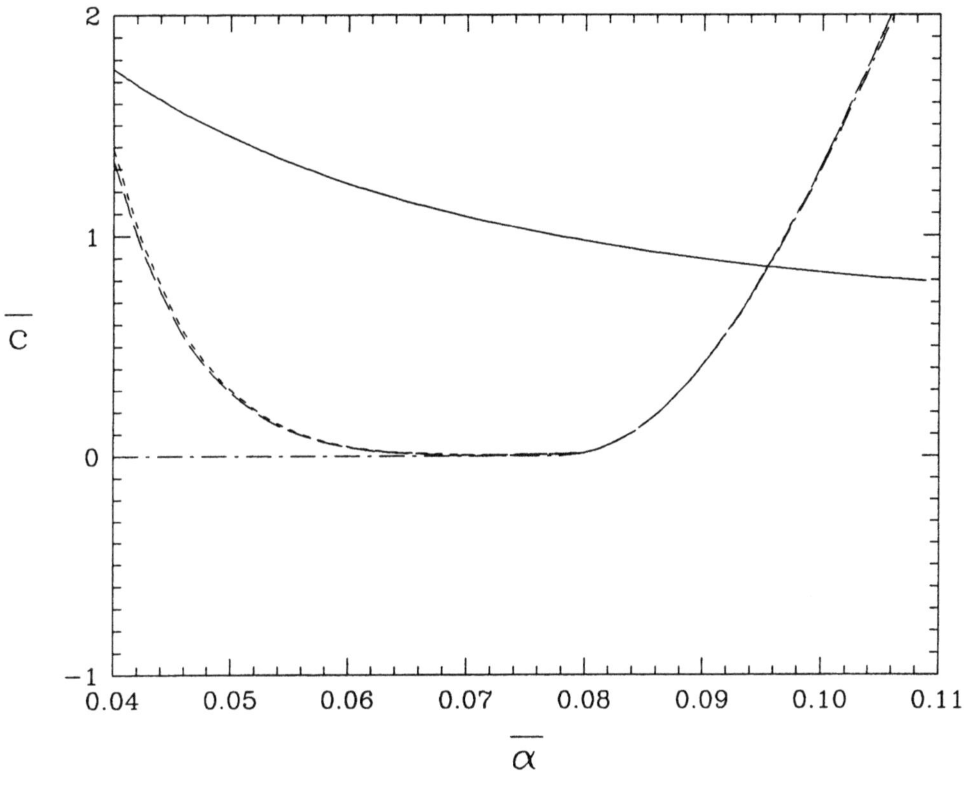

Fig.6. Dispersion curves for boundary-layer flow over a double-layer compliant wall for the fundamental (downstream-travelling) mode. Data as for Fig. 5 with $Re_{\delta^*} = 3600$. ——, $\bar{c}_R$ from full solution (all terms of Eqns. 17,18); — —, $\bar{c}_I \times 10^3$ from full solution; — · —, $\bar{c}_I \times 10^3$, inviscid-shear layer only; - - - $\bar{c}_I \times 10^3$, fluid shear-stress suppressed.

Figure 6 is a dispersion curve which illustrates the various effects of the fluid-forcing terms on a double-layer wall. Wall properties are identical to those used in Fig. 5. $U_\infty = 6\ m/s$ and the kinematic viscosity, ν, is $10^{-6}\ m^2/s$; a Reynolds number of 3600 has been used to generate these results. Note that only the fundamental downstream-travelling mode has been plotted – since this has the lowest shear wave speed, it is the most susceptible to TWF, the instability occuring where the imaginary part of the wave speed, $\bar{c}_I$, is positive. The results of Fig. 6 suggest that the presence of the fluid shear stresses is mildly stabilising but that the effect of the wall viscous layer is generally destabilising. However, neither of these mechanisms has an important influence at the onset ($\bar{\alpha} \approx 0.076$) of TWF or thereafter, where the principal source of irreversible energy transfer is provided by the essentially inviscid part of the boundary layer.

Throughout the above analysis, displacements and stresses have been couched in a form (Eqns. 5) which only admit temporal instabilities. Since TWF is known to be a convective instability, a more appropriate formulation has that perturbations are proportional to:

$$\exp\left(i(\alpha x - \omega t)\right)$$

where α is a complex wavenumber and ω is a (real) radian frequency of oscillation. The above solution procedure which employed the equivalent form: $\exp\left(i\alpha(x - ct)\right)$ may be adapted by allowing α to become a complex quantity along with c and finding solutions subject to the constraint that $\alpha c = \omega$, where ω is wholly real. The equivalent eigenvalue problem to that summarised by Eqn. 19 is:

$$\mathcal{E}(\bar{\alpha}, \bar{\omega}, Re_{\delta^*}) = 0 \tag{20}$$

where $\bar{\omega} = \omega\delta^*/U_\infty$. In the results that follow, it is this spatial representation of unstable waves which has been used.

Attention is now briefly turned to the other two instabilitities – namely divergence and TSW – which might exist in the interacting wall/flow system. Divergence first occurs at the flow speed for which the upstream-travelling wave seen in Fig. 5 turns to travel downstream under the action of the hydrodynamic forces. Exactly at the onset flow speed, where $U_\infty = U_d$, $\bar{c} = 0$ for the divergence waves. Since divergence essentially occurs through the action of the conservative hydrodynamic forces, the value of U_d can be found by solving the system equation, 10, with:

$$P_v = C_{m2}(1 - \bar{c})^2 \qquad\qquad P_h = T_v = T_h = 0$$

and considering the asymptotic behaviour of the characteristic equation as $c \to 0$. This procedure yields the following (dimensional) result:

$$\rho_f U_d^2 = \frac{E_s W(\alpha d, \kappa)}{(1 + \nu_s)} + \frac{E_p (\alpha b)^3}{12(1 - \nu_p^2)} \tag{21}$$

where ρ_f is the fluid density, $\kappa = c_T^2/c_L^2$ and the function W is defined by:

$$W \equiv \frac{1 + (\alpha d)^2(1 + \kappa^2) + (1 - \kappa^2)\sinh^2(\alpha d)}{(1 + \kappa)\sinh(\alpha d)\cosh(\alpha d) - (1 - \kappa)(\alpha d)}$$

Note that in the limit of an incompressible substrate, $\kappa \to 0$ and thus $W \equiv W(\alpha d)$. The right-hand side of Eqn. 21 can be evaluated for a variation of α to find the minimum value of U_d and the wave number of the critical divergence mode.

The model described above is ideally suited to the investigation of TWF and divergence. In fact, the type of TWF calculations presented in §4 could not have been performed using the conventional Orr-Sommerfeld approach to the problem because of the difficulties posed by the continuous spectrum. However, this model cannot be used to predict TSW because the existence of these waves is fundamentally dependent upon viscous effects in the boundary layer. To investigate the TSW, the Orr-Sommerfeld equation has been solved, imposing continuity of velocity and stress as the interfacial boundary boundary conditions. The solution employs a spectral Chebyshev-tau method which has been coded for use on a highly parallel computer system based an AMT DAP-510, the Orr-Sommerfeld differential operator being approximated by a matrix operating on the vector of Chebyshev coefficients. A non-linear eigenvalue equation results, which can be solved using an iterative technique. This method allows a large number of eigenvalues to be obtained comparatively rapidly. Further details of the numerical methods are given in Dixon and Lucey (1991).

4. Optimization Procedure and Results

The theoretical and numerical methods discussed above are now used as tools appropriate to the prediction of each of the main instabilities that may be encountered in the wall/flow system. The compliant wall can now be optimised so that the accumulated growth rates of TSW are kept within pre-transition limits whilst always avoiding both TWF and divergence instability. Here, in order to illustrate the procedure, we consider the case of a double-layer compliant panel of prescribed (substrate) depth, d. Variations to the upper-layer thickness, b, are entertained, but, given that this is a very thin layer, these do little to alter the overall dimensions of the combined wall. The results presented below are for $d = 0.015\ m$ and $U_\infty = 6\ m/s$. Of course, suitable scalings could be used to modify the present results for a different applied flow speed without a complete repetition of the optimization. We present wall parameters in dimensional form so that a clear physical feel is engendered for the realisitic walls that we propose. The aim of the optimization is to determine the maximum length of the compliant panel seen in Fig. 1 so that the transitional Reynolds number of the boundary-layer flow coincides with the Reynolds number of the panel trailing edge. We use a conservative value of $n = 7$ in the e^n calculations to determine the transition point. This value is generally recognised as corresponding to the limits of the linear regime of transition. Usually n is taken to be 9.5 for transition prediction. The overall strategy adopted is to target a Reynolds number and optimise the wall properties for minimum accumulated growth of the TSW at that point whilst ensuring that $n < 7$ for all lower Reynolds numbers. If the value of n at the target Reynolds number is less than 7, then the target Reynolds number is incremented and the procedure repeated. The point at which $n = 7$ for the target Reynolds number yields the optimal wall properties for transition delay.

In the methodology that follows, the wall damping is, perhaps, the key variable in the wide range of parametric investigations which can be carried out. As discussed in §1, the dissipation of wall energy is beneficial to the attenuation of TWF but detrimental to the suppression of TSW. For the case of a single-layer wall, Dixon & Lucey (1991) have shown that there exists an optimal value of wall damping which gives the lowest TSW growth rates whilst eliminating TWF. These results show that the inclusion of wall damping allows a lower value of wall elastic modulus to be used; in consequence the softer wall shows an improved performance with regard to TSW. A similar strategy was found to work in the experimental study of drag reduction conducted by Chung & Merrill (1984). It is also remarked that the useful inclusion of wall damping separates the present work from the optimization of spring-backed flexible-plate compliant walls carried out by Carpenter (1987,1991) in which the inertial effects of the plate proved to be the key factor for the elimination of TWF. The approach adopted in the present optimization of double-layer compliant walls follows from that developed for the single-layer viscoleastic walls discussed above. A useful feature of having wall damping as a primary variable is that the onset flow speed of divergence instability remains unaffected by changes to this variable.

Turning our attention, therefore, to double-layer compliant walls, typical TWF neutral curves for different values of substrate damping, γ_s, in the ω-Re_{δ^*} plane are shown in Fig. 7. These have been carried out for given values of substrate elastic modulus, E_s, and plate flexural rigidity, B_p. If $Re_{\delta^*} = 3600$ is set as the target Reynolds number for a first optimization, then the value of γ_s which places the "nose" of the TWF region at $Re_{\delta^*} = 3600$ is found. This criterion for the determination of required damping is, perhaps, a little pessimistic since it implies the complete removal of TWF throughout the Reynolds number range up to the target. However, little would be gained by permitting *some* TWF growth since, as shown by Lucey *et al.* (1991), once underway this instability is very powerful. Moreover, the elimination of TWF within the target range precludes the possibility of the powerful modal-coalescence instability found by Carpenter *et al.* (1983), Carpenter & Garrad (1985) and Sen & Arora (1988). The combination of the values of E_s, B_p and the critical γ_s found from Fig. 7 then contributes a single point to the summary plot of Fig. 8 which is a record of all such combinations of these properties for which the wall is marginally stable with respect to TWF at the target Reynolds number. Also included in this figure are divergence criteria. These are marked as thick vertical lines drawn at the value of E_s (together with B_p) for which the divergence-onset flow speed, as given by Eqn. 21, is exactly equal to the design flow speed, $U_\infty = 6\ m/s$. Thus any combination of E_s, B_p and γ_s that falls to the right of the divergence cut-off is permissible. At this point in the optimization procedure, damping in the upper layer has been excluded. The distribution of damping throughout the two-layer wall is discussed later.

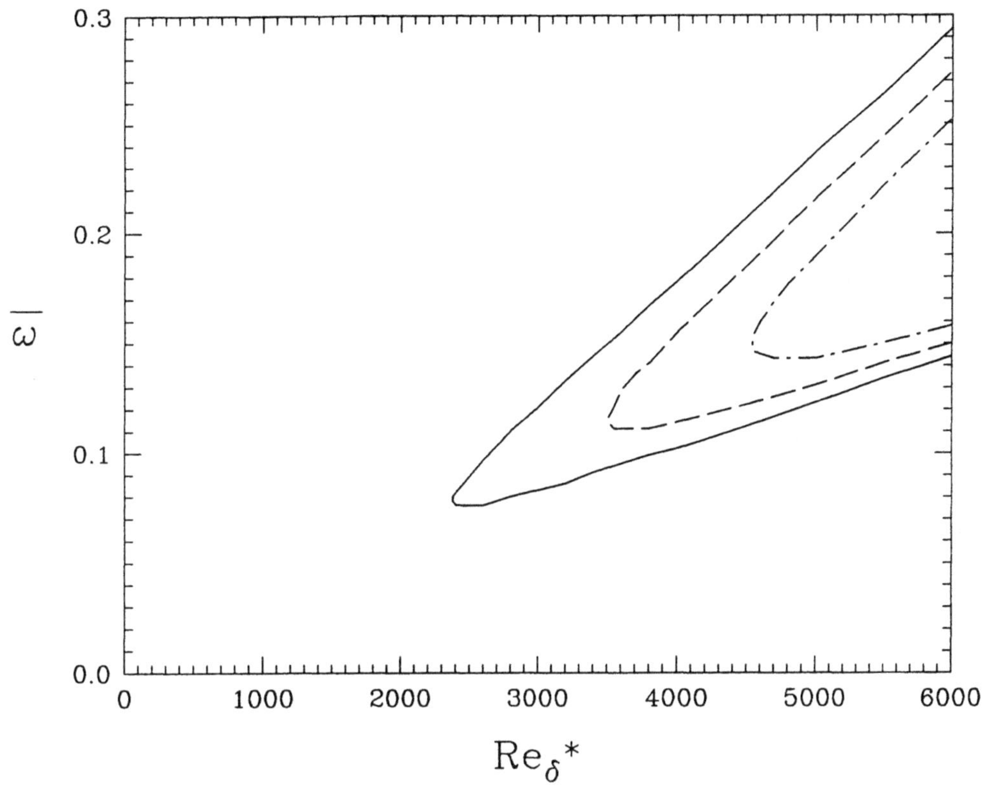

Fig.7. TWF neutral-stability curves for different values of substrate damping, γ_s. Double-layer wall with properties as for Fig. 5. ——, $\gamma_s = 0.010$; — —, $\gamma_s = 0.015$, — · —, $\gamma_s = 0.020$.

Working from Fig. 8, e^n calculations for TSW are carried out along each line representing a suitable combination of wall properties and such that divergence instability is avoided. The integrations of the TSW growth rates are carried out from the onset of instability either through to exit from the instability loop or up to the target Reynolds number, whichever is the greater. It has been found that for a given plate flexural rigidity, the minimum of accumulated TSW growth at the target Reynolds number occurs at the divergence limit. In other words, where the substrate elastic modulus is at its lowest value (and, thus, γ_s at its highest value) for a given B_p-line in Fig. 8. By repeating the TSW calculations for each B_p-line, it is possible to determine the optimal value of B_p for the target Reynolds number. Figure 9 shows the accumulated TSW growth rates obtaining from such an optimization carried out for $Re_{\delta^*} = 3600$. For the purposes of comparison, equivalent results for an optimal single-layer wall and a rigid wall are included in this figure. It is noted that the double-layer wall fares better in the reduction of the TSW growth rates. It is also remarked that neither the single- nor the double-layer wall has been optimised for maximum transition delay since $n < 7$ at $Re_{\delta^*} = 3600$.

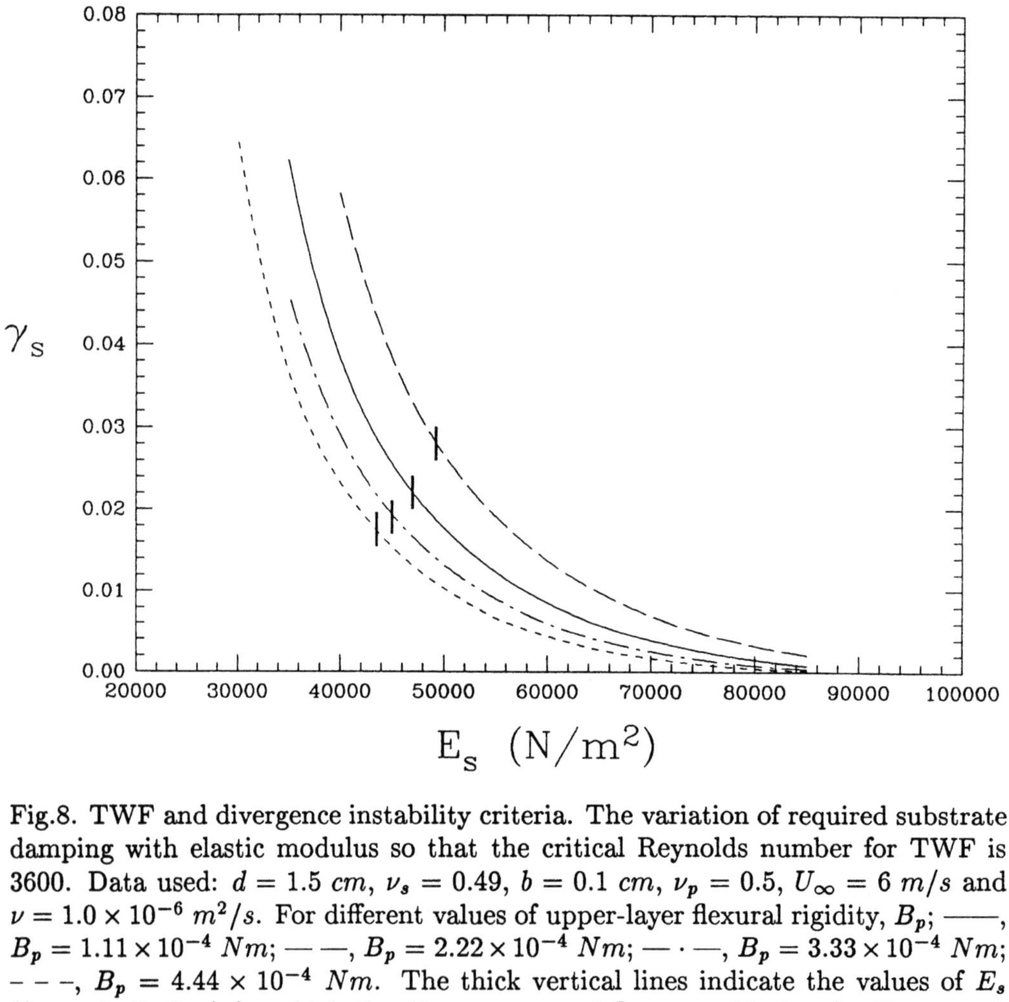

Fig.8. TWF and divergence instability criteria. The variation of required substrate damping with elastic modulus so that the critical Reynolds number for TWF is 3600. Data used: $d = 1.5$ cm, $\nu_s = 0.49$, $b = 0.1$ cm, $\nu_p = 0.5$, $U_\infty = 6$ m/s and $\nu = 1.0 \times 10^{-6}$ m^2/s. For different values of upper-layer flexural rigidity, B_p; ——, $B_p = 1.11 \times 10^{-4}$ Nm; — —, $B_p = 2.22 \times 10^{-4}$ Nm; — · —, $B_p = 3.33 \times 10^{-4}$ Nm; - - -, $B_p = 4.44 \times 10^{-4}$ Nm. The thick vertical lines indicate the values of E_s (for each B_p-line) for which the divergence-onset flow speed is 6 m/s. Data used: $d = 1.5$ cm, $\nu_s = 0.49$, $b = 0.1$ cm, $\nu_p = 0.5$, $U_\infty = 6$ m/s, $\rho_p = \rho_f = 1000$ kg/m^3 and $\nu = 1.0 \times 10^{-6}$ m^2/s.

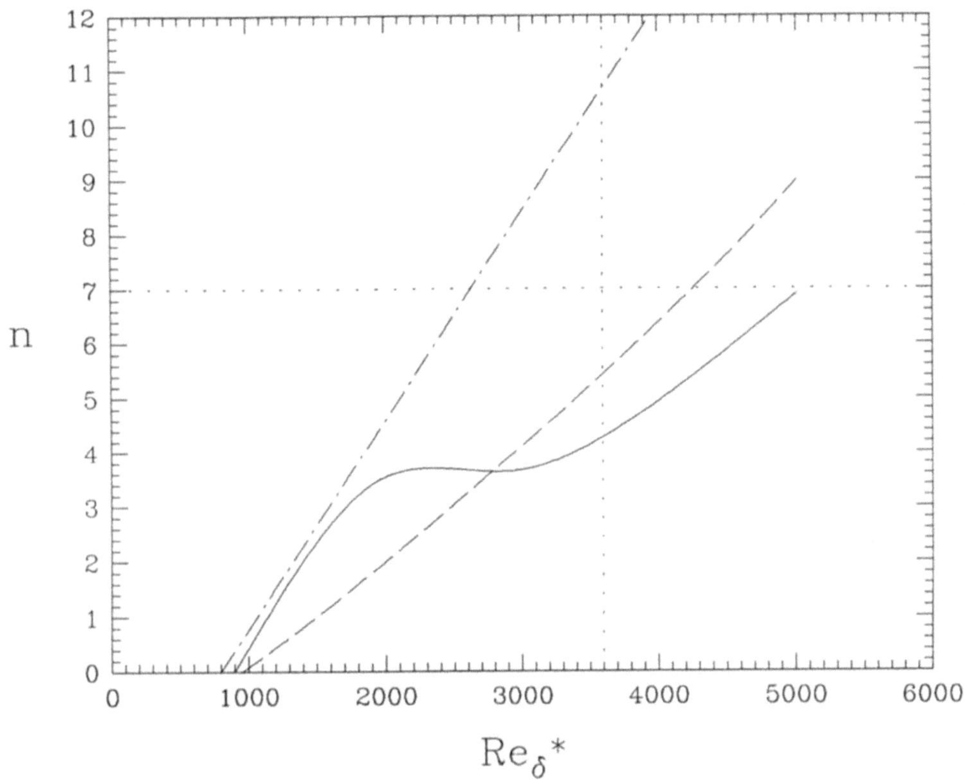

Fig.9. Instability growth-rate envelopes for TSW using the e^n method. — · —, rigid wall; — —, single-layer wall; ——, double-layer wall. The latter two have been optimised at $Re_{\delta^*} = 3600$.

To optimise for transition delay, the target Reynolds number is now increased and the procedure described in the two preceding paragraphs repeated. Figure 10 shows a set of accumulated growth-rate results for TSW where the target Reynolds number is 5700. The five curves pertain to different values of the upper-layer flexural rigidity. (Four of these curves constitute a family of results at $\gamma_p = 0$ whilst the fifth is an optimised result when $\gamma_p = 0.075$ which is discussed in the next paragraph.) In each case the TSW calculations have been carried out for the combination of E_s, B_p and γ_s such that E_s takes its lowest value whilst the combination of E_s and B_p yield a divergence speed of 6 m/s. This figure clearly illustrates how the flexural rigidity of the upper layer takes an optimal value. The strategy, therefore, is to find the value of B_p which gives the local maximum (found at the lower Reynolds numbers) a value of $n = 7$ since this stiffest permissible plate can continue to exercise influence at the higher Reynolds numbers where the most dangerous disturbances have longer wavelengths and lower frequencies. In Fig. 10, it is noted that the result for $B_p = 5.0 \times 10^{-4}$ Nm shows the local maximum (at $Re \approx 3000$) having $n = 7$, together with the accumulated growth rate at the target Reynolds number of 5700 showing a similar value for n. This, therefore, is the optimal configuration for transition delay.

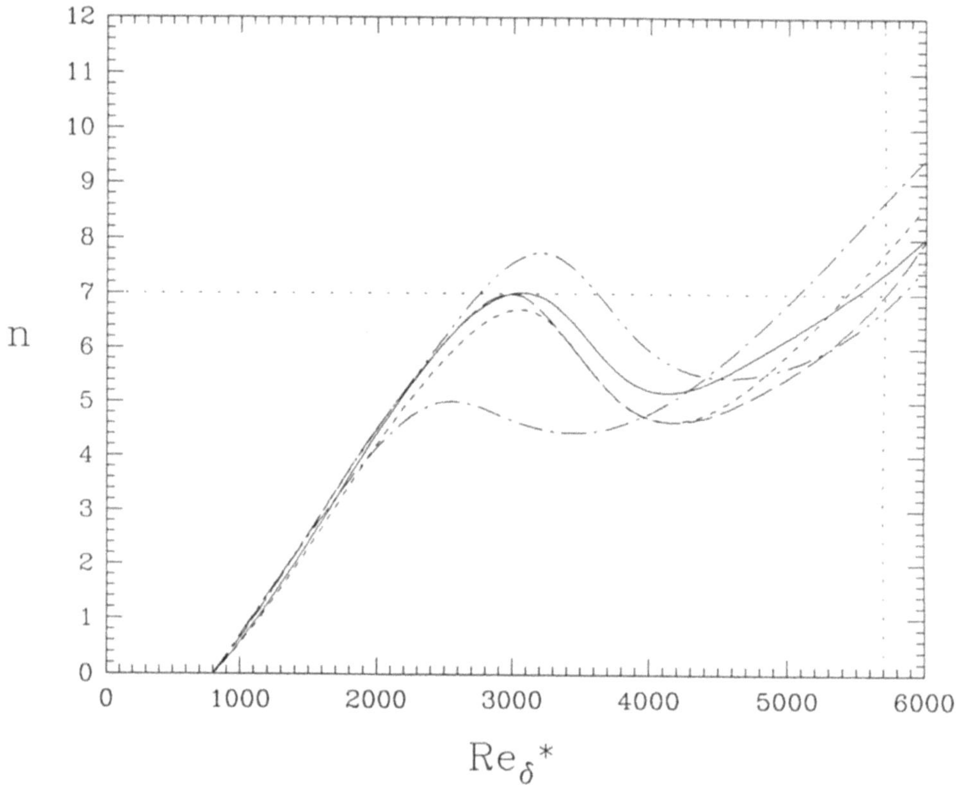

Fig.10. Instability growth-rate envelopes for TSW. Optimization at $Re_{\delta^*} = 5700$.
(i) No plate damping, $\gamma_p = 0$: — · —, $B_p = 2.22 \times 10^{-4}$ Nm, $E_s = 47000$ N/m^2; - - -, $B_p = 4.44 \times 10^{-4}$ Nm, $E_s = 43500$ N/m^2; ——, $B_p = 5.00 \times 10^{-4}$ Nm, $E_s = 42750$ N/m^2 **(optimum result)**; — · · —, $B_p = 6.66 \times 10^{-4}$ Nm, $E_s = 41000$ N/m^2.
(ii) With plate damping, $\gamma_p = 0.075$: — —, $B_p = 3.80 \times 10^{-4}$ Nm, $E_s = 44500$ N/m^2. **(optimum result)**.
Other data as for Fig. 8.

In the above procedure the upper-layer damping has been set to zero. Further results have considered the inclusion and variation of this parameter. Similar optimizations, after first assigning γ_p a value, have been undertaken; the results for an optimised wall when $\gamma_p = 0.075$ are included in Fig. 10. The result is very similar to the optimised case when $\gamma_p = 0$ although the new optimal value of B_p is reduced (accompanied by a rise in E_s to prevent divergence instability). Similar comparisons have shown that the location of the damping – whether it be in the upper or lower layer – has only a marginal effect on the wall performance. In design cases where a choice exists, it is preferable to locate the damping in the upper layer.

Throughout this work, we have chosen to use the flexural rigidity, B_p, as the primary upper-layer variable. Since $B_p = E_p b^3/12(1-\nu_p^2)$, where the elastic modulus, E_p, and the plate thickness, b, could be varied, the present approach might seem to pose something of an unreasonable restriction. In fact, these types of variation are used in the optimization of spring-backed flexible-plate compliant walls. Thus, we have investigated the dependence of the above optimizations upon the manner in which a given B_p is constituted from E_s and b. From a wide variety of such calculations, we can conclude that it is sufficient to characterise the upper layer solely by its flexural rigidity. From the discussions of Carpenter & Garrad (1985,1986), a high value of E_s with a relatively low value of b is favourable for TWF, which, in the present work, allows a lower value of wall damping to be used. A lower damping is, in turn, favourable for the reduction of TSW growth rates. However, the decrease in plate inertia (through decreasing b) is detrimental to the the suppression of TSW. It would seem that these counter-effects on the TSW almost exactly cancel each other for the types of wall being studied here. We may therefore take the optimal wall properties as being those indicated in Fig. 10 which give a transitional Reynolds number of 5700.

5. Conclusions

Efficient methods have been developed for the prediction of TWF and divergence instability for boundary-layer flow over a compliant wall essentially comprising two viscoelastic layers in which a relatively stiff thin layer is supported by a much softer thicker substrate. These methods, together with a rapid solver for TSW eigenvalues, have been used to establish a systematic procedure for the optimization of such compliant walls for transition delay. The most important findings in the optimization are as follows:

- Substrate damping is desirable for, although in itself detrimental to TSW suppression, the reductions to the substrate stiffness that it permits (whilst maintaining neutral stability for TWF) results in an overall benefit for the reduction of TSW growth rates.

- The optimum transition-delaying performance of the compliant wall is largely insensitive to the location of the wall damping, although the optimal structural parameters of both upper and lower layers are dependent on the distribution of damping between the two layers. It may be marginally more desirable to locate as much of the wall damping as possible in the upper layer.

- The key parameter for the upper-layer characterisation is the flexural rigidity.

For a given flexural rigidity, different combinations of elastic modulus and plate thickness have little effect on the TSW growth-rate envelopes.

The optimization of a double-layer compliant wall of prescribed thickness has been carried out. The maximum transitional Re_{δ^*} is found to be 5700 as compared with 2700 for a rigid wall. This corresponds to a transitional Reynolds number (Re_x) about 4.5 times that for a rigid wall. This result is very close to that found for the single-panel compliant walls comprising a spring-backed flexible plate and discussed in Carpenter (1991).

What sort of drag reductions are feasible with the present optimum compliant walls? The Reynolds-number ranges for certain marine applications are approximately as follows: 5 to 20 × 10^6 for hydrofoils; 20 to 50 × 10^6 for small submarines; and 40 to 70 × 10^6 for torpedoes. Using standard methods for calculating skin-friction drag coefficient, C_{Df}, for a mixed laminar-turbulent boundary layer over a flat plate the results shown below in Table 1 were obtained for selected Reynolds numbers in the ranges cited above.

Case	Re_L	$(Re_x)_t$	C_{Df}
Rigid Wall	10×10^6	2.25×10^6	0.00252
	20×10^6		0.00248
	50×10^6		0.00227
Optimal Compliant Wall	10×10^6	10.36×10^6	0.00042
	20×10^6		0.00158
	50×10^6		0.00197

Table 1: Values of C_{Df}.

It can be seen that very considerable benefit is predicted in terms of drag reduction when compliant walls are used for $Re_L \leq 20 \times 10^6$ It should be noted, however, that the estimates used for predicting the transitional Reynolds number are based on the conservative value of $n = 7$. This may well be an unduly restrictive assumption in view of the recent results of direct numerical simulations by Metcalfe *et al.* (1990,1991) showing that the benefits of wall compliance extend well into the nonlinear regime of transition. If the value of n used were to be increased to 9 or 10 a more than tenfold increase in transitional Reynolds number would be possible, leading to a considerable further improvement in the achievable drag reduction.

Furthermore, it is emphasised that the optimization carried out in §4 is for a wall of prescribed depth (at a given flow speed) which itself may not necessarily be optimal. It is noted in Fig. 10 that a steep increase in TSW growth rates occurs at the higher Reynolds numbers. This rate of increase would be reduced by using a thicker substrate. Of course, any anticipated improvements to the TSW would have to be offset against more severe restrictions on the wall structural properties demanded by the prevention of TWF and divergence. Nevertheless, an optimal substrate depth could be determined but it is doubtful that it would yield a significant increase to the maximum transitional Reynolds number found above. A more attractive proposition is to adapt the ideas of Carpenter (1991) to design an optimised two-panel compliant wall. The present work suggests that the second compliant panel would have its leading edge at $Re_{\delta^*} \approx 3500$ and that it should have a comparatively thicker (or softer) substrate combined with a stiffer upper layer than the first panel.

Acknowledgements

This work is part of a research programme at the University of Warwick which is supported by the Ministry of Defence (Procurement Executive).

References

Benjamin, T.B. 1960 Effects of a flexible boundary on hydrodynamic stability. *J. Fluid Mech.* **9**, 513-532.

Benjamin, T.B. 1963 The threefold classification of unstable disturbances in flexible surfaces bounding inviscid flows. *J. Fluid Mech.* **16**, 436-450.

Carpenter, P.W. 1985 The optimization of compliant surfaces for transition delay. *Univ. of Exeter, Sch. of Engineering* , Tech. Note 85/2.

Carpenter, P.W. 1987 The optimization of compliant surfaces for transition delay. In *Proc. IUTAM on Turbulence Management & Relaminarisation, Bangalore, India* (ed. H.W. Liepmann & R. Narasimha), pp. 305-313. Springer.

Carpenter, P.W. 1990 Status of transition delay using compliant walls. In *Viscous Drag Reduction in Boundary Layers* (ed. D.M. Bushnell & J.N. Heffner), New York: AIAA, pp. 79-113.

Carpenter, P.W. 1991 The optimization of multiple-panel compliant walls for delay of laminar-turbulent transition. *AIAA Paper 91-1772*.

Carpenter, P.W. & J.S.B Gajjar 1990 A general theory for two- and three-dimensional wall-mode instabilities in boundary layers over isotropic and anisotropic compliant walls. *Theoretical & Computational Fluid Dynamics*, **1**, 349-378.

Carpenter, P.W. & A.D. Garrad 1985 The hydrodynamic stability of flows over Kramer type compliant surfaces. Part 1. Tollmien-Schlichting instabilities. *J. Fluid Mech.* **155**, 465-510.

Carpenter, P.W. & A.D. Garrad 1986 The hydrodynamic stability of flows over Kramer type compliant surfaces. Part 2. Flow-induced surface instabilities. *J. Fluid Mech.* **170**, 199-232.

Carpenter, P.W., M. Gaster & G.J.K. Willis 1983 A numerical investigation into boundary layer stability on compliant surfaces. In *Numerical Methods in Laminar and Turbulent Flow*, pp.166-172. Pineridge.

Carpenter, P.W. & P.J. Morris 1990 The effect of anisotropic wall compliance on boundary-layer stability and transition. *J. Fluid Mech.* **218**, 171-223.

Chung, K.H. & E.W. Merrill 1984 Drag reduction by compliant surfaces measured on rotating discs. Presented at *Compliant Coating Drag Reduction Program Review*, Office of Naval Research, Washington, D.C., Oct. 1984.

Daniel, A.P., M. Gaster & G.J.K. Willis 1987 Boundary layer stability on compliant surfaces. *British Maritime Technology Ltd., Teddington, U.K.*, Final Rept. No. 35020.

Dixon, A.E. & A.D. Lucey 1991 The application of parallel algorithms for the optimisation of single- and double-layer compliant walls for transition delay. To appear in *Numerical Methods in Laminar & Turbulent Flow*, Pineridge Press, Swansea.

Duncan, J.H., A.M. Waxman & M.P. Tulin 1985 The dynamics of waves betwen a visco-elastic coating and a fluid flow. *J.Fluid Mech.* **158**, 177-197.

Fraser, L.A. & P.W. Carpenter 1985 A numerical investigation of hydroelastic and hydrodynamic instabilities in laminar flows over compliant surfaces comprising one or two layers of visco-elastic material. In *Numerical methods in Laminar and Turbulent Flow*, pp. 1171-1181. Pineridge.

Gad-el-Hak, M., R.F. Blackwelder & J.J. Riley 1984 On the interaction of compliant coatings with boundary-layer flows. *J. Fluid Mech.* **140**, 257-280.

Garrad, A.D. & P.W. Carpenter 1982 A theoretical investigation of flow-induced instabilities on compliant coatings. *J. Sound Vib.* **85**, 483-500.

Gaster, M. 1987 Is the dolphin a red herring? In *Proc. IUTAM on Turbulence Management & Relaminarisation, Bangalore, India* (ed. H.W. Liepmann & R. Narasimha) pp. 285-304. Springer.

Gyorgyfalvy, D. 1967 Possibilities of drag reduction by the use of flexible skin. *Journal of Aircraft*, **4**, 186-192.

Joslin, R.D. & P.J. Morris 1989 The sensitivity of flow and surface properties to changes in compliant wall properties. *J. Fluid and Structures* **3**, 423-432.

Landahl, M.T. 1962 On the stability of a laminar incompressible boundary layer over a flexible surface. *J. Fluid Mech.* **13**, 609-632.

Lucey, A.D. & P.W. Carpenter 1991a A numerical simulation of surface behaviour in the compliant-wall/inviscid-flow interaction. To be published.

Lucey, A.D. & P.W. Carpenter 1991b A numerical simulation of the three-dimensional inviscid-flow/compliant-panel interaction using parallel computing. To appear in *Numerical Methods in Laminar & Turbulent Flow*, Pineridge Press, Swansea.

Lucey, A.D., P.W. Carpenter & A.E. Dixon 1991 The role of wall instabilities in boundary-layer transition over compliant walls. *Proc. Royal Aeronautical Society Conf. on Boundary Layer Transition and Control, Cambridge, U.K.*, pp. 35.1-35.10.

Metcalfe, R.W., F. Battistoni & J. Ekeroot 1990 The effect of a compliant wall on delay of transtion to turbulence. Bull. Amer. Phys. Soc. **35**, 2291.

Metcalfe, R.W., F. Battistoni, S.A. Orszag & J. Ekeroot 1991 Evolution of boundary layer flow over a compliant wall during transition to turbulence. *Proc. Royal Aeronautical Society Conf. on Boundary Layer Transition and Control, Cambridge, U.K.*, pp. 36.1-36.14.

Riley, J.J., M. Gad-el-Hak & R.W. Metcalfe 1988 Compliant coatings. *Annual Review of Fluid Mechanics* **20**, 393-420.

Sen, P.K. & D.S. Arora 1988 On the stability of laminar boundary-layer flow over a flat-plate with a compliant surface. *J. Fluid Mech.* **197**, 201-240.

Willis, G.J.K. 1986 Hydrodynamic stability of boundary layers over compliant surfaces. Ph. D. Thesis, University of Exeter.

Yeo, K.S. 1988 The stability of boundary-layer flow over single- and multi-layer viscoelastic walls. *J. Fluid Mech.* **196**, pp. 359-408.

Nonlinear evolution of modes in the flow over compliant surfaces

J. S. B. GAJJAR

University of Manchester
Manchester, U.K.

K.-S. Choi (ed.), Recent Developments in Turbulence Management, 223–239.

Abstract

The paper discusses the nonlinear stability of the Travelling Wave Flutter modes in the boundary layer flow over a compliant surface. The work of Carpenter & Gajjar (1990) is extended to include nonlinear effects and a set of equations describing the spatial/temporal evolution of the disturbance is obtained. It is shown that there is a strong similarity between the nonlinear evolution of the Travelling Wave Flutter modes and the nonlinear evolution of the inviscid modes in the stability of hypersonic boundary layers. The evolution equations are solved numerically and several results are presented.

1. Introduction

The study of the flow over flexible or compliant surfaces has received much attention in recent years. Many of the earlier investigations were motivated by the experimental results of Kramer (1957,1960,1961,1962) who reported that significant drag reductions could be obtained on the surfaces of bodies fitted with compliant coatings. Benjamin (1960,1963) and Landahl (1962), amongst others, were the first to study the stability of the flow over compliant surfaces using the ideas of hydrodynamic stability theory. They concluded that whilst some types of compliant surfaces could reduce drag, by increasing the critical Reynolds number and hence delaying the onset of transition, this was unlikely to be true for the Kramer type surfaces. This last conclusion is not however supported by more recent numerical studies of the linear stability problem.

The many interesting and fascinating aspects of the flow over compliant surfaces and in particular a discussion of the early and more recent experimental and theoretical work, is given in the comprehensive reviews by Bushnell *et al* (1977), Gad-el-Hak (1986,1987), Riley *et al* (1988), and Carpenter (1990).

The early theoretical work showed that the stability properties were strongly influenced by the fluid substrate interaction. From an analysis of the energies associated with a wave travelling over a compliant wall, Benjamin and Landahl were able to identify three distinct types of instabilities possible in the flow over a compliant surface. They classified these as Class A, B and C type waves and each instability responded in a different manner according to the substrate properties. The Class A wave was destabilised by damping, the Class B wave was stabilised by damping and the Class C wave was unaffected by damping. The Tollmien-Schlichting instability similar to that occurring in the boundary layer flow over a rigid wall but modifed by wall compliance, belongs to the Class A type of wave. The Travelling Wave Flutter, (and hereafter referred to as TWF), is an inviscid instability with the waves travelling at almost free-stream speed, belongs to the Class B category. An example of the third type of wave is the 'static divergence' instability observed in the experiments of Gad-El-Hak (1984). Whereas the Tollmien-Schlichting and the TWF instabilities are convective instabilities the static divergence instability is believed to be an absolute instability.

Whilst the linear stability properties have been studied in great detail, there has been little or no work dealing with the nonlinear aspects. Some exceptions to this include the work of Thomas (1991a,b) who has looked at the resonant triad interactions, and also extended the triple-deck work of Rothmeyer & Hiemcke (1988) and Mackerrell (1988) to include weakly non-linear effects. He finds that if the wall is sufficiently compliant the Tollmien-Schlichting mode can be subcritically unstable but otherwise it is supercritically stable. Domaradzki & Metcalfe (1987) and Metcalfe *et al* (1986,1991) have conducted numerical simulations of the full Navier Stokes equations with a compliant wall and with a temporally growing boundary layer. Their results indicate that secondary instabilities can occur and which can trigger some linearly stable modes to become unstable. They also showed that substantial transition delay could be achieved with suitably chosen wall parameters. However the extrapolation of these results to a spatially growing boundary layer is unclear. In addition questions concerning the use of linearised wall conditions but with a fully nonlinear set of equations for the fluid motion require further study. For the TWF modes this may very well be appropriate but it is more questionable for the Tollmien-Schlichting mode which has a very different structure and properties. The Tollmien-Schlichting mode is strongly influenced by the wall motion as many of the rigid wall triple-deck studies for the flow with humps, corners, slot injection, for example, indicate.

Despite these studies there are still many unanswered questions concerning the nonlinear stability of the boundary layer flow over compliant surfaces. Of particular interest are questions concerning the evolution of the different modes, do they saturate nonlinearly or not, the nonlinear interactions between the modes of which the resonant triad interaction is just one of several possible, and so on. These questions are not just of academic interest since much valuable information can be obtained about the behaviour of the various modes present. In addition the nonlinear theory can provide a useful check for the large scale numerical simulations. One of the aims of the present work is to look at the nonlinear evolution of the TWF mode in the flow over isotropic compliant walls. The analysis presented here also extends to the anisotropic case in a straightforward fashion but for simplicity we have restricted ourselves to the isotropic case only. Further modifications to consider modal interactions are also possible.

A suitable starting point for the nonlinear theory is the work of Carpenter & Gajjar (1990) (hereafter referred to as GC) who obtained the linear asymptotic description of the TWF mode. They considered the two-dimensional boundary layer flow over isotropic and anisotropic compliant surfaces with the wall motion modelled by the simple spring-plate equation, see equation (2.2) in their paper, which for the isotropic wall reduces to

$$b\rho_m \frac{\partial^2 \eta'}{\partial t^2} + B\frac{\partial^4 \eta'}{\partial x^4} + K_E \eta' = -p'_w. \tag{1}$$

Here η' is the small vertical displacement of the wall, see Fig. 1 , p'_w is the fluctuating pressure at the wall, and the remaining parameters describe the properties of the wall with b being the plate thickness, ρ_m is the density of the plate, B is the flexural rigidity of the plate, and K_E is an equivalent spring stiffness. The work in GC was based on the assumptions that the wavelength of the TWF was greater than the boundary layer thickness and that the growth rates were small. An expansion for the disturbances combined with matched asymptotic expansions were then used to obtain asymptotic expressions for various quantities such as the p'_w/η'. This combined with similar expressions obtained from the membrane equation were then used to calculate the dispersion curves, and it was found that the predicted wavenumbers and growth rates were in excellent agreement with those calculated numerically using the Orr-Sommerfeld equation.

Given the linear theory as described in GC, the next question concerns the inclusion of the nonlinear terms. There are several ways in which this can be done although in the present context unsteady nonlinear critical theory is the most appropriate. This is because the growth rates of the most unstable TWF mode are small and in addition there is a distinct critical layer in the flow with the phase speed of the disturbance less than than the freestream speed, see Fig. 9 of GC. This situation is similar to that arising in many shear and boundary layer flows where unsteady nonlinear critical theory has been used succesfully to describe the spatial/temporal evolution of the linearly unstable modes, see for example the papers by Gajjar & Smith (1985), Goldstein & Leib (1988), Goldstein & Hultgren (1988). In shear flows comparisons of theoretical predictions for some flow quantities were found to be in excellent agreement with experimentally observed values, see Hultgren (1991).

We do not intend to go into an extended discussion of nonlinear critical theory and the interested reader is referred to the excellent review articles by Stewartson (1981), and Maslowe (1986). The main effect of a linear critical layer is to produce a jump of '$i\pi$' in the disturbance velocities across the critical layer. This result is used in the numerical calculations of the linear stability equations in the more familiar form as an indented contour around the critical point. It also contributes to a jump in the Reynolds stress across the critical point and which is directly related to the growth rate of the instability wave. With a nonlinear critical layer the jump in the disturbance velocities is no longer constant and depends nonlinearly on the amplitude of the disturbance wave. As the wave evolves downstream the jump changes and is in fact directly coupled to the evolution of the vorticity inside the critical layer.

Below we consider the spatial/temporal evolution of the TWF mode on slow time and length scales, and use a multiple scale analysis together with unsteady nonlinear critical layer theory to obtain a set of equations describing the evolution of the TWF mode. Regarding the scalings, the results of GC show that if the wavenumber is $O(\alpha)$ then the growth rates are $O(\alpha^2)$ and the frequency of the disturbance is $O(\alpha)$. Using these scalings an unsteady nonlinear partial differential equation is obtained which describes the evolution of the vorticity inside the critical layer, and this is coupled to an amplitude equation. It is interesting to note that the equations obtained in the physical coordinates are very similar to those arising in hypersonic flow in the work of Goldstein & Wundrow (1990), although as in the latter paper, with a suitable renormalisation these reduce to the standard unsteady nonlinear critical layer equations.

Our results show that the main effect of nonlinearity on the TWF mode is to cause the growth rate to decrease as the TWF mode evolves downstream, and to cause a roll up of vorticity inside the critical layer with the generation of harmonics. Allied with the comment made earlier the jump in the Reynolds stress across the critical layer is driven to zero as the disturbance evolves downstream These results are typical of those occurring in other contexts involving unsteady nonlinear critical layers.

In the following sections (x, y, t) denote the non-dimensional Cartesian coordinates and time, (u, v) the nondimensional velocities, p is the non-dimensional pressure, and R the Reynolds number, based on the distance from the leading edge, is taken to be large throughout.

2. Problem Formulation

We consider the two-dimensional boundary layer flow over an isotropic compliant wall with the fluid/substrate interaction modelled by the simple spring membrane equation described earlier, see Fig. 1 also.

The basic boundary layer flow is given by

$$\underline{u} = \underline{U}_B(x, Y) = (U_B(x, Y), R^{-\frac{1}{2}} V_B(x, Y)) + \ldots, \quad p = p_B,$$

with $Y = R^{\frac{1}{2}} y$ being the boundary layer coordinate.

Following the discussion in the Introduction we also introduce the multiple-scaling

$$\frac{\partial}{\partial x} = R^{\frac{1}{2}}[\alpha_0 h \frac{\partial}{\partial \xi} + h^2 \frac{\partial}{\partial X_1}],$$
$$\frac{\partial}{\partial t} = R^{\frac{1}{2}}[-\alpha_0 c_0 h \frac{\partial}{\partial \xi} + h^2 \frac{\partial}{\partial T_1}],$$

where α_0, c_0 are taken to be real and $0 < c_0 < 1$, so that a critical layer exists. The wavenumber is $\alpha_0 h$ and the scaling parameter h with $h << 1$ is introduced because we will be considering long waves. The $O(h^2)$ terms above allow for the spatial/temporal growth of the wave on the (X_1, T_1) scales.

The nonlinear theory and the derivation of the amplitude equation

The brief details given below serve only to emphasize the important points in the derivation of the equations describing the nonlinear evolution of the TWF mode. The linear theory is described in detail in GC and the full details of the nonlinear theory will be described elsewhere.

Disturbances to the basic flow of size δ are introduced with $\delta << 1$, so that in the main part of the boundary layer we have

$$\begin{aligned} u &= U_B + \delta(u_0 + h u_1 + \ldots), \\ v &= \delta(h v_0 + h^2 v_1 + \ldots), \\ p &= p_0 + \delta(h p_0 + h^2 p_1 + \ldots) \quad . \end{aligned} \tag{2}$$

The disturbance size δ is taken to be $O(h)$ so that the nonlinear terms contribute nontrivially in the critical layer. This can be deduced from a balance of terms inside the critical layer.

The expansion (2) when substituted into the Navier-Stokes equations give the leading order solutions

$$u_0 = A_0(X_1, T_1)U_{BY}E + c.c, \quad v_0 = -i\alpha_0 A_0(X_1, T_1)(U_B - c_0)E + c.c,$$

$$p_0 = P_0(X_1, T_1)E + c.c,$$

where $E = \exp(i\xi)$ and $c.c$ denotes the complex conjugate. The functions A_0, P_0 are slowly varying amplitude functions. At the next order we obtain

$$v_1 = -i\alpha_0 P_0 E(U_B - c_0)\int_0^Y \frac{dy}{(U_B - c_0)^2} - (A_{0T_1} + c_0 A_{0X_1})E + A_1^{\pm}(U_B - c_0) + c.c. \tag{3}$$

The integral in (3) is singular at the critical point $Y = Y_c$ where $U_B = c_0$ and the $\pm$ terms are there to account for the jump across the critical layer.

The boundary conditions are that

$$v = \eta_t \quad \text{at} \quad Y = \eta = h(\eta_0 + h\eta_1 + \ldots),$$

to match the normal velocity fluctuation with that of the compliant wall $Y = \eta(x, t)$, and

$$v \to 0 \quad \text{as} \quad Y \to \infty.$$

In addition to these there is also the membrane equation which is rewritten in nondimensional form as

$$-p_w = h^{-1}B_0\eta_{tt} + h^{-3}B_1\eta_{xxxx} + hB_2\eta, \tag{4}$$

where the constants B_0, B_1, B_2 can be related to the original parameters in equation (1). Also p_w is the pressure evaluated at the wall $Y = \eta$.

Application of the boundary conditions and matching, and setting $\eta_0 = \bar{\eta}_0 E + c.c$, leads to the dispersion relations

$$P_0 = \alpha_0(1 - c_0)^2 A_0, \quad \bar{\eta}_0 = -A_0$$

and

$$\alpha_0(1 - c_0)^2 = -B_0\alpha_0^2 c_0^2 + B_1\alpha_0^4 + B_2,$$

for determining the wavenumber and phase-speed. It can be seen that A_0 also represents the amplitude of the leading order surface displacement. The second order matching and boundary conditions yields the amplitude equation

$$\frac{\partial A_0}{\partial T_1} + c_g \frac{\partial A_0}{\partial X_1} + \frac{\sigma_2}{\sigma_0} A_0 = -\frac{2\alpha_0^3(1-c_0)^4 \lambda_2 A_0(\phi^+ - \phi^-)}{\lambda_1^3 \sigma_0}, \tag{5}$$

where $c_g = \frac{\partial(\alpha_0 c_0)}{\partial \alpha_0}$ is the group velocity of the wave,

$$\lambda_1 = \left(\frac{\partial U_B}{\partial Y}\right)_{Y=Y_c} \qquad 2\lambda_2 = \left(\frac{\partial^2 U_B}{\partial Y}\right)_{Y=Y_c},$$

and σ_2, σ_0 are constants dependent on α_0, c_0 and the wall parameters and given by

$$\sigma_0 = -2i\alpha_0(1-c_0) + 2iB_0\alpha_0^2 c_0,$$

$$\sigma_2 = -\alpha_0^3(1-c_0)^4 \int_0^{*\infty} \frac{dy}{(U_B - c_0)^2} + \alpha_0^4 \int_0^{*\infty} (U_B - c_0)^2 \, dy.$$

In addition the $\phi^\pm$ term in equation (5) are related to the $A_1^\pm$ terms by

$$(A_1^+ - A_1^-) = -\frac{2i\alpha_0 P_0 \lambda_2}{\lambda_1^3}(\phi^+ - \phi^-)E + h.h.t + c.c,$$

with $h.h.t$ denoting the higher harmonic terms in $A_1^\pm$. The notation $\int^*$ means that the Hadamard finite part of the integral is to be taken. Note that σ_2 is real and σ_0 purely imaginary. The equation (5) describes the nonlinear evolution of the TWF wave with the nonlinearity coming in via the $(\phi^+ - \phi^-)$ jump term. This jump is unknown at this stage but is obtained from solving the critical layer problem described below. In the linear theory $(\phi^+ - \phi^-)$ is the familiar '$i\pi$' jump, but in the nonlinear theory $(\phi^+ - \phi^-)$ depends on the amplitude of the wave.

In the critical layer with $Y = Y_c + h\bar{Y}$ we have

$$U_B = c_0 + h\lambda_1 \bar{Y} + h^2 \lambda_2 \bar{Y}^2 + \dots,$$

$$u = c_0 + hU_0 + h^2 U_1 + \dots,$$

$$v = h^3 V_0 + \dots, \quad p = p_B + h^2 P_0 + \dots,$$

and after some analysis it is found that

$$\Omega_1 = U_{1\bar{Y}}$$

satisfies

$$\frac{\partial \Omega_1}{\partial T_1} + c_0 \frac{\partial \Omega_1}{\partial X_1} + \alpha_0 U_0 \frac{\partial \Omega_1}{\partial \xi} + V_0 \frac{\partial \Omega_1}{\partial \bar{Y}} = 0. \tag{6a}$$

Here Re denotes the real part and

$$U_0 = \lambda_1 \bar{Y} + 2\lambda_1 \mathrm{Re}(A_0 E)$$

$$V_0 = -2\alpha_0 \lambda_1 \bar{Y} \mathrm{Re}(iA_0 E) + \bar{V}_0(X_1, T_1),$$

where $\bar{V}_0$ is a function of the slow coordinates and related to A_0 by

$$\bar{V}_0 \lambda_1 = -(i\alpha_0 P_0 + \lambda_1 A_{0T_1} + \lambda_1 c_0 A_{0X_1})E - i\alpha_0 \lambda_1^2 A_0^2 E^2 + c.c.$$

Together with (6) we have the boundary conditions

$$\Omega_1 \sim 2\lambda_2(\bar{Y} + 2\mathrm{Re}(A_0 E)) + \frac{4\lambda_2 \mathrm{Re}(P_0 E)}{\lambda_1^2 \bar{Y}} \quad \text{as} \quad \bar{Y} \to \pm\infty. \tag{6b}$$

The jump $(\phi^+ - \phi^-)$ is determined from

$$\frac{2\lambda_2 \alpha_0 (1-c_0)^2 A_0}{\lambda_1^2}(\phi^+ - \phi^-) = \frac{1}{2\pi} \int_0^{2\pi} \int_{-\infty}^{*\infty} (\Omega_1 - 2\lambda_2 \bar{Y} - 2\lambda_2 \mathrm{Re}(A_0 E))\ e^{-i\xi}\, d\bar{Y}\, d\xi. \tag{7}$$

The equations (6,7) describe the evolution of the disturbance vorticity inside the critical layer, and (5,7) show how this is coupled to the evolution of the disturbance amplitude via the jump and the amplitude equation. As mentioned in the Introduction (6,7) are not the standard critical layer equations but of a form very similar to those obtained by Goldstein & Wundrow (1990) in their study of the stability of hypersonic boundary layers.

3. Results and Discussion

The linear results of GC are retrieved by setting $(\phi^+ - \phi^-)$ equal to $i\pi$ in the amplitude equation. If we set $\partial/\partial X_1$ to zero and $A_0 \sim \tilde{A}_0 e^{\kappa T_1}$ then from (5) the (scaled) linear growth rate is given by

$$\kappa_r \equiv \mathrm{Re}(\kappa) = -\frac{2\alpha_0^2(1-c_0)^4 \lambda_2 \pi}{\lambda_1^3 [2B_0 \alpha_0^2 c_0 - 2\alpha_0(1-c_0)]},$$

which agrees with that obtained by GC. Comparisons of the linear asymptotic results with numerical calculations from the Orr-Sommerfeld equation are given in GC.

For the nonlinear problem the set of equations (5-7) has to be solved numerically. These can be cast into a more standard form by first changing variables to $(\xi, \tilde{Y}, X_1, T_1)$ where $\tilde{Y} = \bar{Y} + 2\mathrm{Re}(A_0 E)$, and then using the renormalisation

$$\begin{aligned} \tilde{Y} = d_1 Z, \quad T_1 = d_2 T + T_0, \quad X_1 = d_3 X + X_0, \\ A_0 = d_4 A e^{i\xi_0}, \quad \xi^* = \xi + \xi_0, \quad \Omega_1 = d_5 \Omega + 2\lambda_2 \tilde{Y}, \end{aligned} \tag{8}$$

where the d_i are given by

$$d_1 = \frac{\kappa_r c_0}{\pi \alpha_0 c_g \lambda_1}, \quad d_2 = \frac{1}{\alpha_0 \lambda_1 d_1}, \quad d_3 = \frac{c_0}{\alpha_0 \lambda_1 d_1},$$

$$d_4 = \frac{\lambda_1^2 d_1^2}{2(1-c_0)^2}, \quad d_5 = \frac{4\lambda_2 (1-c_0)^2 d_4}{\lambda_1^2 d_1}.$$

This then gives

$$\frac{\partial \Omega}{\partial T} + \frac{\partial \Omega}{\partial X} + Z \frac{\partial \Omega}{\partial \xi^*} - \mathrm{Re}(iAe^{i\xi^*}) \frac{\partial \Omega}{\partial Z} - \lambda \frac{\partial^2 \Omega}{\partial Z^2} = \mathrm{Re}(iAe^{i\xi^*}), \tag{9a}$$

$$\Omega \sim \frac{\mathrm{Re}(Ae^{i\xi^*})}{Z} \quad \text{as} \quad Z \to \infty, \tag{9b}$$

together with

$$\frac{c_0}{c_g} \frac{\partial A}{\partial T} + \frac{\partial A}{\partial X} = -\frac{i}{\pi} \int_0^{2\pi} \int_{-\infty}^{*\infty} \Omega e^{-i\xi^*} \, dZ \, d\xi^*. \tag{9c}$$

The additional viscous term (multiplied by λ) in (9a) has been included for the viscous problem, and

$$\lambda = \frac{R^{-\frac{1}{2}} h^{-4}}{\alpha_0 \lambda_1 d_1^3}.$$

If h is sufficiently small, $h = O(R^{-\frac{1}{8}})$, then this term cannot be neglected. For h much larger than this ordering this term is effectively zero. The constants ξ_0, X_0, T_0 in (8) can be chosen to match the initial conditions so that, for instance, for the spatial problem $A \to e^{\pi X}$ as $X \to -\infty$. Note that in (9c) we have, for convenience, taken σ_2 to be zero which just affects the real part of the wavenumber. This is equivalent to redefining ξ.

The set of equations (9), with $\frac{\partial}{\partial T}$ set equal to zero, were solved numerically using the spectral method as described in Gajjar (1991). Results for various values of λ are given in Figures 2,3. In Figure 2 the scaled growth rate $\text{Re}(A_X/A)$ is plotted as function of the scaled slow streamwise coordinate X, and in Figure 3 $\ln|A|$ is plotted against X. It can be seen that the growth rate follows the linear value (equal to π with the renormalisation), and then rapidly starts to decrease as the nonlinear terms start to exert their influence. Downstream the growth rate reaches a minimum and then starts to oscillate with the oscillations becoming more pronounced with decreasing values of λ. These oscillations represent a continual exchange of energy between the mean flow and the disturbance. In contrast to the shear flow results, the growth rate oscillates about a non-zero mean implying that the wave amplitude, and hence the surface displacement also, is still growing downstream. The rapid adjustment from the linear to the nonlinear growth rate value takes place within a fairly short X lengthscale. The nonlinearity in the critical layer equation (9a) causes the generation of the higher harmonics and smaller and smaller scales. This is evident in the plots of the disturbance vorticity in the shear layer problem, see Goldstein & Leib (1988), Goldstein & Hultgren (1988). These papers also show how the vorticity rolls up as it evolves downstream with the appearance of the familiar Kelvin cat's eyes pattern. This is in the transformed coordinates but in the physical coordinates the picture is somewhat distorted although regions of thin and intense shear layers are still present, see Goldstein & Wundrow (1990). Similar trends also occur here.

In summary we have obtained a set of nonlinear equations which describe the spatial/temporal evolution of the TWF mode in the flow over isotropic compliant walls. The analysis extends readily to the flow over anisotropic compliant walls, and more work needs to be done to investigate the effects of anisotropy, etc, on the nonlinear results. Also the implications of some of our results as far as numerical simulations and experimental observations are concerned is far from clear, and it would be useful to make comparisons with relevant work. It has been shown that there is a close correspondence between the governing equations and results here and those in the stability of hypersonic boundary layers and this is further highlighted when the limit of the phase speed approaching the freestream value is considered. This and the full details of the nonlinear theory will be addressed elsewhere.

Acknowledgement

The author would like to thank the S.E.R.C. of U.K. for the AMT DAP-510 parallel computer bought with the grant GRE/7072.6 and used for the computations reported here, and also for an earlier grant GR/E 5114 for computer time on the Cray -XMP facility at Rutherford. One of the referees is also thanked for his helpful comments.

References

Benjamin, T. B., (1960), "Effects of a flexible surface on hydrodynamic stability," *Jour. Fluid Mech.,* **6**, 513.

Benjamin, T. B., (1963), "The threefold classification of unstable disturbances in flexible surface bounding inviscid flows," *Jour. Fluid Mech.,* **16**, 436.

Bushnell, D. M., Hefner, J. N. and Ash, R. L., (1977) "Effects of compliant wall motion on turbulent boundary layers," *Phys. Fluids,* **20**,S31-S48.

Carpenter, P. W., (1990), "Status of transition delay using compliant walls," in *Progress in Astronautics and Aeronautics* **123** (eds. D. M. Bushnell and J. F. Hefner,) AIAA Washington DC.

Carpenter, P. W. and Gajjar, J. S. B., (1990), "A general theory for two- and three-dimensional wall-mode instabilities in boundary layers over isotropic and anisotropic compliant walls," *Theor. Comp. Fluid Dyn.,* **1**, 349.

Domaradzki, J. A. and Metcalfe, R. W., (1987), "Stabilization of laminar boundary layers by compliant membranes," *Phys. Fluids,* **30**, 695.

Gad-El-Hak, M., (1984), "An optical technique for measuring the flow-induced motion of a compliant wall," *Symp. Flow Induced Vib.* **5** ed. M. P. Paidoussis, A. J. Kalinowski, 9-22, ASME, New York.

Gad-El-Hak, M., (1986), "Boundary layer interactions with compliant coatings: an overview," *Appl. Mech. Rev.,* **39**, 511.

Gad-El-Hak, M., (1987), "Compliant coating research: a guide to the experimentalist," *Fluids Struc.,* **1**, 55.

Gajjar, J. S. B., (1991), "Nonlinear evolution of modes in compressible boundary layers," in *Proc of the Royal Aero. Soc. Meeting on Boundary Layer Transition and Control*, Cambridge, 12.1.

Gajjar, J. & Smith, F. T., (1985), "On the global instability of free disturbances with a time-dependent nonlinear viscous critical layer," *Jour. Fluid Mech.,* **157**, 295.

Goldstein, M. E., and Hultgren, L. S., (1988), "Nonlinear evolution of a spatially excited instability wave in a free shear layer," *Jour. Fluid Mech.,* **197**, 295.

Goldstein, M. E., and Leib, S. J., (1988), "Nonlinear roll-up of externally excited free shear layers," *Jour. Fluid Mech.* **191**, 481.

Goldstein, M. E., and Wundrow, D. M., (1990), "Spatial evolution of nonlinear acoustic mode instabilities on hypersonic boundary layers," *Jour. Fluid Mech.,* **219**, 585.

Hultgren, L. S., (1991), "Nonlinear spatial equilibration of an externally excited instability wave in a free shear layer," submitted to *Jour. Fluid Mech.*

Kramer, M. O., (1957), "Boundary layer stabilization by distributed damping," *J. Aero. Sci.,* **24**, 459.

Kramer, M. O., (1960), "Boundary layer stabilization by distributed damping," *Nav. Eng. Jour.,* **72**, 25.

Kramer, M. O., (1961), "The dolphin's secret," *Nav. Eng. Jour.,* **73**, 103.

Kramer, M. O., (1962),"Boundary layer stabilization by distributed damping," *Nav. Eng. Jour.,* **74**, 341.

Landahl, M. T., (1962), "On the stability of a laminar boundary layer over a flexible surface," *Jour. Fluid Mech.,* **13**, 609.

Mackerrell, S. O. (1988), "Hydrodynamic instabilities of boundary layer flows," Ph. D. Thesis, Univ. of Exeter.

Maslowe, S., (1986), "Critical layers in shear flows," *Ann. Rev. Fluid Mech.,* **18**, 405.

Metcalfe, R. W., Rutland, C. J., Duncan, J. H. and Riley, J. J., (1986), "Numerical simulation of active stabilization of laminar boundary layers," *AIAA J.,* **24**, 1494.

Metcalfe, R. W., Battistoni, F. and Ekeroot, J., (1991), "Evolution of boundary layer flow over a compliant wall during transition to turbulence," in *Proc of the Royal Aero. Soc. Meeting on Boundary Layer Transition and Control,* Cambridge, 36.1.

Riley, J. J., Gad-El-Hak, M. and Metcalfe, R. W. (1988), "Compliant Coatings," *Ann. Rev. Fluid Mech.* **20**, 393.

Rothmeyer, A. P. and Hiemcke, C., (1988), Unpublished report, Dept. Aero. Eng., Iowa State Univ.

Stewartson, K., (1981), "Marginally stable inviscid flows with critical layers," *I.M.A.J. Appl. Math.,* **27**, 133.

Thomas, M. (1991a), "On the resonant triad interaction in flows over rigid and flexible boundaries," submitted to *J. Fluid Mech.*
Thomas, M. (1991b), "On the nonlinear stability of flows over compliant walls," submitted to *J. Fluid Mech.*

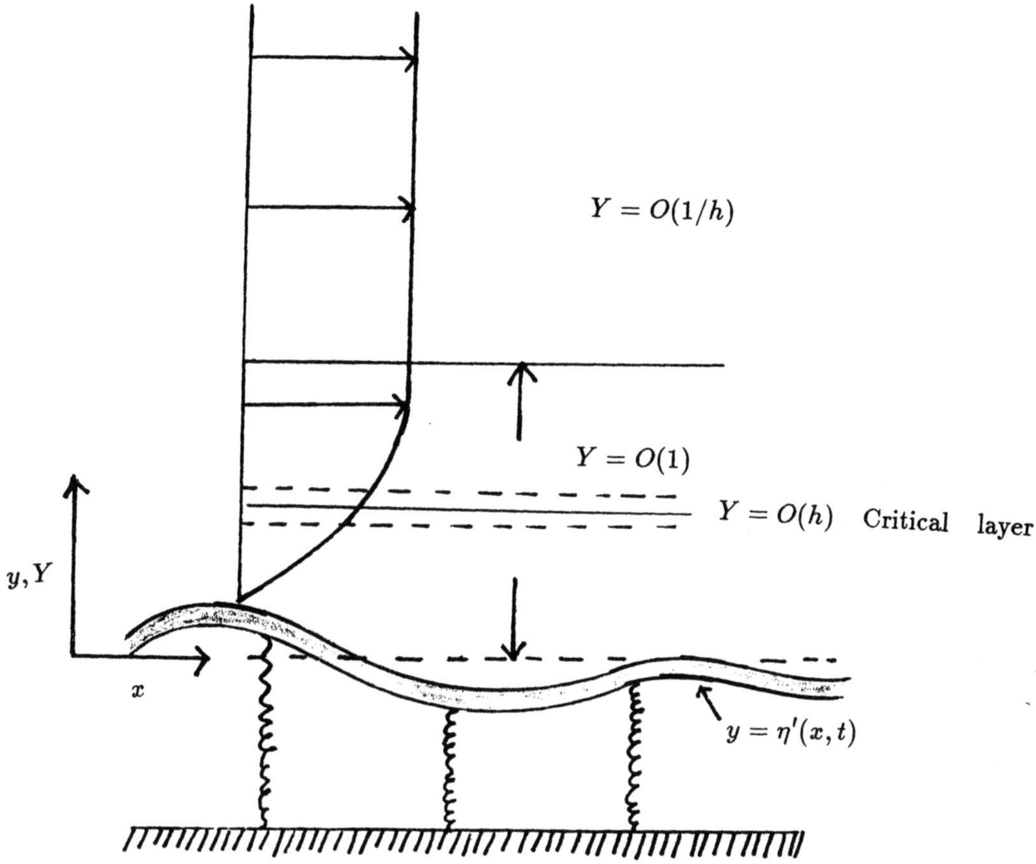

Figure 1. A schematic diagram of the boundary layer flow over an isotropic compliant wall showing the critical layer region and the outermost potential flow region. The normal coordinate is y and Y is the non-dimensional boundary layer coordinate. The small vertical displacement of the wall is η' which satisfies the model spring plate equation (1).

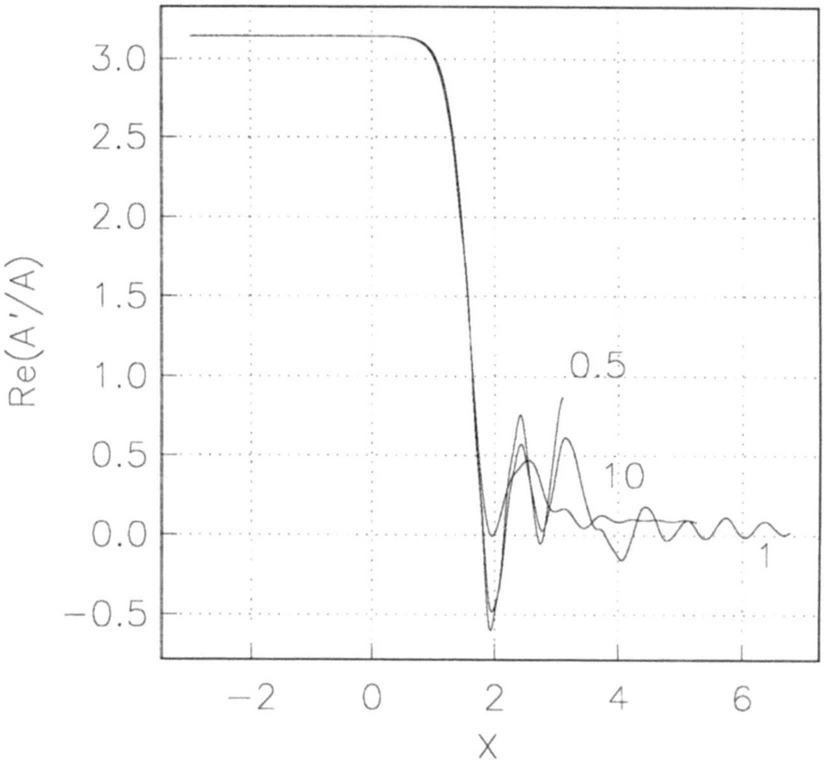

Figure 2 A plot of the scaled growth rate $\mathrm{Re}(\frac{A_X}{A})$ versus the scaled slow streamwise coordinate X, for several values of the viscous parameter λ, $\lambda = 0.5, 1$, and 10.

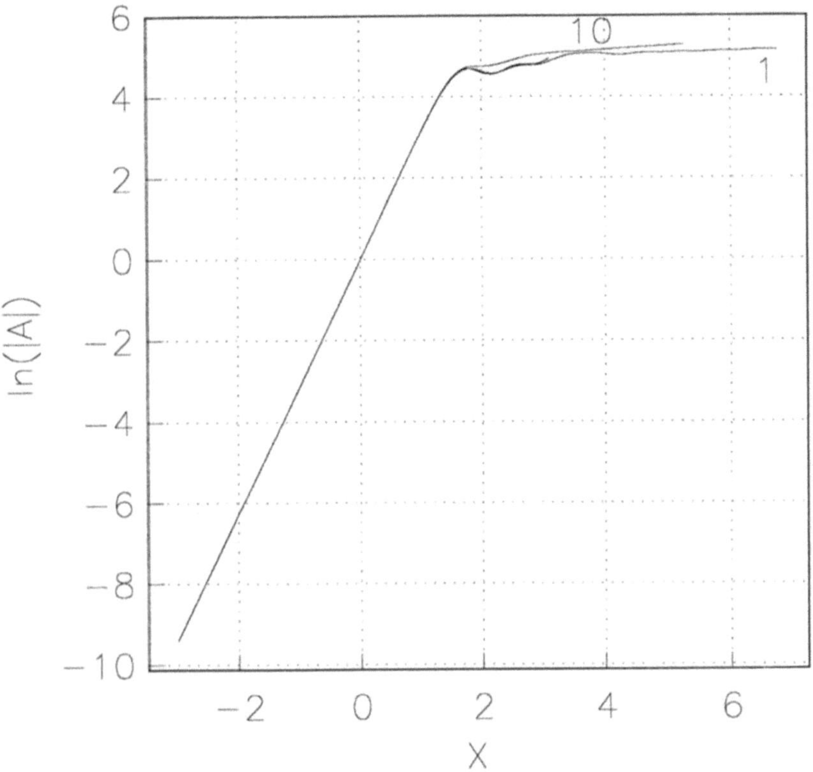

Figure 3 A plot of $\ln|A|$ against X for the same values of λ as in Figure 2, although the $\lambda = 0.5$ curve is not marked.

On conditions of modelling and choice of viscoelastic coatings for drag reduction

B. N. SEMENOV

Institute of Thermophysics
USSR Academy of Sciences, Siberian Branch
Novosibirsk, USSR

K.-S. Choi (ed.), Recent Developments in Turbulence Management, 241–262.

Abstract

The experiments of various authors show a critical situation of the coating vibration properties choice for the drag reduction security. The Kramer hypothesis of the energy absorption isn't explain a cause of facts of the friction increase. According to the interference theory,the action of viscoelastic boundary is critical. It can lead to decrease or increase of the turbulent energy generation in dependence of the wave properties of a surface. Modelling parameter is complex compliance of a boundary, which is conditioned by amplitude and phase of the boundary displacement relative to the turbulent pressure pulsation. Conditions for a choice of the vibration properties are written on the base of the interference analysis of the viscoelastic boundary action on near-wall turbulence. The first condition is the requirement of quick attenuation or absence of free vibrations of coatings. The second condition is limitation of the coating compliance from the condition of hydraulic smoothness. They follow as necessary conditions at the statement of a problem on interaction of viscoelastic boundary with viscous sublayer. The third condition of drag reduction is a choice of natural frequency according to the interference theory. The way of the material search is shown on the example of one-layer coating from material with spacious plateau of viscoelastic properties.

1. Introduction

The experiments of various authors show the possibility of the turbulent friction reduction till 50-60% using the elastic coatings. However, coatings were chosen by chance and no all experimenters could reach drag reduction, and the hydrodynamic friction increase was registrated in a series of cases (see, for example, Bushnell *et al.* 1977). This testifies on a critical situation of the coating vibration properties choice for the drag reduction security.

That is why, the main problem of study is a revealing of their correlation with the near-wall turbulence characteristics in order to choice the viscoelastic coatings for drag reduction.

Two forms of wavy action of the viscoelastic boundary on near-wall turbulence caused by specific properties of compliant coating are considered for the modelling of a phenomenon. Firstly, it is absorption,dissipation (inside viscoelastic coating) of the energy of pressure pulsations deforming a wall. This hypothesis was offered by Kramer (1957) for the laminar flow stabilization. Thát was used by Semenov (1976), Voropaev and Babenko (1978), Korobov and Babenko (1983) modelling the action on near-wall turbulence. But a priori it is clear that the Kramer hypothesis of the energy absorption can't explain a cause of facts of the friction increase. Kulik (1986) analysed experimental data for one-layer coatings and determined that generation and dissipation of turbulent energy in turbulent boundary layer is greater by many fold than absorption and dissipation of the pulsation energy by damping coating. That is why, this factor of the action on near-wall turbulence isn't essential. Moreover, there are the experimental data contradicting to the hypothesis. The coatings lose the ability to reduce turbulent friction and even increase it with the coating energy absorption increase.

Another form of the compliant boundary action was analysed by Semenov (1971, 1976, 1978) on the base of linear harmonic solution on its kinematic-dynamic interaction with viscous sublayer of turbulent near-wall flow. He used the near-wall turbulence model of Sternberg (1962). This simplified linear model allows to obtain a solution in the analytical form. This is important for an analysis of phenomenon. The field of the pulsation velocities as well as the Reynolds stresses are formed as a result of superposition of two waves: 1) going out of turbulent core, being powerful stable generator of long-wave perturbations, and 2) reflexed from a wall. Addition of compliance to boundary leads to the variation of amplitude and phase of induced wave and, respectively, to a change of the interference picture of the turbulence generation. Its action can lead to decrease or increase of turbulent friction in dependence upon the wave properties of flowed surface. The next solutions were obtained by Skripatchev (1977) for the near-wall turbulence model of Schubert and Corcos (1967) and by Trifonov (1978, 1986) and Kereyko (1990) for monoharmonic model of Goldshtik and Shtern (1977). These calculations on the base of more exact models confirm previous main conclusions resulting from simplified theory, which can't prognosticate data of drag reduction, but can prognosticate vibration properties, which are necessary for its security.

Comparison with the experimental data testifies the validity of the interference approach. That is why, the term "interference compliant coatings" should be used instead of the term "damping coatings".

Viscoelastic coatings are made from polymer materials being in high-elastic state for operating range of temperature and frequency. The description of their vibrations even by the simplest Kelvin and Maxwell models foresees three parameters - reduced mass m_c, stiffness C_o, viscosity ν_c. In general case, these parameters are functions of time, wave length, temperature, pressure. Therefore, three conditions are necessary, at least, for the parameter choice determined by specifity of turbulent flow near viscoelastic boundary. They were formulated by Semenov (1976, 1978, 1981) for three schemes of coatings shown in fig.1.

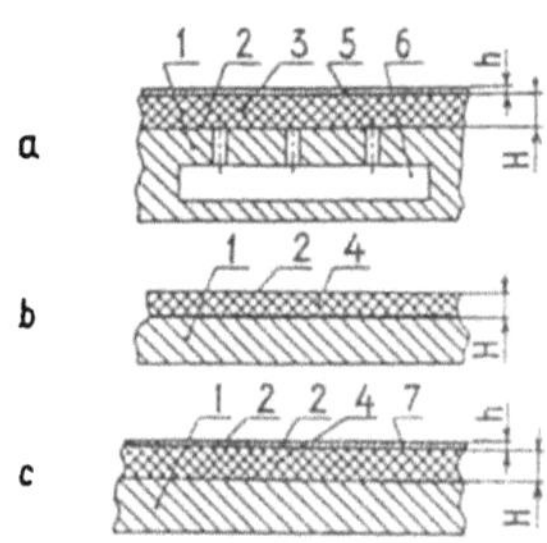

1-rigid basis
2-the adhesion layer
3-elastic foam
4-vulcanizate
5-smooth film
6-reservoir with fluid
7-the layer of lacquer

Fig.1. Skin construction. a - the laboratory scheme. b - the one-layer skin. c - the two-layer skin.

Thin facings, the deflection vibrations of which can be considered in the first approximation in accordance with a notation on viscoelastic base, were studied.

2. Two restrictions for the bringing up of the problem

Following conditions of modelling and choice of viscoelastic coatings for drag reduction are written on the base of theoretical solution on interaction of viscoelastic boundary with viscous sublayer of turbulent flow. A choice of this region is explained by its main role in the near-wall turbulence generation. Now it is impossible to solve the problem completely. That is why, it is neccessary to introduce restrictions for the bringing up of the problem. And besides, following conditions have an important physical sense. They take into consideration a reality of the turbulence properties in modelling approach.

2.1. The requirement of absence or quick attenuation of free vibrations

Aperiodic properties of turbulent pulsations, wide spectrum of their frequencies allows to state the problem on induced vibrations of coatings in a flow correctrly only in the case of absence of free vibrations. Here, forced vibrations of viscoelastic boundary are taken into consideration. Only such approach permits to get single- valued results. Solution is grounded physically, if the free surface waves don't radiate in all directions from every local pressure pulsation, but only the surface deformations follow forcedly to the stochastic changing pressure. I.e. the meaning restriction is given beforehand, but, in principle, one can assume the problem consideration at enough quick attenuation of free vibrations. However, in this case required restriction must be determined from an experiment.

For a laboratory scheme of coating (see fig.1,a) this condition of absence of free vibrations is written

$$n/\omega_o \geq 1 \tag{1}$$

here $2n = \nu_c / m_c$, $\omega_o^2 = C_o / m_c$, n is damping factor of free vibrations, ω_o is natural circular frequency of the first form of purely elastic deformations of skin. However, already at $n/\omega_o = 0.6$ the damping is so great, that the amplitude at the end of period is by factor of 10^2 less than the initial one. Apparently, the restriction $n/\omega_o > 0.4$ is quite sufficient.

The damping factor increase leads to decrease of compliant ability of coating acording to the formula:

$$K_d = [(1 - \omega^2/\omega_o^2)^2 + 4n^2\omega^2/\omega_o^4]^{-1/2} \qquad (2)$$

K_d is dynamic coefficient of oscillations, i.e. ratio of amplitude of forced oscillations with frequency ω to the displacement under action of the same, but static pressure. That is why, it can search optimal damping factor. Kramer (1960) had proposed as an optimum $n/\omega = 0.7$ on the base of his experiments.

2.2. Limitation of the coating compliance

The facing compliance restriction was written by Semenov (1976) from the condition of the principle observance of hydraulic smoothness of oscillating surface. This restriction follows as a necessary condition at the statement of the problem on interaction of viscoelastic boundary with viscous sublayer, so far as a notion of viscous sublayer exists only near a smooth wall.

Here it is important to emphasize a physical role of this limitation. Waves and eddies are guilty of the near-wall turbulence generation near smooth surface. The wave action role decreases as a result of the roughness increase. Viscoelastic coatings respond to the pressure pulsation waves. That is why, the viscoelastic boundary action loses a physical sense as a result of high roughness.

Carrying out the analogy with the effect of distributed roughness on friction, it was supposed that until the surface deflections under the action of pressure pulsations are less than critical height of roughness, they don't lead to the friction increase. According to classical condition of hydraulic smoothness (see, for example, Schlichting 1951) it is required

$$|\zeta_o| U/\nu \leq 70 \qquad (3)$$

Here, $|\zeta_o|$ is the RMS surface deflection, U, ν are the velocity and kinematic viscosity coefficient of a flow.

$$\zeta_o = K_d p/C_o \qquad (4)$$

I .e. the surface deflection is proportional to the pressure fluctuation, which has the highest value for low frequencies (or for low wave numbers). According to the condition (1) is $K_d \leq 1$. Moreover, it is realized for $n/\omega_o \geq 0.7$. That is why, it was taken for a calculation of the RMS pressure fluctuation on hard smooth plate

$$|p| = \rho U^2 Re_o^{-0.3}/2 \tag{5}$$

ρ is density of flowing fluid. Reynolds number Re_o is written on abscissa of the skin beginning.

As a result the required stiffness of coating is

$$C_o \geq 0.007 \rho U^3 Re_o^{-0.3}/\nu \tag{6}$$

For more general case it can be written

$$C_o \geq 0.007 K_d \rho U^3 Re_o^{-0.3}/\nu \tag{6,a}$$

Thus, it should be noted this formula was written using the experimental results only for hard smooth plate for $Re < 2 \cdot 10^6$. But these didn't used from tests for compliant surfaces. The analogous approach gives for large Reynolds numbers

$$C_o \geq 0.003 K_d \rho U^3 Re_o^{-0.21}/\nu \tag{7}$$

The experimental results for compliant coatings , which were tested in water flow at velocities from 0.5 m/s till 21 m/s and in the air flows for velocities from 5 m/s till 70 m/s for $10^5 < Re_o < 8 \cdot 10^6$ were analysed. The analysis has shown that the conditions (6) and (7) allow correctly to determine permissible compliance value. Some examples are given in figs 2, 3, 4, 5.

Kramer (1962) gave experimental data on the drag coefficient of the models with "ribbed coating" at towing speed 32 knots as a function of the coating stiffness (fig.2). $x_o = 0.47$ m. The limiting stiffness, which was calculated according to (7), corresponds to minimal drag coefficient for the third variant. But it is displaced a little at left from minimum for variants 1 and 2.

We and many other investigators had a opportunity to vary wide towing speed or flow velocity, but had only a small collection of materials. Therefore, it is worth while to analyse the limiting velocity following from formulas (6) and (7):

$$U_{lim} \leq (C_o \nu^{0.7} x_o^{0.3} / 0.007 \rho K_d)^{1/2.7} \quad (8)$$

or

$$U_{lim} \leq (C_o \nu^{0.79} x_o^{0.21} / 0.003 \rho K_d)^{1/2.79} \quad (9)$$

In the sixties most of researchers studied the action of viscoelastic coatings of laboratory scheme (fig.1,a) on near-wall turbulent flow. A layer of elastic foam impregnated by a damping fluid and covered by thin smooth film was held on rigid base. This scheme was very convenient for laboratory studies, in so far as it allows a wide modification of properties of viscoelastic coating: ν_c varies due to viscosity of liquid filling agent, m_c varies by the elastic foam thickness H and thickness h, the density ρ_2 of the film, C_o varies by the change of H:

$$m_c \simeq \rho_o H/3 + \rho_2 h, \qquad C_o = E_o H(1-\mu_o^2) \quad (10)$$

Here ρ_o, E_o, μ_o are density, modulus of elasticity, Poisson's ratio for filled elastic foam.

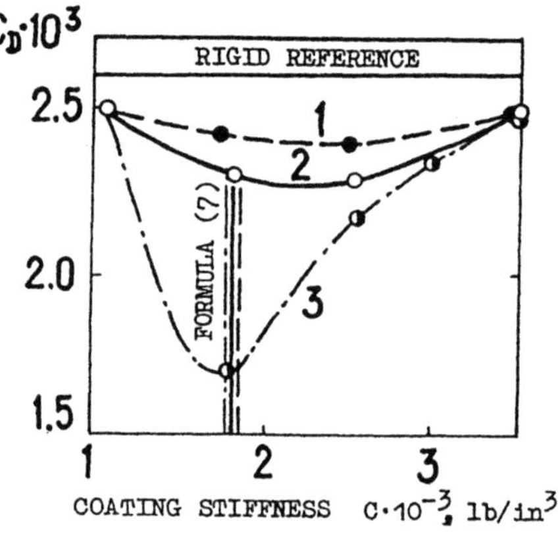

Fig.2. Comparison of the calculated limiting stiffness with experimental data of Kramer (1962) on drag coefficient of the models with "ribbed coating" at 32 knots towing speed and 10.000 centistoke viscosity of the damping fluid as a function of stiffness. Thickness of outward diaphragm - 0.05 in. (1), 0.02 in. (2), 0.03 in. (3).

In our analysed tests (fig.3, 4) polyurethane foam with small open pores was used as a material of main layer. H = 3 mm. $C_o = 1.7 \cdot 10^7$ N/m^3. The foam layer was covered by thin sheet of polyvinylchloride. Drag of the model section from 0.36 m till 1.43 m was measured (fig.3).

$$\Psi = 100\% \, (1 - R_c/R_o)$$

R_c is drag of the section with compliant coating, which was glued on length from 0.40 m till 1.37 m; R_o is drag of a hard reference with polished surface. A difference between mean pressures at the beginning and at the end of this section was controlled in the tests. It was equal to zero. A flow about a measuring part of the model had no pressure gradient (see fig.3,a). Temperature of water flow was 11.7...12,5 C°. Relative error of the drag measuring was 3.5 %.

Other experiments were carried out with the model coming to the surface (fig.4,a). Stiffness of compliant coating (variant 11) was C_o = $1.5 \cdot 10^7$ N/m³. The foam layer (thickness 3 mm) was covered by thin polyethylene film (thickness 0.06 mm) and was impregnated by sea water. These tests had very high accuracy. Relative error of the drag measuring was ≃ 0.1%. Efficiency of total drag reduction for whole model Ψ_D is shown in fig.4,b. Of course, it is less than local friction reduction so far as covered surface is only half of total surface of the model.

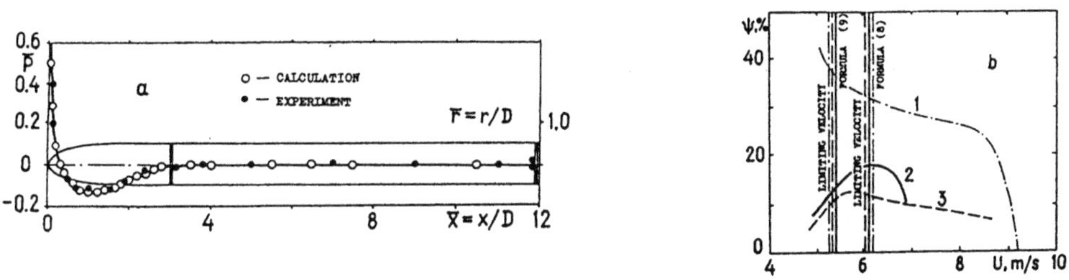

Fig.3. a-the hydrodynamic pressure distribution along the model. b-comparrison of the calculated limiting speed with our experimental data on the drag reduction by compliant coatings of laboratory scheme as a function of towing speed. Thickness of outward diaphragm - 0.23 mm (1), 0.48 mm (2,3). Damping fluids - distillate (1,2), glycerine (3).

These experimental data show a typical dependence of drag reduction by compliant coating on the flow velocity: at an existence of positive efficiency the friction coefficient decreases as velocity increases, arrives at minimum and then increases exceeding friction coefficient of smooth hard reference. The drag coefficient minimum (or the drag reduction maximum) corresponds approximately to the calculated limiting velocity. Apparently, a principal cause of difference is an action of Reynolds number, so far as formula (8) is written for moderate numbers, but formula (9) corresponds to very large values. Formula (8) predicts exceeded limit of velocity for the studied Reynolds number range but formula (9) is on the contrary.

Analogical example is in fig.5 for tests in air flow (Blick et al. 1969). Compliant coatings of the laboratory scheme were used. $x_o = 0.8$ m. Marks of coatings in this figure are from quoted article. $C_o = 5 \cdot 10^5$ N/m^3 for the 27PPI-PVC coating and $C_o = 1.6 \cdot 10^6$ N/m^3 for the 40PPI-PVC coating. As a result it is important to note that considered coatings are deformed a little. For example, the displacement amplitude of compliant boundary must be less than 3.5 μm at the water flow velocity 20 m/s. This corresponds to the relative deformation value 10^{-3} for the facing with thickness of 3.5 mm.

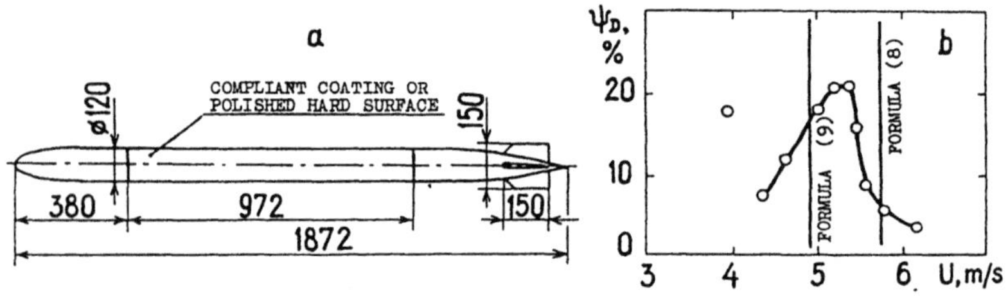

Fig. 4. a-scheme of the model, b-comparison of the calculated limiting velocity with experimental data on total drag reduction by coating (variant 11) as a function of the model velocity.

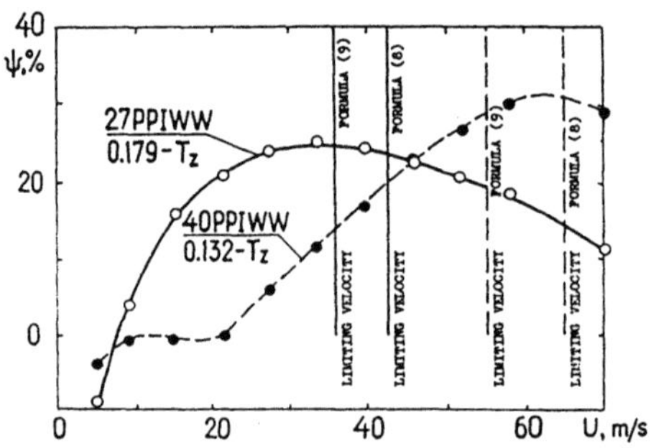

Fig. 5. Comparison of the calculated limiting velocities with experimental data of Blick et al. (1969) on the friction reduction as a function of the air flow velocity.

3. On boundary conditions for compliant boundary and on the criterion of modelling

The pressure pulsations in viscous fluid flowing along a plate with velocity U have the form of running waves, which spread in the flow direction with the phase velocity U_w (fig.6). They induce running bending waves of viscoelastic surface, which spread with the phase velocity U_c. For important particular case of the free vibration absence (U_c=0) the bending displacement of a boundary ζ follows forcedly to the pressure wave with the phase delay θ. In this case

$$\zeta = \mathrm{Real}\{(-K_d p/C_o)\exp[i(k_x x + k_z z - \omega t + \theta)]\} \qquad (11)$$

k_x, k_z are wave numbers in the direction of axis x, z, correspondly.

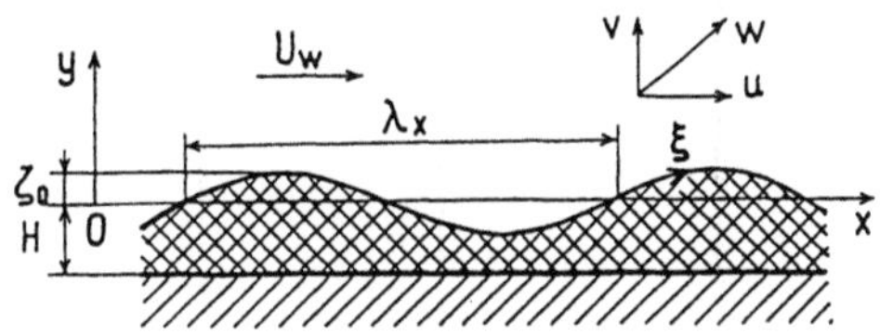

Fig.6. Coordinate system.

According to the condition of the flow absence and sticking on boundary, its deformation rate and the fluid velocity must be equal. The velocity component which is perpendicular to a boundary, is

$$v_\zeta = \partial\zeta/\partial t = \mathrm{Real}\{(i\omega K_d p/C_o)\cdot\exp[i(k_x x + k_z z - \omega t + \theta)]\} \qquad (12)$$

The longitudinal component of the velocity pulsation (along the wave surface) is

$$u_\xi = \partial\xi/\partial t = (\partial\zeta/\partial x)\cdot(\partial\zeta/\partial t)/[1+(\partial\zeta/\partial x)^2]^{1/2} \qquad (13)$$

By analogy, the boundary condition is written for the transversal component of the velocity pulsation.

For the next analysis we used two scales: the friction velocity v_d and viscous scale ν/v_d ($v_d^2 = \tau_w/\rho$, τ_w is friction tension on a wall).

$$k_1 = k_x\nu/v_d, \quad k_3 = k_z\nu/v_d, \quad X = U_w/v_d,$$
$$q = p/\tau_w, \quad Y = \rho v_d^3 K_d/(\nu C_o), \quad y^+ = y v_d/\nu. \qquad (14)$$

The bondary condition for complex amplitude of normal component of the velocity pulsation follows from (12):

$$\hat{v}_\zeta = i k_1 X q Y \cdot \exp(i\theta) \qquad (15)$$

The limitation of the coating compliance was written earlier (section 2.2), according to which one must consider compliance

$$Y \le 1.15|q| \qquad (16)$$

$|q|$ is the known coefficient of Kraichnan (1956). Different authors inform $|q| = 1.8...3.5$. That is why, one can conclude from (15) that normal component of pulsation velocity on compliant wall can have an order of the dimensionless frequency magnitude.

For example, $k_1 X \simeq 0.07$ for monoharmonic theory of Goldshtick and Shtern (1975) takes place. This value corresponds to the frequency range, which is responsible for a main generation of near-wall turbulence (see Hinze 1959). These fluctuations have linear dimensions (in the flow direction) by far more than viscous scale. Monoharmonic generation of turbulence is realized at $k_1 = 0.002$.

Now we can evaluate amplitudes of longitudinal and transversal components:

$$\hat{u}_\zeta \simeq k_1 \hat{\zeta}_o \hat{v}_\zeta, \quad \hat{w}_\zeta \simeq k_3 \hat{\zeta}_o \hat{v}_\zeta.$$

k_1 and k_3 have the values of the same order. As a result, amplitudes of the velocity pulsation components have the next order for limiting compliance:

$$\hat{v}_\zeta \simeq 1, \quad \hat{u}_\zeta \simeq 10^{-3}, \quad \hat{w}_\zeta \simeq 10^{-3}.$$

From the continuity equation we have: $(d\hat{v}/dy^+)_\zeta \simeq 10^{-6}$.

According to these estimations, we can believe for $y = \zeta$:

$$\hat{u}_\zeta = \hat{w}_\zeta = (d\hat{v}/dy)_\zeta = 0 \tag{17}$$

Conditions (12), (17) are written on movable boundary, but arisen difficulties for solution are very essential. Therefore, the boundary conditions are transfered to y=0. This operation has the base. The Taylor expansion near boundary with the linear terms keeping gives at every fixed time for normal component of the pulsation velocity $\hat{v}_o = \hat{v}_\zeta + (d\hat{v}/dy^+)_\zeta \cdot \zeta^+_o$ and according to (17): $\hat{v}_o = \hat{v}_\zeta$.

Then analogous expansion near y=0 gives $(d\hat{v}/dy^+)_o = 0$. According to the continuity equation: $k_1 \hat{u}_o + k_3 \hat{w}_o = 0$. Therefore, it is quite correct for simple longitudinal fluctuations ($k_3 = 0$): $\hat{u}_o = 0$. For the three- dimensional fluctuations it is necessary to make additional estimations.

The Taylor linear expansion for longitudinal component near y=0 gives at every fixed time $\hat{u}_\zeta = \hat{u}_o + \zeta^+_o \cdot (d\hat{u}/dy^+)_o$, but from (17) it follows $\hat{u}_\zeta \simeq 0$. According to experimental data of Bellhouse and Schultz (1966) it is $(d\hat{u}/dy^{+})_{o} \simeq 10^{-1}$, i.e. $\hat{u}_{o} \simeq 10^{-1}$. It is by a factor of 10 less than maximal value of the velocity pulsation in viscous sublayer. Therefore, one can believe $\hat{u}_{o} \simeq 0$. Mean velocity of a flow U at y=0 is equal to zero too.

A relative error of the region calculation for the equation integration is determined by a ratio of the displacement amplitude and of the calculated sublayer thickness l. This error is estimated 10^{-2}.

The next system of the boundary conditions for y=0 was written as a result:

$$U = \hat{u} = \hat{w} = d\hat{v}/d\hat{y} = 0; \quad \hat{v} = ik_{1}XqY\exp(i\theta) \tag{18}$$

The action of viscoelastic boundary on near-wall turbulence is determined by complex compliance $Y\exp(i\theta)$, which is a main criterion of the modelling for this problem. The restriction of the parameter modulus was discussed on principle in the section 2.2. The role of the parameter modulus was shown by formula (18) for normal component. Here it is important to note that Y is proportional to U^{3} but it is in linear dependence on density and inversely proportional to viscosity of medium (see formula (14)). That is why, Y can be invariable at the transition from a water to an air, if the motion velocity will approximately twenty two times as large for constant Reynolds number. Of course, the similarity condition can be satisfied only then, if complex parameter $Y\exp(i\theta)$ will be invariable. The role of the phase delay θ will be discussed in the next section.

4. The interference analysis

The boundary conditions (18) were used by Semenov (1971) for the solution of the problem on an interaction of viscoelastic boundary with viscous sublayer of turbulent near-wall flow. This interaction has the interference form. The field of the pulsation velocities is formed as a result of superposition of two waves: 1) the one going out turbulent core and 2) the second wave reflected from a wall. The boundary flexibility leads to the change of amplitude and phase of induced wave and to the change of the interference picture of the turbulence generation.

4.1. On a role of the phase shift

The obtained solution shows the restriction of the region $\theta(\omega)$ for positive action of viscoelastic boundary (for drag reduction). The positive action of compliant coating on turbulent friction is connected with decrease of the turbulence generation. For fixed frequency it is necessary the generation change of the turbulence energy should be

$$-\int_0^l [\langle uv\rangle (dU/dy) - \langle uv\rangle_c (dU/dy)_c] dy \geq 0 \qquad (19)$$

Here, index "c" corresponds to compliant boundary. The interference action for fixed frequency ω is neutral, if this integral is equal to zero. After the near-wall turbulence model of Sternberg (1962) calculated viscous sublayer thickness l is connected with the pulsation frequency as follows

$$l \simeq 5/(\omega/2\nu)^{1/2} \quad \text{or} \quad l v_d/\nu = 7/(k_1 x)^{1/2} \qquad (20)$$

For the neutral action variant, the mean velocity profile U(y) is written according to the experimental data for hard smooth wall. This variant is an opportunity so far as it can be used for the analysis of the experimental data on the turbulence structure and its role near hard boundary.

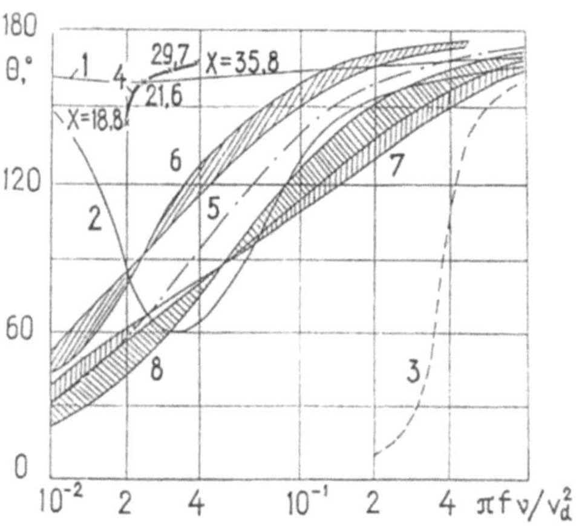

Calculations:

1 – $k_3 = 0$ $\quad U^+ = 16[1-\exp(-y^+/16)]$

2 – $k_3 = k_1$

3 – $k_1 = k_3$ $\quad U^+ = y^+$

4 – $k_1 = 0.002125$ $\quad U^+ = 16[1-\exp(-y^+/16)]$

Phase characteristics of tests:

5 – Ψ_D = 50%, Kramer (1960)

6 – Ψ_T = 25%, Blick et al. (1969)

7 – ψ = 42%, var.1, Semenov (1981)

8 – ψ = 44%, var.11, ibid.

Fig.7. Phase-frequency diagram.

Four examples of the neutral line are shown in fig.7. The phase delay θ of the coating boundary displacement relative to acting pulsating pressure is on the ordinate. Its cyclic frequency f ($\omega = 2\pi f$), which is dimensionless by kinematic coefficient of the motion medium viscosity ν and dynamic velocity v_d, is on the abscissa. The region from the left or above neutral line corresponds to the turbulent friction decrease, from the right or below – to increase.

The first calculated variant and the fourth one in fact also correspond to the plane oscillations. Calculations testify that positive action region of compliant coating is narrow for these pulsations. Moreover, it will be more narrow for the phase velocity increase.

But experimental investigations of the near-wall turbulence testify that oscillations have three-dimensional form mainly, so $k_3/k_1 = 0.8-1.5$. Therefore, the cases 2 and 3 are more close to real conditions of near-wall turbulence among the given calculated variants. The second and the third variants are different in the form of the mean velocity profile. The second case corresponds to turbulent flow near smooth hard wall at $y^+ \leq 50$. This profile was written by Schubert and Corcos (1967). The third calculated variant shows the profile linearization increases the positive action region. The results of the third variant for $lv_d/\nu \leq 10$ were shown in fig.7.

4.2. The choice of natural frequency

The phase delay $\theta = 90^\circ$ corresponds to the pulsation frequency, which is egual to natural frequency of compliant coating. Therefore, the described calculations of the phase-frequency region of positive action are taken as a principle of the choice of natural frequency.

Here it is important to compare the calculations with the experimental data. The circumstance that the phase shift increases monotonically with the frequency growth having the limit of 180° for skins of laboratory scheme is of particular interest. This means that there exists a conceptual possibility to obtain by the choice of parameters of a facing, that phase characteristics of skin were in the positive region of action relative to the neutral line 3 (fig.7). Or, at least, it is also possible for frequencies only of main range of the turbulence generation, but already relative to more critical curve 2. Monotonical phase change in this case (at the frequency variation of forced oscillations) allows to correlate uniquely the phase-frequency characteristics of tested coatings with calculated neutral curves. In so far as the action of viscoelastic boundary on turbulent friction depends on two basic parameters (compliance and phase shift) according to the interference theory, a comparison of the phase diagrams is carried out for the compliance values close to the above-described optimal ones (section 2.2).

Here (see fig.7) the experimental data of Kramer (1960), Blick et al. (1969) and the author (1981) (the experiments were above described) are used.

Comparing the phase characteristics of viscoelastic coatings having in described experiments the great effect of turbulent friction decrease (Ψ_τ is for the local one, Ψ is for the surface region), with neutral curves (fig.7), one can conclude the following.

- The given characterictics of facings are completely in the region of positive action relative to neutral line 3 without any interaction with it. Therefore, the top frequency can be written in following way:

$$\pi f_o \nu / v_d^2 < 0.37 \tag{21}$$

- Considering at estimations that dissipative characteristics n/ω_o are possibly determined unexactly, one should note a high accuracy of the determination of ω_o for given analysis, and hence of the value of $\pi f \nu / v_d^2$, for which complies the phase shift $\theta = 90^\circ$ .The fact is of interest that at variation of stiffness of compliant coatings by a factor of 10^2 tested under very different conditions the range of dimensionless natural frequencies of facings having high friction decrease effect is very narrow and lies in the zone of the pit of neutral curve 2:

$$0.02 < \pi f_o \nu / v_d^2 < 0.06 \tag{22}$$

5. Two notes

5.1. On the equation linearization for a flow near compliant wall

In the sixties and seventies new models of near-wall turbulence were suggested by Schubert and Corcos (1967), Chi and Stuart (1969), Kader (1970), Goldshtik and Shtern (1977). All the theories as well as the

Sternberg theoretical model (1962) foresee quasistationary pulsations, a possibility of their harmonic expansion and the further analysis of separate harmonics using the linearized Navier-Stokes equations. A neglect by the nonlinear component action is necessary for the problem solution. It is permissible near hard wall so far as all velocities become zero at y=0. This assumption leads to an error, which increases as distance y increases. The nonlinear component action can be significant already near y=0 for compliant boundary. The equation linearization is permissible keeping the condition $d\langle \rho uv/\tau_w \rangle/dy^+ \ll k_1 q$. Therefore, it is possible for a flow near compliant wall if the next condition is carried out at y=0

$$\chi \cdot |R_{uv}| \cdot |d\hat{u}/dy^+| \ll 1/Y. \tag{23}$$

Here, $|R_{uv}|$ is the correlation coefficient for the pulsation velocity components u, v.

This condition restricts the coating compliance more than the above written condition (16). As a result it is permissible to consider only slightly deformed coatings.

5.2. Restriction of the coating thickness

It was already noted in introduction that thin coatings are considered here. It was proposed that in the first approximation the coating vibrations can be considered in accordance with a notation on the viscoelastic pillar deformation. Here, it is necessary to write a restriction of the coating thickness according to used conception of thin coating. The problem on the plane bending of a plate with thickness h (elastic modulus E_2, Poisson's ratio μ_2), which lies on elastic base with thickness H (elastic modulus E_o, Poisson's ratio μ_o) was considered. A bending is excited by statical sinuous loading with wave length λ_x. The solution is written after Vlasov and Leontyev (1960):

$$C_B/C_o = 1 + \frac{H}{2L_B}\cdot\left(\frac{2\pi L_B}{\lambda_x}\right)^4 + \frac{1-\mu_o}{3}\cdot\left(\frac{H}{L_B}\right)^2\cdot\left(\frac{2\pi L_B}{\lambda_x}\right)^2 \tag{24}$$

$$L = h\cdot[E_2(1-\mu_o^2)/6E_o(1-\mu_2^2)]^{1/3}$$

Here C_B is real bending stiffness and C_o is the coating stiffness according to formula (10). It is clear that $C_B \rightarrow C_o$ at $\lambda_x \rightarrow \infty$. If $h \rightarrow 0$, $C_B \simeq C_o$ takes place already at $2\pi H \leq \lambda_x$. But ratio C_B/C_o increases sharply as L_B increases. Elastic unit of length L_B is proportional to the plate thickness h. For laboratory scheme of coating this plate is outward diaphragm. Its thickening decreases the coating compliance for the frequency region of main generation of near-wall turbulence. But in this case the limiting compliance is restricted as usually by boundary displacement for the long wave pulsations. Therefore, the diaphragm thickening can lead to the effectiveness decrease (see fig.2, 3). However, the role of the diaphragm thickness is more complicated so far as it has the action on the frequency characteristic.

6. Use of the interference calculations for prognosticated appraisement of compliant coatings

The above written conditions and parameters of modelling are a basis for prognosticated appraisement of viscoelastic coating goodness for drag reduction. Convincing of the validity of the above estimations on the instance of the laboratory scheme facing one can analyse other schemes. The simplest technological variant of viscoelastic skin is one monolithic layer of vulcanized rubber fixed on rigid base (fig.1,b). The solution algorithm for free and forced oscillations of the one-layer facing is written by Semenov (1978).The problem is solved in two-dimensional statement. In the quoted calculations density ρ_o and dynamic modulus of elasticity E_o' of material are taken constants independending on time and thickness.

As regards the third parameter of material (viscosity) two cases are considered. In the first case, the material viscosity is given as a constant one. Another case, analysed in detail by Semenov (1981) is consideration of oscillations of the one-layer skins from materials characterized by constant angle φ of losses. From mechanics of polymers (see, for example, Ferry 1961) it is known that many materials in a high-elastic state have an extensive plateu of viscoelastic properties. In wide range of frequencies and temperatures the dynamic elastic modulus and angle of losses are constant or vary weakly. This variant, which is interesting in a pactice, is analysed further.

The requirements to the wave motion of compliant boundary written on the base of the interference analysis, establishment of algorithm of its connection with properties of materials for concrete schemes of skins allow to prognosticate constructive parameters and to hold the material search for prescribed hydrodynamic conditions or to determine the range of hydrodynamic conditions, in which the turbulent friction decrease is possible. Let's consider the way of the solving of the first problem as an example.

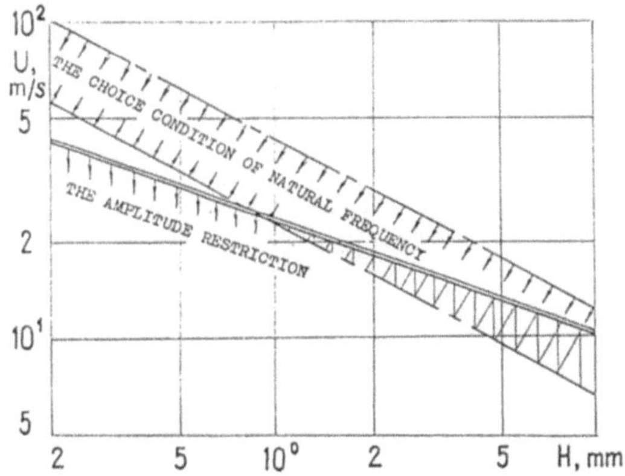

Characteristics of material:
$E_\infty = E'_o = 1.8 \cdot 10^6\ N/m^2$
$tg\varphi = 0.4$
$\rho_o = 1150\ kg/m^3$
$\mu_o = 0.5$

Conditions of a flow:
$x_o = 0.8\ m$
$\nu = 10^{-6}\ m^2/s$
$\rho = 1000\ kg/m^3$

Fig.8. Calculation of the zone of positive action of one-layer coating (of region of decrease of the turbulence generation).

Taking into account the above-made analysis on the first stage of the material search we'll choose the materials having sufficiently high angle of losses ($tg\varphi \geq 0.2$). On the following stage having an information about properties of samples: ρ_o, E_∞, μ_o, $E'_o(\omega)$, $tg\varphi(\omega)$ with their temperature dependencies on the base of conditions (8) or (9) and (22) we can determine possible field of the coating using (or its absence) for drag reduction. In fig.8 the example of determination of this region is carried out graph-analitically in coordinates: thickness of skin – velocity of flow. As it is seen, only narrow "wedge" (which is shown in figure by double hatching), cut off by the line of the amplitude limitation of oscillations after condition (8) from the band determined from the condition (22) of the choice of natural frequency, is the zone of prognosticated positive action (i.e. the region of the turbulence generation decrease and drag reduction).

The next step is calculation of drag reduction in the "wedge" region which can be carried out according to Trifonov (1986). The analysis by Kulik (1986) of hydrodynamic tests of 10 one-layer monolithic skins confirmed these calculations. The interference analysis explains the causes of a series of found peculiarities. For example, in the tests of Kulik, Poguda and Semenov (1984) skins NN 7, 8 studied at two temperatures of a flow, difference of which is only two degrees, show extremally various, even opposite effects. From the interference analysis follows the explanation on prognosticated diagram H − U : experimental conditions correspond to the "wedge" top approximately.

Hence this important conclusion can be followed. It is necessary to choice thickness of coatings in a wide part of prognosticated "wedge" for stabilized positive results.

References

Bellhouse, B.J. & D.L. Schultz 1966 Determination of mean and dynamic skin friction, separation and transition in low-speed flow with a thin-film heated element. *J. Fluid. Mech.* 24, pt 2, 379-400.

Blick, E.F., R.R. Walters, R. Smith & H. Chu 1969 Compliant coating skin friction experiments. *AIAA Paper*, N69 - 165.

Bushnell, D.M., J.N. Hefner & R.L. Ash 1977 Effect of compliant wall motion on turbulent boundary layers. *Phys. Fluid* 20, N 10, pt 2, 31-48.

Chi, J.M.H. & E.B. Stuart 1969 An analytical model for the viscous region in wall turbulence *AIAA Paper* N 69-163.

Ferry, J.D. 1961 Viscoelastic properties of polymers. New York - London.

Goldshtik, M.A. & V.N. Shtern 1977 The monoharmonic theory of near-wall turbulence. In *Turbulent flows*, Moscow, 102-110, in Russian.

Hinze J.O. 1959 Turbulence, New York-Toronto-London.

Kader, B.A. 1970 A turbulence in the viscous sublayer near wall. In *Turbulent flows*. Moscow, 69-73, in Russian.

Kereyko, G.V. 1990 On an interaction of near-wall turbulence with compliant surface. *Izv. AN SSSR. M.J.G.* N 4, 67-72, in Russian.

Korobov, V.I. & V.V. Babenko 1983 On a mechanism of interaction of elastic wall with flow. *J. Eng. Phys.* **44**, N 5, 730 - 733.

Kraichnan, R.H. 1956 Pressure fluctuations in turbulent flow over a flat plate. JASA **28**, N 3, 379-390.

Kramer, M.O. 1957. Boundary layer stabilization by distributed damping. *J. Aeron. Sci.* **24**, N 6, 459-460.

Kramer, M.O. 1960 Boundary layer stabilization by distributed damping. *J. Amer. Soc. Nav. Eng.* **72**, N 1, 25-33.

Kramer, M.O. 1962 Boundary layer stabilization by distributed damping. *Nav. Eng. J.* **74**, N 2, 341-348.

Kulik, V.M., I.S. Poguda & B.N. Semenov 1984 Experimental study of the effect of one-layer viscoelastic coatings on the turbulent friction and pressure pulsations at the wall. *J. Eng. Phys.* **47**, N 2, 189-196.

Kulik, V.M. 1986 The analysis of interaction of one-layer monolithic damping covers with turbulent flow. *J. Eng. Phys.* **51**, N6, 959-965.

Schubert, G. & G.M. Corcos 1967 The dynamics of turbulence near wall according to a linear model. *J. Fluid Mech.* **29**, pt 1, 113-135.

Semenov, B.N. 1971 Interaction of an elastic boundary with the víscous sublayer of a turbulent boundary layer. *Zh. Prikl. Mekh. Tekh. Fiz.* N3, 58-62, in Russian.

Semenov, B.N. 1976 The effect of elastic covers on a turbulent boundary layer. In *Investigations of boundary layer control.* Novosibirsk, 92-101, in Russian.

Semenov, B.N. 1978 On interferent form of the influence of viscoelastic boundary on wall turbulence. In *The influence of polymer additives and surface elasticity on wall turbulence.* Novosibirsk, 57-74, in Russian.

Semenov, B.N. 1981 Analysis of deformation characteristics of viscoelastic coatings. In *Hydrodynamics and acoustics on near-wall and free flows.* Novosibirsk, 57-76, in Russian.

Skripatchev, V.V. 1978 The theoretical analysis of viscoelastic boundary action on near-wall turbulence. In *Proc. conf. on drag reduction.* Moscow, in Russian.

Sternberg, J. 1962 A theory for viscous sublayer of a turbulent flow. *J. Fluid Mech.* **13**, N 2, 241-271.

Trifonov, G.F. 1978 Theoretical investigation of the influence of elastic boundary characteristics on turbulent flow. In *The influence of polymer addittives and surface elasticity on wall turbulence*. Novosibirsk, 75-85, in Russian.

Trifonov, G.F. 1986 Self-consistent model of interaction between turbulent flow and deformable wall. In *Thermogasdynamics of turbulent flows*. Novosibirsk, 5-16, in Russian.

Vlasov, V.Z. & N.N. Leontyev 1960 Beams, flagstones and covers on an elastic basis. Moscow, in Russian.

Voropaev, G.A. & V.V. Babenko 1978 The turbulent boundary layer on elastic surface. In *Hydrodynamics*. Kiev, N 38, 71-77, in Russian.

Experimental investigation of one-layer viscoelastic coatings action on turbulent friction and wall pressure pulsations

V. M. KULIK, I. S. POGUDA and B. N. SEMENOV

Institute of Thermophysics
USSR Academy of Sciences, Siberian Branch
Novosibirsk, USSR

K.-S. Choi (ed.), Recent Developments in Turbulence Management, 263–289.

Abstract

The purpose of the paper is an experimental study of the one-layer viscoelastic coatings action on turbulent friction and the wall pressure pulsations jointly with measurement of mechanical characteristics of these coatings under conditions, which are equivalent to the coating work in a flow. Complex carried out study, allowed to analyze the viscoelastic coating work in turbulent flow. This analysis testifies a validity of interference theory of viscoelastic boundary action on near-wall turbulence.

1. Introduction

The possibility of turbulent drag reduction by the surface deformations attracts an attention of researchers for a long time. The first arisen based on the bionics assumption on the interaction of the dolphin skin with a flow. This idea was successfully realized by Kramer (1957, 1960, 1962), who had achieved double friction reduction. Alluring application perspectives of a new effect gave a rise in the peak of studies next years, in general, of the experimental ones. At present it is possible to carry out the classification of these studies based on different approaches to design of deformable surface (fig.1).

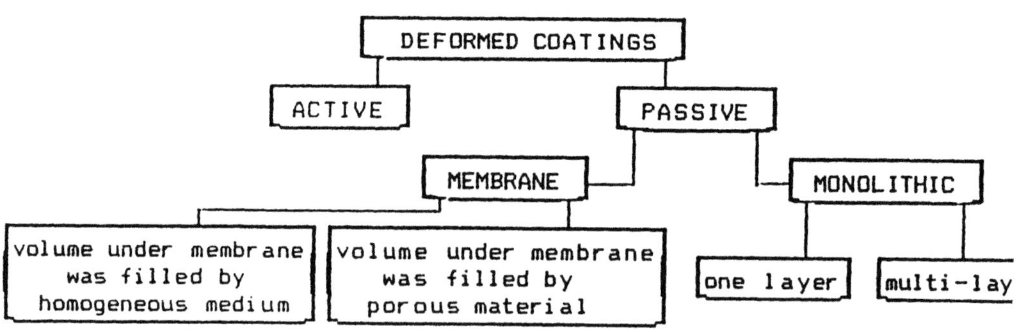

Fig.1. Classifying of compliant coatings.

The majority of studies was carried out with passive coatings, the surface deflection of which is the result of action of the pressure pulsations of turbulent flow. Active coatings are studied significantly less. Deformation of their surface is realized according to prescribed by the experimenter program with the energy action outside, for example, the experiments of Kendall (1970) and Loof (1974). Unfortunately, in these works is no information on effectiveness of this phenomenon from the point of view of the energy expenditure.

Passive coatings can be the membrane and monolithic ones. The membrane coating design is the following. The volume is closed by thin smooth film with regulated stresses along flow and transversal direction. This volume can be free or filled by liquids of different viscosity or porous materials, which also can be impregnated by viscous liquids. This scheme is suitable for the laboratory tests (this is why, it is known as a laboratory skin), since allows to vary widely the coating properties, however, can't be used in a practice. The film layer deforming under action of shear stresses forms the ring hillock. The filling liquid strengthens these defects at a reflow. And besides, at pressure gradient along a surface these defect deformations are very large that prevents to the idea realization. The attempts to remove them by the film tension lead to the factor appearance of resonant action, which doesn't give a possibility for the wave deformation of boundary outside of resonant frequency. May be, the zero result in the experiments of McMichael, Klebanoff and Mease (1979) is explained by this factor.

Practically, monolithic coatings, which are weakly deformable, durable, usable for the facing of surfaces flowed with pressure gradient are the most interesting. Additionally, the technology of their making is simpler and cheaper.

Chronologically it was so that the first stage of the studies was in the main connected with the chance search of coatings which were able to reduce the friction drag. Experimenters used all-possible materials and the coating schemes in order to obtain any drag reduction. And besides, a specific attention was not paid on description of viscoelastic properties of the used coatings. More often trade mark of material was simply shown. In the papers of this period (see reviews of Blick (1974), Fischer *et al.* (1975)) such data are not even given. Most of the experimental works was carried out with the use of the membrane coatings. However, as far as the choice of coatings was chance, the experimental results were different.

The first physically full studies were carried out in the Oklahoma university by the group of Blick (Fischer and Blick (1969), Looney and Blick (1966), Blick and Walters (1967), Smith and Blick (1965)). The experiments were carried out in aerodynamic tunnel with the membrane coatings. Drag reduction of 40% was obtained, when as a dampening medium was used motor oil. The authors give only quasi-equilibrium modulus of elasticity of materials and the design data.

Great study of integral and pulsation characteristics of turbulent boundary layer in water flows near elastic coatings of the membrane type was carried out in the Kiev Institute of hydromechanics (Voropaev *et al.* (1978), Korobov *et al.* (1981), Kanarsky *et al.* (1982)). The existence of correlation of the turbulence suppression and the drag reduction effectiveness was confirmed, and also decrease of longitudinal component of pulsation velocity in the case of drag reduction was found. In this case, near viscoelastic boundary the transition region (intermediate layer) from viscous sublayer to turbulent core decreases, and thickness of viscous sublayer increases.

However, in the carried out studies viscoelastic properties of the coating materials were insufficiently measured: only for resonant frequency (260 Hz) or for enough narrow frequency band (20 - 400) Hz. Besides, vibration characteristics of the coatings were not calculated (dynamic coefficient, phase angle between the coating surface displacement and applied strain, spectral density of energy dissipation and so forth). In given case this was very difficult problem since used coatings were the multi-layer ones. This is why, given by the authors results are seen unfit for theoretical analysis.

In the seventies years studies of compliant walls were begun in the Langley Research Center (Fischer *et al.* (1975)). The main attention was paid to study of "rigid" coatings which were suitable for the application to the flight devices. From requirements of the hydrodynamic smoothness and the obtaining of some minimal level of the surface displacements were chosen viscoelastic polymers with low modulus (polyuretane foams and polyvinilchloride (PVC) plastics). Coating from polyuretane foams covered over milar membrane showed comparatively small drag reduction of 16 %. The PVC cover provided the local friction decrease of 61 %. However, significant lack of the PVC was a strong dependence of its modulus of elasticity and coefficient of losses on temperature which was insufficiently controlled in the tests that made a correct analysis difficult. Besides the above-mentioned works are more than hundred of published data on compliant walls which show drag reduction. The papers exist also, in which one can't obtain drag reduction, and in the some ones obtained even drag increase (see, for example, Kawamata *et al.* (1973), Blick (1974)). Unsuccessful attempts to repeat successful experiments in the other laboratories, apparently, can be explained by the difference in viscoelastic properties of coatings which were not measured in the majority of the works. Non-accordance of the attempt results can be explained by the size decrease (downstream) of the used compliant coatings, and also by the difference in the turbulent flow parameters.

Since the first reason of the action of viscoelastic coating on the flow characteristics is the surface deformation by turbulent pressure pulsations, for the explanation of the mechanism of the flow interaction with a wall it needs to know the wall deformation value. The one can be determined experimentally (Kanarsky *et al.* (1979), McMichael *et al.* (1979), Gad-el-Hak (1984); *et al.*). This direct method is very cumbersome and labour input both on applied apparatus and because of the huge volume of information which is necessary for a description of accidental value (frequency spectra, spacious and time correlations and so on).

Another way is calculation of these deformations. In dependence on the damping degree, two approaches to a description of the coating surface deformations are possible. The first one (Landahl (1962), Merkulov (1981)) is valid at low damping and considers the action of running waves of the coating deformation on a flow. In the second approach, which is valid at high damping, elastic coating is considered as a local deformable, i.e. its deformations in arbitrary point depend only on the pressure fluctuations in the same point. The vibration characteristics for the simplest case of one-layer homogeneous monolithic coating made from material with the elasticity modulus E and coefficient of losses η indepenent of frequency, are determined by Semenov (1978) on the base of exact solution of the problem on steady forcing vibrations. Resonant character of dynamic coefficient variation and wave-like for phase delay of deformation on stress θ_d (fig.5) were noted. Analogous characteristics were obtained by Semenov (1981) for the coatings having two monolithic layers.

Below-described series of experiments was carried out:

- with the purpose of an acquisition of the experimental information, enough rich for the analysis of nature of the viscoelastic boundary action on near-wall turbulence;
- with taking into account above-shown experience of preceding studies and theoretical prognosis following from the interference theory of Semenov (1971, 1976, 1978);
- in the tendency to approach scientific investigations to the practice conditions.

This is why, one-layer scheme of monolithic coatings, which is the simplest in technology, having calculated algorithm of the connection of the surface vibrations with the material properties was chosen.

New technique of the viscoelastic property measurement for high - elastic materials was worked out and the unique measuring complex was made.

A search for the coating materials was carried out among polymers having a plateau of viscoelastic properties in the operating frequency - temperature range.

The special attention was paid to the technology simplicity, the use of industrially produced materials, the coating design having good operating properties providing additional advantages (for example, the anticorrosion firmness).

2. Measurement of viscoelastic properties of materials

The base for calculations of vibration characteristics are viscoelastic properties of material: density ρ, Poisson coefficient μ, equilibrium (or quasi-equilibrium) modulus of elasticity E_o, complex modulus of elasticity, real part of which E' is modulus of elasticity, and the imaginary one is E'', phase-shifted on 90°, is modulus of losses ($E''/E' = \eta$ is coefficient of losses). In general case, these parameters are functions of time, temperature, pressure, frequency and amplitude of deformation.

Typical peculiarity of measurements of mechanical properties of polymer materials for compliant coatings is the requirement of the small-amplitude deformations, which must be in the range, correspondent to the coating deformations in a flow. It was shown by Semenov (1976) that to ensure the hydrodynamic smoothness the boundary displacement must be no more than $70\nu/U$, i.e. for instance, at the water flowing (viscosity $\nu=10^{-6}$ m^2/s) with velocity 20 m/s it is less, than 3.5 μm that corresponds to the deformation value $|\varepsilon|\simeq 10^{-3}$ for the facing with thickness of 3.5 mm. That is why, the standard techniques being in rubber industry can not be used here. This is connected with the fact that such standard measurements are carried out at deformations by the factor of 10^2-10^3 more than the coating deformations in flows, and it is well-known that properties of polymer materials are essentially dependent on the deformation value.

Since the reason of the compliant coatings action are the turbulent pressure pulsations, it is necessary to sertificate viscoelastic properties of coatings in all for the main frequency range of pressure pulsations. It is determined by $f\delta^*/U \leq 1$ according to Willmarth and Roos (1965) and the frequency band is up to 20 kHz, for example, at $U = 20$ m/s and at the driving thickness $\delta^* = 1$ mm.

Complete set for the sertification of viscoelastic properties of the coating materials (see Kulik and Semenov, 1983) was made with taking into account the shown considerations in Institute of Thermophysics of Siberian Branch of the USSR Academy of Sciences.

The first installation was intended for measurements of Young's modulus and Poisson's coefficient at the plane sample tension under quasi-equilibrium conditions. The length and wide of the sample part between marks at various loadings is measured with the help of optical comparator. In all experiments maximal relative deformation of the sample is $\varepsilon_l < 0.1$. Minimal value of deformation is always less, at least, by an order of magnitude than the maximal one. In the case of non-linear law of deform ability Young's modulus is determined by extrapolation at $\varepsilon_l \to 0$. Dynamic viscoelastic properties of materials in a wide range of frequencies and temperature at small amplitudes of deformations are measured in the second installation. The material sample on which above is put on the loading mass (fig.2), is fixed on the hard plate of vibrator generating vertical harmonic vibrations of prescribed frequency. The ratio of the acceleration amplitudes A and also their phase displacement θ are measured in the experiment.

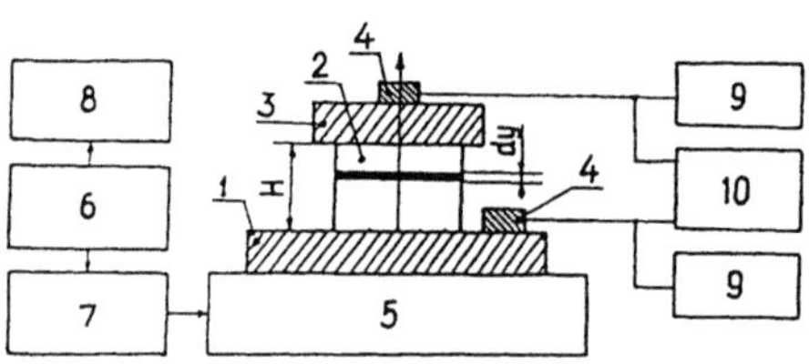

1 – baseplate
2 – sample
3 – loading plate
4 – accelerometer
5 – vibrator
6 – generator
7 – amplifier
8 – frequency meter
9 – voltemeter
10– phase meter

Fig.2. Block-diagram of the installation for the measuring of viscoelastic properties of material.

For the important in practice case, where the deformation wave length is more great than the sample height, the following formulas are valid:

$$E' = M\omega^2 H(A\cos\theta - A^2)/(2A\cos\theta - 1 - A^2); \qquad \eta = \sin\theta/(A - \cos\theta).$$

The exact formula of the material property connection with measured parameters is given by Kulik and Semenov (1986), and correct coefficients for the here shown formulas for general case are given in the work of Kulik (1990).

The installation was made from the lot production instruments. As a vibrator was used the vibration electrodynamic stand VEDS-100B. The sample for study and the vibration table were placed in the thermostat camera connected with the cooler and heat elements. The frequency range is from 60 Hz to 10 kHz. Powerful vibrator mounted on massive base allows to use the loading plates with mass up to 20 kg at the amplitude acceleration of the vibration table up to 50 m/s^2. Therefore, in this installation can be measured after described technique viscoelastic properties of polymers in all characteristics zones including the glassiness state (when $E' \cong 10^{10}$ N/m).

The measuring method is absolute since calibration is not necessary and used parameters can be measured with high accuracy (by weighting and measuring of linear dimensions). The quality of the method is also the fact that the vibrator properties don't affect on measurements. The installation doesn't require mechanical adjustment and is simple in the work.

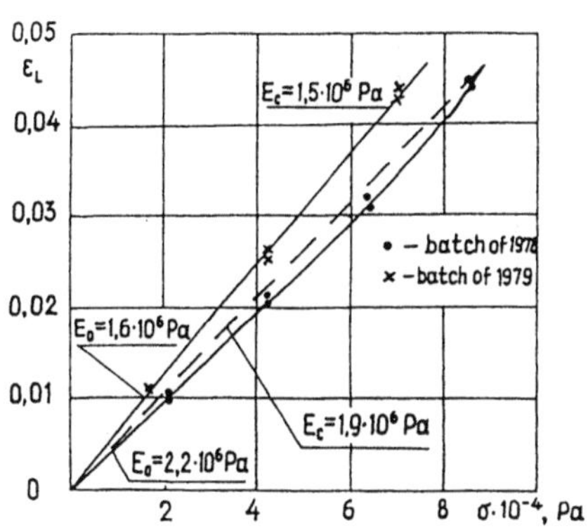

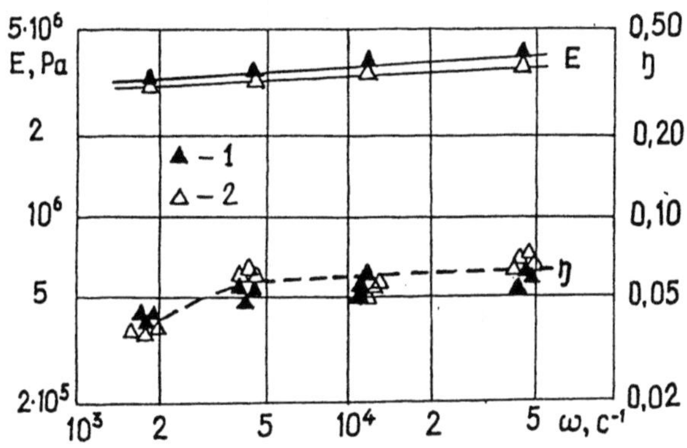

Fig.3. Dependence of deformation on tension at static loading for vulcanizate KLT-30A.

Fig.4. Dependence of elasticity modulus and coefficient of losses on frequency of forced vibrations for KLT-30A. 1-$|\varepsilon|$=0.1%; 2-$|\varepsilon|$=(0.2-0.3)%.

The analysis for the technique has shown that instrumental errors allowed to determine dynamic modulus with the accuracy of 2 %, and coefficient of losses is 3 % for the most of wide-spread materials with coefficient of losses, which is less than 0.6. As an example, in figs.3, 4 are given the measurement results made for vulcanizate of Si-organic compound "KLT-30A". Practical linearity of deformations remains up to $\varepsilon_l \leq 0.03$ for given low-molecular compound with small concentration of fillers. As it is seen from the example, $E_c = \sigma/\varepsilon_l$ at $\varepsilon_l = 0.045$ is already less than E_o by 6-14 %. The experiments were carried out at two temperatures 17 °C and 25 °C. Differences in quasi-equilibrium deform ability were not observed. The Poisson's coefficient is $\mu = 0.46$. The given figure shows also considerable scatter of mechanical characteristics of vulcanizates of industrially lots of compounds.

The results of dynamic measurements carried out at small deformations are given in fig.4. For comparison, the values measured at $|\varepsilon| = 0.05$, temperature 20°C and frequency $\omega = 150\ s^{-1}$ are $E' = (3.0\text{-}3.1)\cdot 10^6\ N/m^2$, $\eta = 0.17\text{-}0.19$. These values are not given in fig.4. Since in this temperature-frequency range is the plateau of viscoelastic properties, it is possible to compare the shown results. From comparison it follows that the deformation amplitude increase up to $|\varepsilon| = 0.05$ didn't lead to significant variation of dynamic modulus, but caused significantly higher losses in material (approximately, 3 times as much). At small deformations $|\varepsilon| = 0.001\text{-}0.003$ viscoelastic characteristics of material are equal practically.

Semenov (1978, 1981) had moved algorithms of the coating vibration characteristic calculation. The main vibration characteristics are phase angle θ_p between pulsating pressure applied to coating and displacement of its surface and also dynamic coefficient K_d (ratio of amplitude of forced surface vibrations to its displacement under action of the same, but static pressure). Wake-like character of the θ_d-variation and resonant variation of K_d is noted in dependence on the vibration frequency. Amplitudes of variation θ_d and resonant value K_d grow at the material loss coefficient decrease.

Difference between wave properties of one-layer monolithic coating (vulcanizate "KLT-30A") calculated in accordance with results of dynamic experiments at small ($|\varepsilon| \leq 10^{-3}$) and comparatively large deformations ($|\varepsilon|$=0.05) are illustrated in fig.5. Here the example of calculations of K_d, θ_d is given versus parameter $H|\delta| = \omega H(\rho/E')^{1/2}/(1+\eta^2)^{1/4}$. Difference between graphs don't need in additional discussion. Hence the validity of the above-written requirements for sertification of viscoelastic properties of the coating materials acting on near-wall turbulence, is practically obvious.

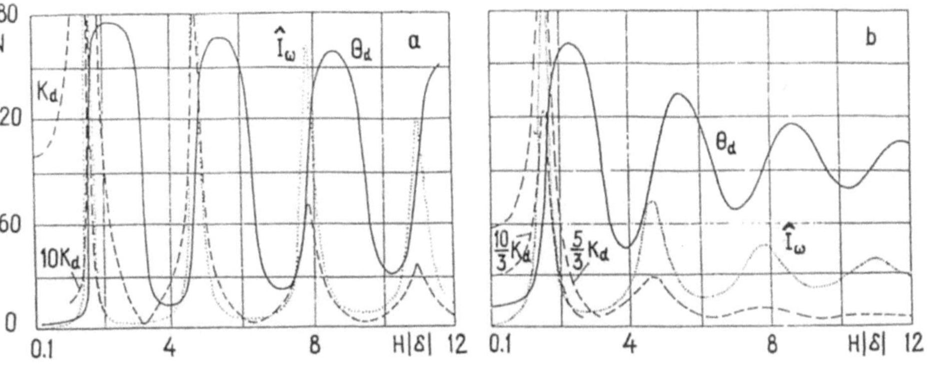

Fig. 5. The amplitude-phase-frequency characteristics on the one-layer coatings. a - η=0.05; b - η=0.2. $\theta_d = N^o$; K_d= N/100; $\hat{I}_\omega$= N/100

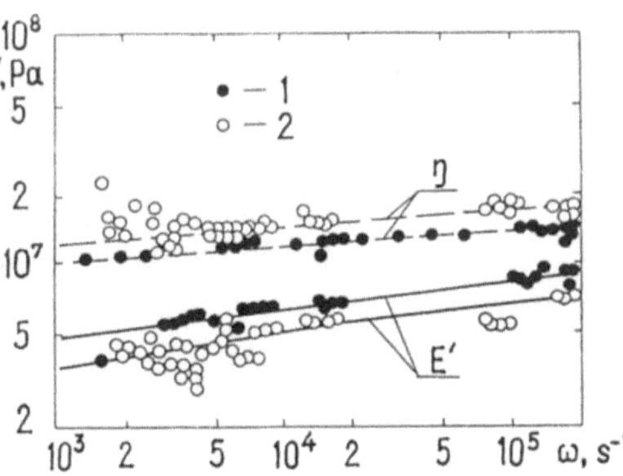

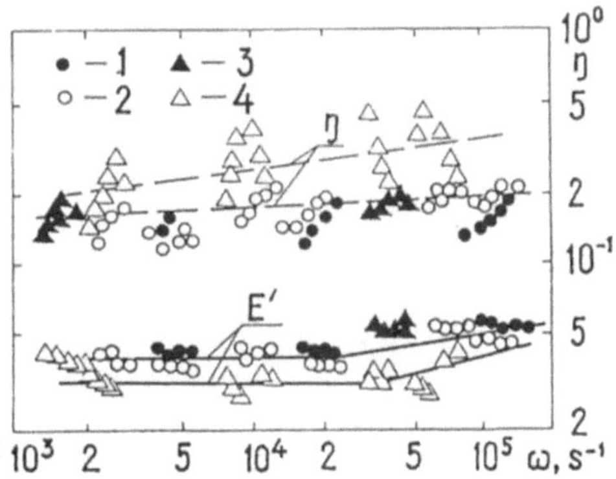

Fig. 6. Dependence of properties of viscoelasticity on frequency for material N1. 1- $|\varepsilon|$=0.1%; 2- $|\varepsilon|$=(0.2-3.2)%.

Fig. 7. Dependence of properties of viscoelasticity on frequency for material N2. 1- $|\varepsilon|$=(0.03-0.04)%; 2- $|\varepsilon|$=(0.08-0.13)%; 3- $|\varepsilon|$=(0.2-0.3)%; 4- $|\varepsilon|$=(0.45-1.3)%.

The measurements for other materials of coatings were carried out after the described technique. Characteristics of materials used in below-given experiments were shown in fig.6, 7 and in the table 1.

Material	$\rho \cdot 10^{-3}$, Kg/m^3	μ	$E_o \cdot 10^{-6}$, Pa
KLT-30A	1.23	0.46	2.2
N1	4.00	0.5	0.47
N2	2.14	0.47	3.7

Table 1. Characteristics of the coating materials.

3. Turbulent friction and wall pressure pulsations

Hydrodynamic tests on viscoelastic coatings were carried out in open basin with natural external turbulence. The model (fig.8) was mounted in front of towing cutter on the depth of 0.8 m. Strain gauge balance of all model and of cylindrical insert (in the length of 640 mm) and the pressure pulsation gauges (made according to the wrought up vibro-compensated scheme) are placed immediately in the model housing. Cylindrical part was made easily prefabricated down with a set of outer accessory cylinders, which were covered by compliant coatings in the length of 590 mm. The insert was assembled with axial clearance 0.5-0.8 mm relatively to the main part of the model. Maximal divergence of butts in direction of normal to a surface was no more than 0.05 mm. The insert and the pressure pulsation gauges were placed in the zone of the gradientless flow at the zero surplus pressure. Calculations show in considered range of the Reynolds numbers (based on mean abscissa of the insert) $Re = (5\text{-}20)\cdot 10^{6}$ near the insert is realized developed turbulent flow. Friction coefficient calculated for mean abscissa $x_{mean} = 1.05$ m is approximately equal to half-sum of friction coefficient determined in onset and in the end of the insert.

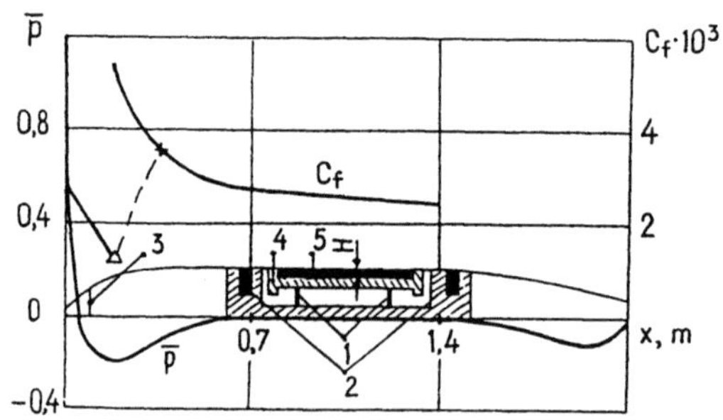

Fig.8. Scheme of the model with hydrodynamics pressure and coefficient of friction. 1-tenzo-weight meter, 2-pressure pulsation gages, 3-injection slot, 4-accessory cylinder, 5-coating.

Careful measuring of friction force and pressure pulsations for the case of hard polished insert surface used for comparison as a standard was carried out. Experiments were made both with the wire turbulizer (D = 1 mm, x = 25 mm) and without it. The proximity of calculated and measured values was obtained and the turbulizer action was not revealed, that characterized turbulent flow near the insert as a developed, self-similar.

Ten one-layer monolithic coatings of various thickness were studied. Characteristics of damping coatings and the results of their study are given in the table 2. Statistical roughness of coatings was measured by the device of the light sections ("Karl Zeiss", Yena). Technical possibilities of the facing making allowed to provide hydraulic smoothness of flowed surface at all velocity regimes in the experiment only for the facings from material 2. Criterion of hydraulic smoothness ($y_s v^*/\nu < 5$) was satisfied only for velocity $U < 10$ m/s for all coatings with the exception of one from KLT-30A by thickness of 1.5 and 2 mm. Coatings NN 8, 10 had the surface defects in the form of separate shells, total area of which didn't exceed 1 %. Defects were filled by putty before experiments and the surface was carefully evened. Variation of the friction coefficient of the facing insert C_f was compared with one of hard smooth boundary C_{fo} (in percents) $\psi = 100\cdot(C_{fo} - C_f)/C_{fo}$ and variation of mean square values of pressure pulsations [P]: $\psi_p = 100\cdot([P]_o-[P])/[P]_o$ was given in the table 2.

Coatings can as increase (minus) as decrease (plus) friction and pressure pulsations on a wall in dependence upon the properties of coatings and the flow velocity. Relative values of their variation coincide practically. This means that the wall pressure pulsation level variation in energy-containing part of spectrum is analogous to the friction variation, i. e. Kraichnan factor keeps its value ($Kr=[P]/\tau_w$) apparently at an introduction of viscoelastic boundary.

Comparative measurements of the pressure pulsation spectra carried out behind the insert with hard polished surface and with coatings confirm this conclusion only partly.

The results of these measurements are given in figures 9 where the values of the spectral density for some regimes are given versus frequency in the range from 500 Hz to 6 kHz. Measured dependencies for hard surface are shown by solid line.

N	Mat	H, mm	y_s, μm	%	6.-6.7	9.-9.3	10.-11.	12.-12.2	13.3-13.5	15.5	16.5
					Flow velocity, m/s						
1	KLT	1.5	41	ψ			-10		-14	-17	
2	-	2.0	48	ψ			+8		+5		
3	-	1.4	12	ψ			-2		-18	-23	
		2.0		ψ_p			0		-10	-12	
4	-	3.0	19	ψ		+6	-3		-7	-14	-17
				ψ_p		0	-7		-12	-12	-8
5	-	4.0	19	ψ			0		-10		
				ψ_p					-8		
6	N1	2.0	20	ψ	-2	-12	-15	-15	-15		
7	N2	2.5	5	ψ	+14		+10		+4	+4	+6
8	-	4.0	16	ψ			+17		+4	+2	-4
				ψ_p			+16		+9	0	-2
9	-	4.4	3	ψ	+17		+9	+10	+3	+3	+5
10	-	7.0	5	ψ	+15	+6	+13	+6			

Table 2. Characteristics of compliant coatings and results of their hydrodynamic testing.

Direct accordance of the spectral density behavior with ψ and ψ_p is seen comparing the results given in these figures and in the table 2. However, in a series of experiments (for example, coating N4 at velocity 15.5 m/s and 16.5 m/s, coating N6 at velocity 10.5 m/s and 13.5 m/s) is observed the spectral density decrease at frequency more then 2-3 kHz even at increase of friction and mean square level of wall pressure pulsations.

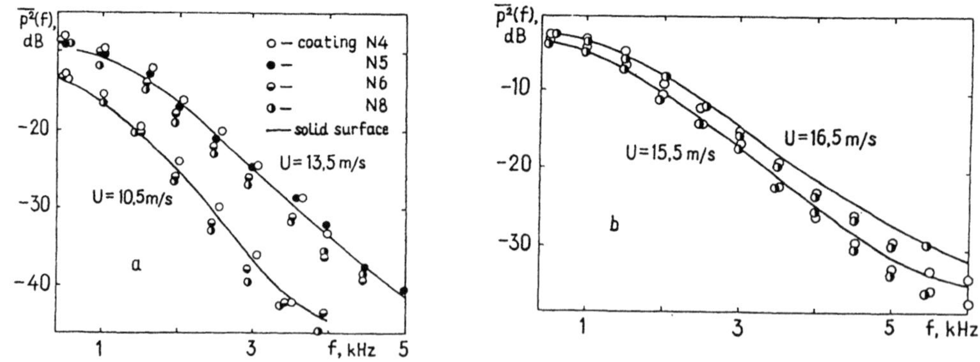

Fig. 9. Spectra of pressure pulsations behind solid body and compliant coatings.

The prognosis of the action of compliant coatings on turbulent friction was made before hydrodynamic tests. The requirement of the keeping of hydrodynamic smoothness of the surface deformations and the conditions restricting the choice of resonant frequency of the surface vibrations give the thickness of compliant coating in dependence upon the motion velocity. In fig.10 the example is shown when the solution is sought by graph-analytical method. The line 4 results from the condition of the hydrodynamic smoothness conservation. The velocity regimes for the turbulent friction reduction are assumed only below this line. The lines 2 and 3 limit the choice of resonant frequency. The diagram region between them (double hatching) recommends expedient velocity regimes in dependence upon the coating thickness from the point of view of interferent form of the compliant boundary action on near-wall turbulence. The region (single hatching) is the region of possible positive action. The line 1 limiting it below was calculated in accordance with the limit neutral curve (Semenov 1976, 1978).

Above mentioned requirements for the prognosis were written for the membrane coatings, the phase and amplitude characteristics of which vary monotonically as frequency grows. As it was shown by Semenov, one- layer monolithic coatings have more complex form of the frequency characteristics. This is why, it was assumed that these prognosis estimations for having coatings have the orientation character.

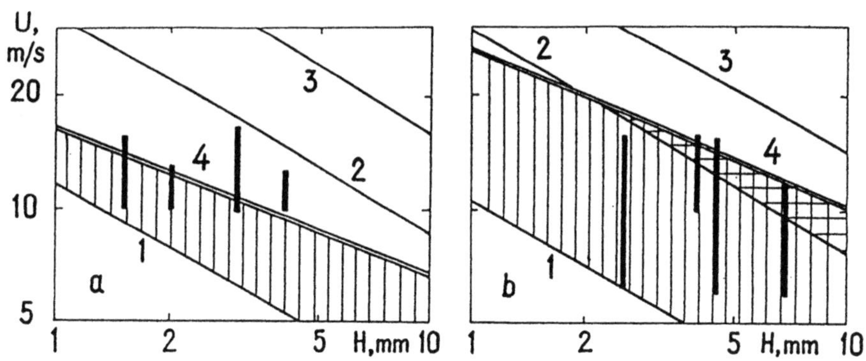

Fig.10. Prognosis of influence on turbulent friction by compliant coatings from material: a-KLT-30A; b-N2.

The prognosis of drag reduction by having compliant coatings is shown in fig.10. Drag reduction can be expected only for thickness 1.5 mm and 2.0 mm for coatings made from KLT-30A. For coatings from material 2, drag reduction can be expected for all the operating velocity regimes, and is most probable for coatings of greater thickness.

Although an application of the prognosis estimations to tested coatings was shown as a restricted one, made experiments satisfy that the estimations allow correctly to show the thickness of thin one-layer compliant coatings, for which realized drag reduction.

4. Analysis of the test results

Since the cause of the compliant coating action on the the flow characteristics is deformation of the facing surface by turbulent pressure pulsations, it is necessary to take into account a dependence on frequency both of active onset (of pressure pulsations) and passive (of damping coating) at the analysis of this phenomenon. In figs. 11-13 various factors acting on interaction of compliant walls with turbulent flow are given as frequency functions.

θ_o is phase angle, at which is provided neutral action of viscoelastic boundary on turbulent friction (Semenov, 1978). Phase angles θ_d above neutral curve can lead to friction reduction, and the ones under the curve can lead to increase.

As it is seen from given figures, the frequency region, where $\theta_d > \theta_o$ and, accordingly, the friction reduction can be expected, takes place at all chosen regimes. However, various frequencies from the region of possible positive action act on friction reduction in different way.

From the condition of the liquid non-flowing through compliant boundary the equality of the surface deflection velocity under action of turbulent pressure pulsations dy/dt and normal component of pulsating wall flow velocity v_o follows. Since value of the Reynolds friction stresses $\langle v \cdot u \rangle$ is proportional to v, the action of compliant coating on characteristics of turbulent flow in the first approximation is determined by dy/dt. This is why, weight factor considering a contribution of some frequencies, is proportional to normal component of pulsating wall velocity $|v_o|_\omega$.

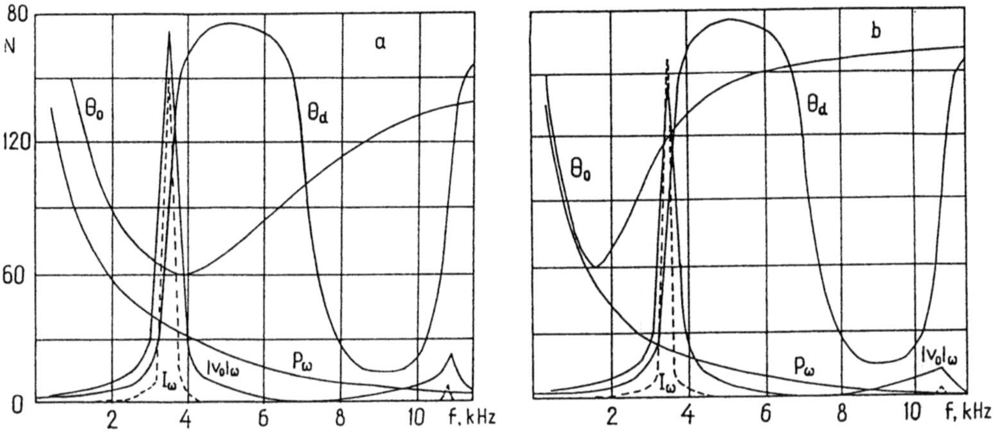

Fig. 11. Dependence of the properties of coatings from KLT-30A with thickness 3 mm on frequency. a – U=15.5 m/s; b – 10.5 m/s. θ°_o = N; θ°_d = N; a – $P_\omega[N^2 s/m^4]$ = N/3, b – N/10; a – $|v_o|_\omega[m/s]$ = N/10^5, b – N/$2\cdot10^5$; a – $I_\omega[J/m^2]$ = N/$5\cdot10^4$, b – N/$3\cdot10^5$.

Taking the pulsation pressure as a set of harmonic components $\varphi_T(\omega,t)$ with spectral density of distribution P_ω we obtain accidental value of the surface deflection consist also of the harmonics set $y_T(\omega,t) = \varphi_T(\omega,t)K_d H/E'$. In figs 11-13, the value $|v_o|_\omega = \omega H K_d (P_\omega)^{1/2}/E'$, determined as a square root of spectral density of the elastic surface deformation rate, is shown. Here evident resonant character of the behavior of these weight factors is seen. Sharpness of the peak (Q-factor) is determined by coefficient of losses of the coating material. The peak value depends on the coating stiffness, resonant frequency and pulsation pressure at this frequency.

Spectral distributions of the energy dissipation rate by a wall per unit of the surface area $I_\omega = \omega\cdot\sin\theta_d\cdot H\cdot K_d\cdot P_\omega/E'$ are also given. For calculations the pressure pulsation spectrum measured by Willmarth and Roose (1965) on flat plate at the ratio $r/\delta \to 0$ is taken. Here r is radius of sensitivity of the pressure pulsation transformer, δ is thickness of boundary layer. Such form of spectrum is close to real, since the averages of pressure pulsations over the gauge area are absent. Calculating the spectral density of pressure pulsations, τ_w, and the displacement thickness δ^* were calculated for coordinate x, which is equal to mean point of measuring insert (x=1.1 m).

Naturally, the presence of compliant wall distorts the pressure pulsation spectrum. This is why, made calculations are valid only for small rate of the action of a coating on a flow. In general case, this can be considered as the first approximation in the chain of direct and indirect action of a flow and coating.

Semenov (1976) and Korobov and Babenko (1983) have determined the condition of non-excess of the limit value, which is equal to critical value of roughness, by the surface deflection. However, in these works the surface deflection value was determined only in integral form.

In accordance with the laws of description of accidental values, with the dynamic roughness value is mean-square value of the surface deflection. Since the surface displacement power spectrum $Py(\omega)=K_d^2 \cdot H^2 \cdot P_\omega/E'$, dispersion of the surface displacement

$$\sigma_y^2 = \int Py(\omega)d\omega = 2\int_0^\infty \left(K_d^2 \cdot H^2 \cdot P_\omega/E'^2\right)d\omega.$$

Hence, mean-square displacement of compliant surface under action of turbulent pressure pulsations has the form:

$$\sigma = \sqrt{\overline{y(t)^2}} = \frac{\sqrt{2}\ H}{E'} \left[\int_0^\omega K_d^2\ P_\omega d\omega\right]^{1/2}.$$

The obtained formula is valid for all the types of compliant coatings, and not only for the one-layer monolithic ones. However, as distinct from well-studied classical types of roughness, for example, the sand one, development of roughness from deflection of compliant surface has characteristic peculiarities. The main difference is this roughness develops only in the coating resonant frequency region. Laws of the action of such roughness on turbulent characteristics of a flow can differ from classical and were not studied at present.

In the table 3 are given calculated data of assumed values of roughness and mean-square values of the surface deflection for described regimes. The assumed value of roughness was determined according to the formula $y_l v_d/\nu = 5$. The value σ_y was obtained by the method of numerical integration. As it is seen from the table, dynamic roughness is less, than assumed static roughness of a surface for all the analyzed regimes. Relative deformation for these regimes is in the range $2\cdot10^{-4}$– $2.6\cdot10^{-3}$ that confirms the validity of the requirements of the coating material sertification at small values of relative deformations.

Now analyze the results of force measurements for the series of compliant coatings with the wall pressure pulsation spectra, measured immediately behind coatings (fig. 9).

Tests of the coating from material KLT-30A in the thickness of 3.0 mm at velocity 10.5 m/s (fig. 11,b) gave some friction increase and the growth of wall pressure pulsations. A possible colculated positive effect can not be significant (minimum of the phase-frequency characteristic θ_o is left the region of resonant interaction) and negative action of technological roughness is higher. Roughness, development of which, in given case, is described by the law of the transition regime ($5 \leq yv_d/\nu \leq 70$), gives the friction $\simeq$ 10 % increase. It was noted the increase of spectral density up to frequencies 3.5 kHz, i. e. up to the onset of the region of the prognosis positive action.

The same coating, but at large velocity U=15.5 m/s (fig.11,a) gave great increase of friction and pressure pulsations. As velocity grows shear to the right of minimum of phase-frequency characteristic of the neutral action curve and its extension on the frequency axis takes place. The velocity variation doesn't act on the frequency band width of resonant interaction and on resonant frequency. Although the phase-frequency conditions for development of positive action became the optimal ones, the dynamic roughness increase and reduction of the assumed level of roughness, apparently, demonstrated a great action. However, the spectral density reduction at frequency above 3.5 kHz (of the first peak of resonant interaction and the onset of the region of the expected positive action) is noted in pressure pulsation spectrum.

In the second region of possible positive action at frequency $\simeq$ 11 kHz, where is the second maximum of phase-frequency vibration characteristic of a coating, rate of the surface deflection and, therefore, action of viscoelastic coating on the flow characteristics is by an order less than at the first frequency of resonant interaction. In the region of the high-frequency part of pressure pulsations the value of their spectral density is small, and this is why the action of compliance is not effective. And besides, it is possible that the mechanism of the compliant coating work in the region of dissipative frequencies differs from the work mechanism in the region of the energy-carrying frequencies.

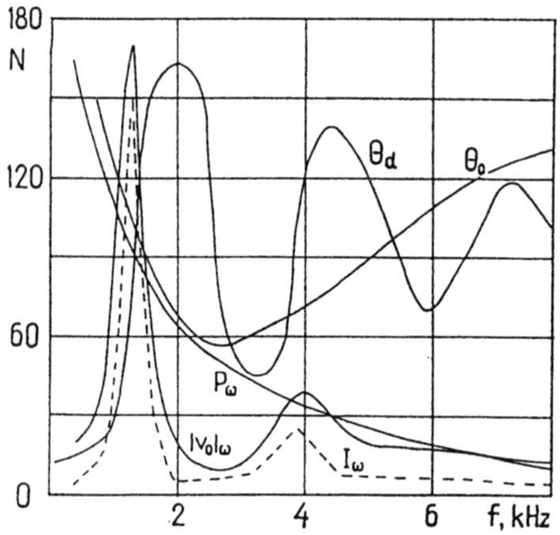

$\theta_o^o = N$

$\theta_d^o = N$

$P_\omega\ [N^2 s/M^4] = 0.3 \cdot N$

$|v_o|_\omega\ [m/s] = N/5 \cdot 10^5$

$I_\omega\ [J/m^2] = 3 \cdot 10^{-6} \cdot N$

Fig.12. Dependence of the propertis of coating from material 1 vith thickness 2 mm on frequency.

The tests of the coating from more soft material ($E_o = 0.47 \cdot 10^6$ Pa) with large coefficient of losses ($\eta = 0.2$) at velocity $U = 13.5$ m/s, characteristics of which are given in fig.12, gave the friction increase of 17% and the pressure pulsation increase of 16%. In given case, minimum of the phase-frequency characteristic of neutral action θ_o is right of the first region of resonant interaction. The region of positive action from the phase-frequency representation is enough large: from 1.5 kHz to 5.5 kHz. All the region of frequency can have affective action, since in it are two peaks of resonant interaction, heights of which are as 4 to 1, and also the frequency region between them is essential. all this was reflected in the behavior of pressure pulsations, spectral density of which is less, than the ones on hard surface in the region of frequencies, which are larger, than 1.5 kHz.

However, the sum value of static and dynamic roughness is 3 times as large, than the assumed value. Total effect of action on friction and pressure pulsations is negative.

Characteristics of coatings from material N2 (with coefficient of losses η=0.2) for thickness 4.0 and 7.0 mm are given in fig.13 at U=10.5 m/s. For the first regime minimum of θ_o is somewhat more left, than the region of resonant interaction. The width of this region determined on half-height of resonant peak is equal to 900 Hz. Even although the roughness value is compared with the assumed one, for this regime was observed decrease of friction and level of the pulsation level of 17% and 16%, respectively. Spectral level of pressure pulsations was lower, than for hard surface at frequencies f>2 kHz analogously to described cases.

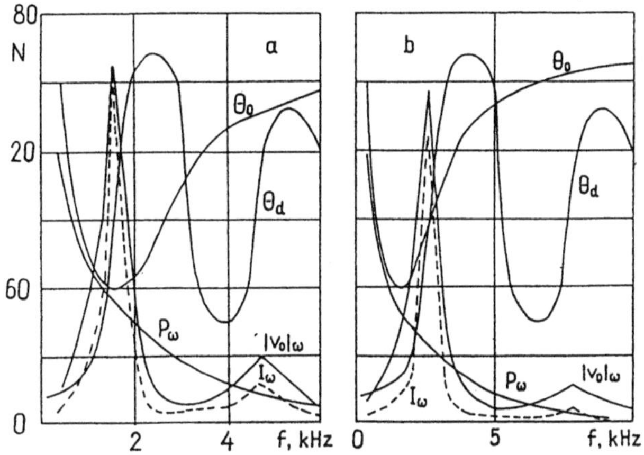

Fig. 13. Dependence of the properties of coating from material 2 on frequency. a - H=7 mm; b - 4 mm. U = 10.5 m/s. θ^o = N; θ^o = N; $P_\omega [N^2 s/m^4]$ = N/10; $|v_o|_\omega$ [m/s] = $N/10^6$; I_ω [J/m^2] = $N/10^6$.

The coating thickness increase from 4.0 mm to 7.0 mm led to decrease of the resonance region width by a factor of $\simeq$ 1.5. And although in this case minimum of the phase-frequency characteristic of neutral action and the resonant interaction peak are at the same frequency, and the sum of static and dynamic roughness don't exceed the limit assumed one, i.e. a coating is hydrodynamic smooth, the test of this coating gave the friction decrease only of 11 %. Apparently, the action frequency band is one of important factors acting on the effectiveness of the compliant coating work.

Coat	U, m/s	y_l, μm	σ_y, μm	A, J/m^2	W, J/m^2	$\frac{W}{A}\cdot 10^4$	ψ, %
4	10.5	13	2.1	84.4	$9.8\cdot 10^{-3}$	1.2	-3
4	15.5	9	5.4	185	$4\cdot 10^{-2}$	2.2	-14
5	10.5	13	4.3	84.4	$1.8\cdot 10^{-2}$	2.1	0
5	13.5	10.4	6.6	140	$3.1\cdot 10^{-2}$	2.2	-10
6	13.5	10.4	5.1	140	$2\cdot 10^{-2}$	1.4	-17
8	10.5	13	0.9	84.4	$7.1\cdot 10^{-3}$	0.84	+17
8	15.5	9	1.8	185	$1.8\cdot 10^{-2}$	0.97	+2
10	10.5	13	1.3	84.4	$4.2\cdot 10^{-3}$	0.5	+11

Table 3. Comparison of specific energy, absorbed and dissipated on viscoelastic coatings.

In the table 3 is compared specific energy dissipated in flow and on the wall section in the length of L. The energy dissipated in flow is equal to $A=\tau_w L$. Absorbed by a wall specific energy is determined by the method of numerical integration according to the formula $W=(L/U_c)\int \tau_w dw$, where L=0.59 m is the coating length, $U_c = 0.8U$ is convective velocity. From the table 3 it is seen that energy absorbed by the wall is only small part of turbulent energy dissipated in flow near a wall: $W/A = (0.5-2.2)\cdot 10^{-4}$.

Comparison of this ratio and the friction decrease effect shows that is no direct connection between them too. Even in opposite, coatings decreasing friction absorb the lesser part of turbulent energy, than coatings, which gave the friction increase. This conclusion confirms a failure of the explanation of the friction decrease effect by the energy absorption wall.

5. Some remarks on practical use

Above-shown comparison with smooth rigid boundary was carried out for the analysis of physical picture of the action of viscoelastic boundary on near-wall turbulence. Considering the question on the coating use in ship building, it is necessary to estimate effectiveness as compared with real strong-colored surface. In accordance with the existing technological norms, roughness of freshly colored metallic surface is 30μk.

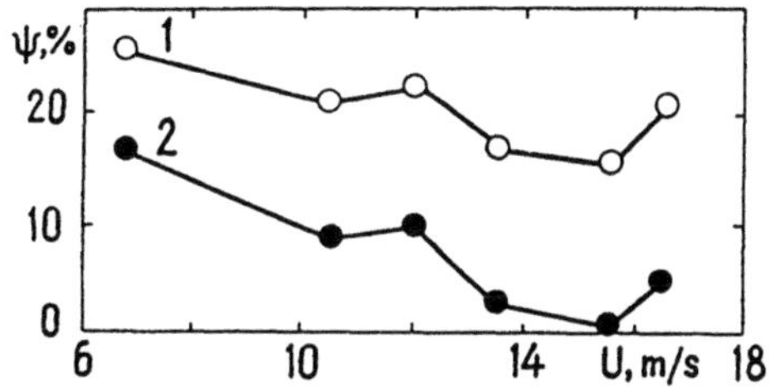

Fig.14. Experimental date of drag reduction for coating N2 as compare with freshly colored metallic surface (line 1) and polished hard surface (line 2).

On fig.14 the values of the friction reduction effectiveness are given as an example. These values were determined by comparison with hydraulically smooth hard surface (line 2) and freshly colored metallic surface (line 1). We see that real approach significantly increases the velocity range of perspectives of the use of worked out coatings (it differs from the conclusions following from purely physical comparison with smooth hard boundary). Made analysis of other experimental data confirms that one-layer monolithic coatings in the thickness of 4.4-7.0 mm from the material N 2 is effective means of decrease of drag and wall pressure pulsations in the motion velocity range from 6 m/s to 17 m/s.

The important quality of described coatings is excellent anticorrosive properties, firmness to the action of microorganisms, physiological inertia helping to the decision of ecological problems in the world.

6. Conclusion

For the effective work of compliant coating in turbulent flow such viscoelastic characteristics of the coating material and such velocity regime must be provided in order the region of positive action obtained from the phase-frequency representation and the frequency band of resonant interaction, which must bee so more wide as it is possible, would be related to the region of the energy-having frequencies. Besides the sum of static and dynamic roughness of coating must be less, than the limit assumed for the providing of hydrodynamic smoothness. The satisfaction of these conditions moves contradictory requirements.

From the carried out consideration of the monolithic one-layer viscoelastic coating work follow the conclusions:

1. Interaction of these coatings with turbulent pressure pulsations has resonant character. The first resonant frequency $f_o = \sqrt{E'/\rho}/4H$ is the most responsible for interaction. The width of the main action band Δf_o is determined by coefficient of losses of the compliant coating material. At constant coefficient of losses $\Delta f_o H/(\rho/E')^{1/2} = \text{const}$ takes place. Hence it follows that for the effect increase it is necessary to choose more hard materials with lesser density and to decrease the coating thickness. The same requirements are necessary for the dynamic roughness decrease.

2. Restrictions of the coating surface deflection rate under action of turbulent pressure pulsations must be. It must not be very small in order to provide the need rate of interaction. Since $|v_o|_v = wHK_d\sqrt{P_v}/E'$, the restriction to the lower requires increase of H and decrease of E' that has a contradiction with 1.

3. The band of the interaction frequencies must be in the region of the energy-carrying frequencies ($\sqrt{E'/\rho}/H < 2U/\pi\delta^*$). That is why, the inequality must be fulfilled that corresponds to the requirement 2.

In made studies this condition could not be satisfied. Its fulfillment (at fixed velocity) requires to increase H and decrease E'/ρ that stimulates reduction of the interaction frequency band and increases the risk of the dynamic roughness development.

4. The condition of the optimum of phase-frequency characteristic of neutral action requires $2\cdot 10^{-2} < \pi\nu f_o/v_d^2 < 6\cdot 10^{-2}$. This, in own turn, requires additional restrictions on the choice of the coating material properties or leads to selective dependence on velocity at the work on concrete coating.

References

Blick, E. 1974 Skin friction drag reduction by compliant coatings. In *Proc. Int. Conf. Drag Reduction*, Cambridge, F2/23-36.

Blick, E. & R. Walters 1967 Turbulent boundary layer characteristics of compliant surfaces. *J. of Aircraft* **5**, 11-16.

Fischer, M. & E. Blick 1966 Turbulent damping by flabby skins. *J. of Aircraft* **3**, 163-164.

Fischer, M., L. Weinstein, R. Ash & D. Bushnell 1975 Compliant wall-turbulent skin-friction reduction research. *AIAA Paper*, N 75-833.

Gad-el-Hak, M. 1984 An optical technique for measuring the flow-induced motion of a compliant surface. In *Symposium on flow-induced vibrations*, **5**, New-Orleans, 9-22.

Kanarsky M., V. Babenko & G. Voropaev 1982 Measurement of kinematic characteristics of turbulent boundary layer on a plate and processing of the obtained information on computer. *Gidromekhanika* **45**, 30-36, in Russian.

Kanarsky, M., V. Babenko & L. Kozlov 1979 Experimental study of turbulent boundary layer on elastic surface. In *Stratified and turbulent flows*, Kiev, 59-67, in Russian.

Kawamata, S., T. Kato, Y. Matsumura & T. Sato 1973 Experimental research on the possibility of reducing the drag acting on a flexible plate. *Theor. and Appl. Mech.* **21**, 507-518.

Kendall, J. 1970 The turbulent boundary layer over a wall with progressive surface waves. *J. Fluid Mech.* **41**, 259-281.

Korobov, V. & V. Babenko 1983 On one mechanism of interaction of elastic wall with a flow. *J. Eng. Phys.* **44**, 730-733.

Korobov, V., V. Babenko & L. Kozlov 1981 Integral characteristics of boundary layer on elastic surfaces. *Doklady AN Ukr. SSR, ser.* A, N 11, 351-352, in Ukrainian.

Kramer, M. 1957 Boundary layer stabilization by distributed damping. *J. Aeronaut. Sci.* **24**, 459-460.

Kramer, M. 1960 Boundary layer stabilization by distributed damping. *J. Amer. Soc. Nav. Eng.* **72**, 25-33.

Kramer, M. 1962 Boundary layer stabilization by distributed damping. *Nav. Eng. J.* **74**, 341-348.

Kulik, V. 1990 Improvement of an accuracy processing the measuring data of viscoelastic properties of materials after two-parametric technique. *Metrologiya*, N 3, 21-24, in Russian.

Kulik, V. 1986 The analysis of interaction of one-layer monolithic damping covers with turbulent flow. *J. Eng. Phys.* **51**, 959-965.

Kulik, V. & B. Semenov 1983 On the certification of wave properties of viscoelastic coatings affecting wall turbulence. In *Hydrodynamic flows and wave processes.* Novosibirsk, 117-129, in Russian.

Kulik,V., I. Poguda & B. Semenov 1984 Experimental study of the effect of one-layer viscoelastic coatings on the turbulent friction and pressure pulsations at the wall. *J. Eng. Phys.* **47**, 189-196.

Kulik, V. & B. Semenov 1986 Two-parametric method of measuring the viscoelastic properties of polymer materials. *Metrologiya*, N 4, 32-38, in Russian.

Landahl, M. 1962 On the stability of laminar incompressible boundary layer over a flexible surface, *J. Fluid Mech.* **13**, 609-632.

Loof, J. 1974 A synthesis on drag reduction experiments at Bertin. From compliant surfaces and gas film to polymers. In *Proc. Int. Conf. Drag Reduction*, Cambridge, F3/37-52.

Looney, R. & E. Blick 1966 Skin friction coefficients of compliant surfaces in turbulent flow. *J. of Spacecraft and Rockets* **3**, 1562-1564.

McMichael, J., P. Klebanoff & N. Mease 1979 Experimental investigation of drag on a compliant surface. In *Techn. Pap. Symp. viscous flow drag reduction*, Dallas, 410-438.

Merkulov, B. 1981 Control by the liquid motion. Novosibirsk, Nauka, in Russian.

Semenov, B. 1971 Interaction of the elastic boundary with the viscous sublayer of a turbulent boundary layer. *Prikladnaya mekhanika i tekhnich. fizika*, N 3, 59-62, in Russian. (TT F-14, 1972, NASA).

Semenov, B. 1976 The effect of elastic covers on a turbulent boundary layer. In *Investigations of boundary layer control.* Novosibirsk, 92-101, in Russian.

Semenov, B. 1978 On interferent form of the influence of viscoelastic boundary on near-wall turbulence. In *The influence of polymer additives and surface elasticity on wall turbulence.* Novosibirsk, 57-74, in Russian.

Semenov, B. 1981 Analysis of deformation characteristics of visco-elastic coatings. In *Hydrodynamics and acoustics of near-wall and free flows.* Novosibirsk, 57-76, in Russian.

Smith, R. & E. Blick 1969 Skin friction of compliant surfaces with foamed material substrate. *J. of Hydronautics* **3**, 100-102.

Voropaev, G. & V. Babenko 1978 Turbulent boundary layer on elastic surface. *Gidromekhanika* **38**, 71-77, in Russian.

Willmarth, W. & F. Roos 1965 Resolution and structure of the wall pressure field beneath a turbulent boundary layer. *J. Fluid Mech.* **22**, 81-94.

V. Polymer Additives

The pulseless injection of polymeric additives into near-wall flow and perspectives of drag reduction

B. N. SEMENOV

Institute of Thermophysics
USSR Academy of Sciences, Siberian Branch
Novosibirsk, USSR

K.-S. Choi (ed.), Recent Developments in Turbulence Management, 293–308.

Abstract

The analysis of experimental data reveals that the polymer consumption is connected with the ship velocity by linear dependence, and that the fuel economy is connected by the cube one. That is why, the velocity range of the profitable use of polymer additives can exist theoretically always. Now, at the existing consumption of polymer injected as a jet through slots into near-wall region, and since polymer is significantly more expensive, than fuel for the present, application of polymer additives can be profitable only of short duration in the high velocity sea transport. For example, "polymeric forcing" allows to increase load-carrying capacity for hydrofoil craft of type "Kometa" of 50% to help to engine at the going out from a water and to use their total power in the cruiser regime of motion. The problem of studies is increase of specific effectiveness ψ/q , which is the main parameter determining a profit. ψ/q is a ratio of drag reduction to dimensionless coefficient of polymer consumption. For the present, $\psi/q < 4\cdot10^7$. Calculations lead to the conclusion that from the point of view of profit it is worth while not to tend to the drag minimization, but to restrict the friction reduction twice. The tests with the wash-out of coatings give the value $\psi/q = 3\cdot10^8$ (for $\psi \simeq 0.2$). The drag reduction method by polymer additives will be profitable at an application for the most part of types of the sea ships at the realization of the given specific effectiveness. Calculations and tests testify the ways of the specific effectiveness increase are the polymer molecular mass increase, preparation of injected solutions and the pulseless injection.

1. Introduction

The first successes in the Toms effect study were so significant that shipbuilders have made trials of sea ship with the injection system of water-poly(ethylene oxide) (PEO) without delay (see for example, Canham *et al.* 1972). However, technical-economical estimation of the obtained results was then not in the high-molecular polymer application favour. Expenses at their injection were higher than economy of expenses at fuel. At last years a situation had changed. Cheap trade-marks of high-molecular polymers having high hydrodynamic effectiveness at small concentrations in a solution (for instance, PEO of trade-marks WSR-301, WSR-701 made in USA, BADIMOL made in Bulgaria) had appeared. The process of the further ratio decrease of cost of polymer and fuel is continued. Injection of polymeric additives in a flow at their consumption decrease is perfected. On given stage it is necessary, firstly, to estimate the perspective of application of polymeric additives in water transport and, secondly, to determine expedient conditions of the use with taking of into account the accumulated experiences.

2. The existence of the velocity region for profitable use of polymer additives

It is important to predict trends and conditions of the given method of hydrodynamic drag reduction (HDR) for the application in shipbuilding, not decreasing a importance of the estimations for concrete types of ships (for example, in the quoted work for trawl ships). Injection of high-molecular polymeric additives in near-wall region of a flow decreases turbulent friction, and for well-flowed bodies (at non-separated flow) decreases analogously the form drag (induced by deviation of the zero stream-line from the surface). The effectiveness of viscous drag reduction $\psi = 1 - \zeta/\zeta_o$, where ζ is coefficient of viscous drag at injection of polymeric additives, ζ_o is the same value without injection. Thereby, power of marine engines, which is necessary for a motion at velocity U, decreases $\Delta N = \psi\zeta_o\rho_o U^3 S/2\eta$ worth. Here ρ_o is water density, S is wet surface of ship; η is total propulsive coefficient. The economized fuel mass is proportional to ΔN. Consumption of dry polymer powder for realization of the HDR method $q_o = q\rho SU$, here ρ is polymer density. On the base of the having data (see, for example, Hoyt 1972) one can think that at keeping the necessary conditions of preparation and injection of polymer solution in near-wall region the effectiveness correlation on dimensionless coefficient of dry polymer consumption q is weakly dependent on velocity (in the range of the ship exploitation velocities). So, at the HDR method realization the polymer consumption is connected with velocity of ship by linear dependence, and the fuel economy is connected by the cube one. The profit criterion is ratio of economized cost of fuel to the polymer cost, i.e. we have the condition:

$$A = 10^{-6}\cdot\frac{q_N}{7.2}\cdot\frac{\zeta_o}{\eta}\cdot\frac{\rho_o}{\rho}\cdot\frac{\psi}{q}\cdot\frac{FC}{PC}\cdot U^2 \geq 1 \tag{1}$$

Here, specific consumption of fuel q_N is written in usual dimension for shipbuilding [kg/kW·hour]. From written condition (1) it is seen if even at large ratios the cost of polymer and fuel (PC/FC), the method is not profitable for the low-velocity ships (type of the above-cited trawl ship), it is profitable for high-velocity ships. $\rho \simeq \rho_o$. According to Voitkunsky *et al.* (1960), carrying out the calculated estimations, it was assumed: $\zeta_o = 2.7\cdot 10^{-3}$; $\eta = 0.6$, $q_N = 0.3$ kg/kW·hour (diesel fuel).

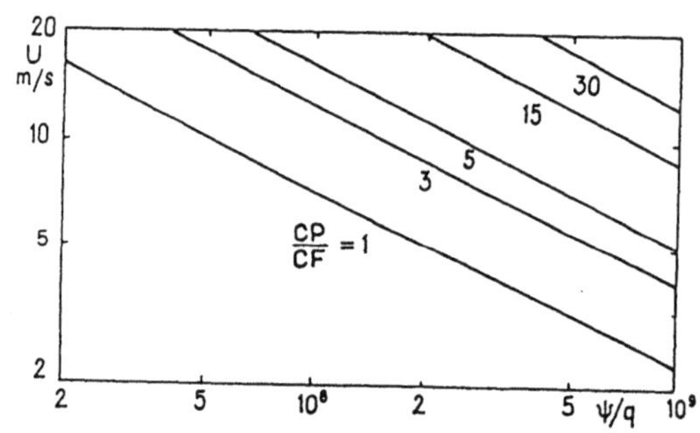

Fig. 1. Lover velocity of profitable application of polymeric additives for the viscous drag decrease of ship in dependence of specific effectiveness (at variation of ratio of cost of polymer and fuel).

The dependence of the exploitation velocity, at which is A=1, on specific effectiveness ψ/q is given in fig.1, i.e. the velocity range (above the calculated lines) of profitable use of the HDR-method is determined. Varied ratio of cost of polymer and diesel fuel corresponds approximately at the present to 3-5 (USA), 12-15 (Bulgaria), 30 (USSR). To a marked degree, it is determined by industrial output of poly(ethylene-oxide), which is the most in USA and the least in USSR. One assumes at further improvement of technology and the polymer output increase the ratio PC/FC will tend to 1.

The written criterion is the upper limit since at the estimation it was not taken into account the possible increase of carrying capacity (analogously and profit) of ship due to the ship power plant decrease (and its mass, respectively), the need fuel supply decrease keeping the required voyage distance. Mass of the polymer solution preparation system, their injection into a flow is small as compare with the engine mass (by analogy, one can consider the ship fire-fighting system at the project estimations). Mass of used polymer, as it is seen from above-given estimation (fig.1) is significantly less than mass of economized fuel (in the FC/PC ratio in given graphs). That is why, in series of cases the given HDR method is technically expedient even for small profit, since the mass economy can turn out a very sufficient factor making the exploitation advantages for some types of ships (for instance, for passenger ships, container ships, ferries), for which with the purpose of rigorous observance for the time-table traffic is foreseen the power augmentation power forcing of the ship power plant two-three times as large in the case of the wave resistance increase under unpropitious meteorology conditions. For this purpose "the polymer forcing" can be used, i.e. no permanent but only temporary injection of polymeric additives decreasing friction and compensating, for example, the wave resistance increase at storm.

Earlier (Semenov *et al.* 1988) the expedience of "polymeric forcing" at the speed-up stage of motion of a hydrofoil craft (HC) is well grounded. Although the part of viscous drag in the total one for HC is significantly less than for ordinary ships "polymeric forcing" allows to increase carrying capacity of HC significantly (by 60 % as large for HC of type "Kometa") due to the engine help at going of a body out of the water and use of their total power on the cruiser regime of motion. In this case specific effectiveness ψ/q is the main parameter too determining profit of the HDR method.

It is analogous to parameter ψ/C introduced by Berman (1978) for the analysis of effectiveness of the hydraulic loss reduction in pipes at the solution flow with homogeneous concentration of polymer in a flow (here and below, the mass part of polymer in solution is considered). Its use gives a possibility both to estimate practical expedience of the HDR method and allows to exclude the consumption parameter at study of multiparametric dependence of the Toms effect.

3. On the drag minimization by polymer additives

It is very interesting to analyze the variant of the friction resistance minimization of smooth plate at injection of polymeric additives in the nose part. It is assumed the wall injection into near-wall region of a flow will be pulseless, that is why, the polymer concentration decrease downstream is solely caused by the processes of turbulent mass transport in boundary layer. It is known (Kutateladze and Leontyev 1972) that at large Reynolds numbers a similarity of the velocity profiles and concentrations is observed in near-wall region of a flow, and the drag reduction is accompanied by the same decrease of mass exchange. Then after the experimental data of Fabula and Burns (1970), one can connect the wall polymer concentration C_w at abscissa (of the plate L-long) with consumption as follows: $C_w(x)=2qL/\delta(x)$. Here, $\delta(x)$ is thickness of dynamic boundary layer.

Apparently, one can take for granted that main deturbulizing action the polymer additives at small concentrations in a flow show in near-wall region. This assumption allows to use having data on the friction reduction in internal flows with constant concentration of polymer C for the problem on outer flowing-around identifying $C_w = C$. From the experiment for the PEO-solution flow between coaxial cylinders (Semenov *et al.* 1988) it is known that at $4 \le CM^{0.85} \le 800$ the plateau of maximal effectiveness drag reduction is observed (M is mean molecular mass, determined after measurements of characteristic viscosity). That is why, for the drag minimization of a plate L-long it is necessary to provide at x=L: $C_w = 4M^{-0,85}$. Finally, one can write calculating the thickness of boundary layer after the Falkner's formula (see Voitkunsky *et al.* 1960) the following dependence of specific effectiveness of maximal drag reduction of a plate on molecular mass of PEO and the Reynolds number of the plate flowing by water ($Re=UL/\nu$, where ν is kinematic coefficient of water viscosity):

$$\psi/q = 5.5\cdot M^{0.85}\cdot Re^{1/7} \qquad (2)$$

In the formula (2) maximal value of drag reduction was taken $\psi max = \psi_{\tau} max = 0.8$ according to experimental data (review of Hoyt 1972). The analysis is given for the self-similarity region of ψ versus shear stress τ. This case when they are significantly greater than the threshold ones for high-molecular PEO, realizes for most regimes of ship motion (thereby, as it is above noted, it is practically expedient to use polymeric additives, first of all, on the high-velocity ships).

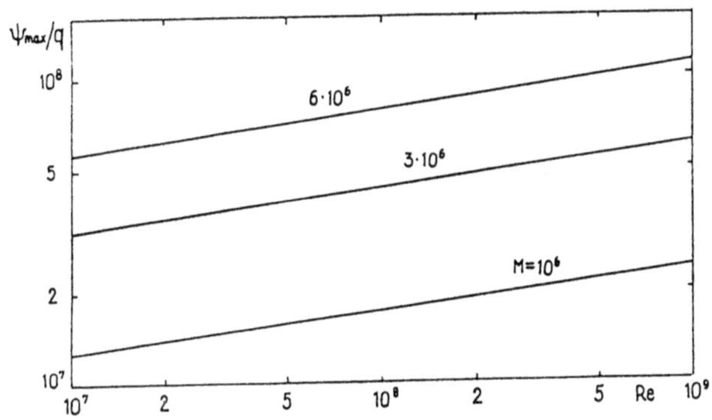

Fig.2. Specific effectiveness of the plate friction minimization by pulseless injection of polymeric additives in dependence on the Reynolds number (at variation of PEO-molecular mass).

Fig.2 illustrates the calculation results. From consideration and comparison of them with the estimation of the application profit of the HDR method in water transport (fig.1) we conclude specific effectiveness increases significantly at increase of the Reynolds number and molecular mass of polymer. However at existing ratio of cost of PEO and fuel it is unlikely expedient to tend to the friction drag minimization.

4. Optimal effectiveness of drag reduction for profit

Berman (1978) had shown that ψ_{τ}/C was decreased as the concentration was on the increase for a flow with constant concentration of polymer additives and was significantly less at the friction minimization than specific effectiveness at moderate values ψ_T. This is connected with nonlinear form of dependence $\psi_T(C)$, its asymptotic behaviour at $\psi_T \to \psi_T max$. For solution of poly(ethylene oxide) it has the form:

$$\psi_T = \alpha_2 \operatorname{arctg}(\alpha_1 CM^{0.85}) \tag{3}$$

Maximal effect $\psi_T max$, which determines the value $\alpha_2 = 2\psi_T max/\pi$, is reached at $\alpha_1 CM^{0.85} \gg 1$. For example, if $\psi_T max = 0.8$, $\alpha_2 = 0.51$ takes place. But if $\psi_T max = 0.69$ (see work of Sedov *et al.* 1974) it should be taken $\alpha_2 = 0.44$. For small values of invariant it is $\alpha_1 CM^{0.85} \simeq \operatorname{arctg}(\alpha_1 CM^{0.85})$. Taking ratio of specific effectiveness of PEO to its value at $C \to 0$, we obtain universal dependence of this ratio on $\psi_T/\psi_T max$ (no depending on value of maximal effectiveness, the Reynolds number, parameters of PEO solution):

$$(\psi_T/C)/\lim_{C\to 0}(\psi_T/C) = 0.5\pi(\psi_T/\psi_T max)/\operatorname{tg}[0.5\pi(\psi_T/\psi_T max)]. \tag{4}$$

In fig.3 it is shown by solid line.

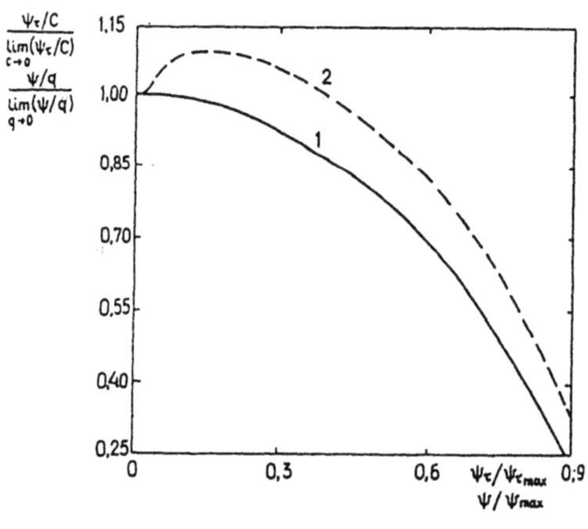

Fig.3. Variation of specific effectiveness of the friction decrease (1) and viscous drag (2) in dependence on the friction decrease level and drag vith respect to maximal effects.

Analogous analysis can be carried out for a flow with variable concentration of polymeric additives in a flow. After the method of Sedov *et al.* (1974) it was taken that injection of polymers into near-wall region of a flow is realized by the pulseless method in the nose section, and polymeric additives don't affect on the friction drag at $x < x_o$. In the point of transient from laminar form a flow to turbulent one (at $x = x_o$) the condition of continuity of lost momentum thickness is written. On its base initial thickness δ_o of turbulent boundary layer is determined, in which was taken the power form of the velocity profile with index 1/11. The system of equations is solved:

$$\psi = \int_{x_o}^{L} \psi_T \tau_o dx \;/ \int_{0}^{L} \tau_o dx; \qquad \psi_T = \alpha_2 \operatorname{arctg}(\alpha_1 C_v M^{0.85});$$

$$C_v = 2qL/\delta_\psi; \qquad \delta_\psi = \delta_o + \int_{x_o}^{x} \frac{(1-\psi_T)\tau_o}{0.0705\rho_o U^2} dx \tag{5}$$

where: $\tau_o = \tau_l = 0.332\rho_o U^2/(Ux/\nu)^{1/2}$ $\qquad 0 \le x \le x_o;$

$\tau_o = \tau_T = 0.0128\rho_o U^2/(Ux/\nu)^{1/7}$ $\qquad x \ge x_o.$

At given Re, x_o we obtain dependence $\psi(q)$, which we normalize relating to calculated maximal effectiveness of hydrodynamic drag reduction ψ_{max}. We normalize the specific effectiveness ψ/q by its value at $q \to 0$. The results of calculation for a plate at $Re = 2 \cdot 10^7$ with initial laminar part $x_o/L = 0.075$ are shown in fig.3 by interrupted line. As distinct from a flow with constant polymer concentration, maximum of specific effectiveness of action of polymeric additives on friction of a plate is replaced to $\psi/\psi_{max} = 0.15$. This is connected with the PEO wall concentration decrease from nose to stern and with the fact that there is initial laminar section at outer flowing along, and turbulent shear stresses in a flow without additives decrease from nose to stern. Accordingly, their contribution in integral drag of a body and also the role of their decrease by the HDR-methods decrease. The analysis leads to practical important conclusion that specific effectiveness of hydrodynamic drag reduction changes comparatively not much at $\psi/\psi_{max} \le 0.6$. That is why, from view of point of profit of this HDR-method it is worth while to get drag reduction approximately twice.

5. Perspectives of the specific effectiveness increase

The normalizing term in (4) is $\lim_{C\to 0}(\psi_T/C) = \alpha_1\alpha_2 M^{0.85}$ but $\lim_{q\to 0}(\psi/q) \approx \lim_{C\to 0}(\psi_T/C)$, that is, specific effectiveness of hydrodynamic drag reduction is proportional to $\alpha_1\alpha_2 M^{0.85}$. Empirical coefficient α_1 characterizes the growth of the drag reduction effectiveness at increase of small polymer concentrations in a flow. Its value, as it follows from our experiments, depends on quality of the solution preparation.

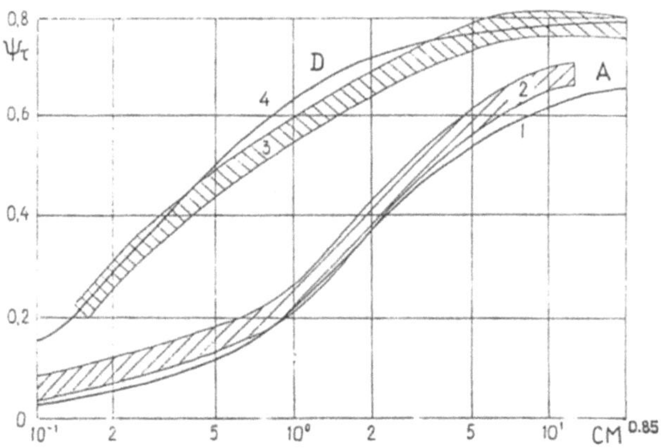

Fig.4. Maximal decrease of turbulent friction of water flows in pipes by PEO-additives. 1 – $\psi = 0.44\mathrm{arctg}(CM^{0.85}/1.8)$; 2 – the data of Sedov et al.; 3 – the data of Ramu and Tullis; 4 – $\psi_T = 0.51\mathrm{arctg}(3CM^{0.85})$.

In the work of Sedov *et al.* (1974) was used the value $\alpha_1 = 1/1.8$. Calculation at $\alpha_1 = 1/1.8$, $\alpha_2 = 0.44$ is illustrated by the line 1 in fig.4. It conforms with experimental data for a flow of PEO-solutions in pipes taken from the work of Sedov *et al.* (1980), but lies significantly below the data from the works of Ramu and Tullis (1976) approximated by calculated dependence (3) at $\alpha_1 = 3.0$ and $\alpha_2 = 0.51$ (shown by line 4 in fig.4). These data correspond to the section ($x/d>40$) of a flow in the pipe (diameter d = 305 mm), on which is realized constancy on cross-section of concentration of the WSR-301 ($M = 4\cdot 10^6$) injected into initial section of the pipe (at $x/d = 3.5$) as a concentrated solution. It is interesting that near the injection location (at $x/d = 12 - 15$) in a zone with variable PEO-concentration on cross-section and length, local drag reduction attains $\psi_T = 0.90$. The more highest data of drag reduction ($\psi_T = 0.92$) was measured by Maksimovic (1985) at the jet injection of the WSR-301 solution into near-wall region of a flow in rectangular channel at a distance 0.6 m from the injection slot. It is possible so large values of the HDR in these experiments are explained by partially (besides action of polymeric additives) by the jet effect (by injection of had been slowing liquid into near-wall region). However, from the analysis of experimental results of Metzner (1977) (in which $\psi_T = 0.965$ was fixed) in the pipe of 2.4 cm in diameter (at the Reynolds number 10^5) for a flow with constant concentration of polyacrylamide and fibres one can think the possibility of the attainment of so great effects and for PEO at C = const might be real. That is, it is evident that correction of coefficients α_1, α_2 to side of increase with the description of their correlation with molecular-mass distribution, conditions of the solution preparation are necessary.

As a result, in the experiments of Ramu and Tullis (1976) specific effectiveness of small polymer additives $\simeq 6.3$ as much, than in the experiments described in the work of Sedov *et al.* (1980). Studies of found-out phenomenon of the growth of drag reduction on initial (on time) section of the Toms effect dynamics (Kulik and Semenov 1986) explained this fact. It was established destruction of supermolecular structures of the colloid particle-type in solutions of high-molecular PEO lead to minimization of turbulent friction (Semenov *et al.* 1988). After the estimations in the cited experiments of Ramu and Tullis the heterophase state of the PEO + water system decreases due to action of turbulent pulsations in the flow zone preceding to the analyzed one. The possibility of sharp decrease of initial time-section of the effect by preliminary thermopreparation of polymeric solution or by addition of low-molecular substances to it was shown by Amirov *et al.* (1986). Increase of specific effectiveness of drag reduction for water solution of polyacrylamide was obtained by vibratory preparation by Bachtiyarov (1987).

For the estimation of the HDR perspectives, action of quality of polymeric additives pulseless-injected from a slot in the nose part of a plate was carried out calculation of $\psi(q)$ for $Re = 2\cdot10^7$, $x_o/d = 0.075$

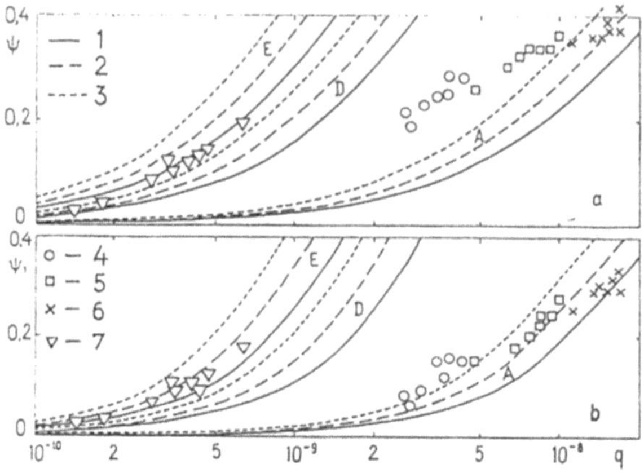

Fig.5. Decrease of total drag of the model (a) and friction of cylindrical section (b) versus PEO-consumption. Calculation in the variants A, D, E: 1 - M=2.2 mln, 2 - 3 mln, 3 - 4 mln. Experiment: the slot injection, M=3.5-4 mln; 4 - $C_i = 10^{-4}$, 5 - $2.5\cdot10^{-4}$, 6 - $5\cdot10^{-4}$; 7 - vashing out of coating, M=2.2 mln.

at variation of molecular mass of PEO and coefficients α_1, α_2. The variant A is $\alpha_1 = 1/1.8$ and $\alpha_2 = 0.44$. The variant D is $\alpha_1 = 3.0$ and $\alpha_2 = 0.51$. The results of calculations are given in fig.5 and compared with experimental data obtained at towing of the model two meters long at velocity 10.2 - 11.2 m/s described in detail in the work of Kulik *et al.* (1984). With the help of strain gauge balance placed inside a model was measured not only its total drag, but and the friction force of cylindrical section of the surface on the extent from $x_1 = 775$ mm to $x_2 = 1430$ mm. Drag reduction on it at injection of polymeric additives is shown in fig.5,b and compared with the calculated effectiveness

$$\psi_1 = \int_{x_1}^{x_2} \psi_T \tau_T dx \Big/ \int_{x_1}^{x_2} \tau_T dx.$$

The results of measurements at injection of the PEO solutions (M = 3.5 - 4 mln) in near-wall region through a slot located in 28 mm from the nose edge of a model are shown in fig.5. Variation of q is realized as due to variation of volumetric consumption of injected solutions as variation of the polymer concentration C_i. It is seen that at increase of C_i corresponding groups of experimental values are shifted from calculated region of the variant D into region of the variant A. By this, specific effectiveness of the HDR decreases. If at $C_i = 10^{-4}$ $\psi/q = 7\cdot10^{7}$, but at $C_i = 5\cdot10^{-4}$ $\psi/q = 3\cdot10^{7}$. As it was above mentioned, possible cause is technology of preparation of injected polymeric solutions used by the majority of investigators and in described experiments. Usually it was prepared basic PEO-solution with concentration 2 %, which was diluted immediately before experiments to operating concentration C_i during $\simeq$ 10 minutes by flowing mixing at temperature of 16-20 °C. According to made study using this technology significant number of supermolecular structures of PEO keeps (especially, for most high-molecular fractions of polymer), which decrease the HDR-effectiveness.

Attracting by simplicity of realization the jet injection of polymeric additives has a series of essential defects a priori. It might lead to interruption of boundary layer due to the taking away of normal (to a wall) velocity component from a wall and due to intensive mass-exchange on boundary of jet and a flow. Although in described experiments were taken measures for decrease of action of these factors (injection velocities were by order less, than velocities of towing of a model, angle of the slot inclination to forming were $\leq 20^{\circ}$), comparison of obtained results with the data of calculation at pulseless injection can have only qualitative character.

For quantitative comparison it was carried out the experiments with the washing out of water-soluble coating (from polymeric composition including PEO with M=2.2 mln, adhesive components and low-molecular substances decreasing intermolecular interaction of PEO) deposited on the nose part of a model from the nose perpendicular to the injection slot. Specific effectiveness of the HDR in this case was equal to $3\cdot10^{8}$. Experimental points (fig.5) are significantly on the left than calculated lines of the variant D. The obtained specific effectiveness of the HDR is twice as much, than the predicted one by the calculations made on the base of the best among having results of Ramu and Tullis in pipes. The data obtained in experiments with use of water-soluble coatings are in good agreement with calculations, if in the formula (3) it is taken $\alpha_1 = 6$, $\alpha_2 = 0.51$ (the variant E). There is a sense to take this variant as a base of the HDR perspective estimation determining the result, to which one must tend using the practically convenient means of injection of polymeric additives in a flow, for example, the slot injection of polymeric solutions.

6. Conclusion

It should be noted here at existing consumption of polymer injected into near-wall region and until polymer is more expensive, than fuel, one can profitably apply polymeric additives for decrease of expenditure of energy only of short duration in high-velocity sea transport.

However, made calculations lead to the conclusion that at realization of drag reduction by pulseless injection of small additives in near-wall region of a flow (according to (3) in the variant E) this method of the HDR will be profitable at the application in the most of type of sea ships. Then at increase of Reynolds number specific effectiveness of the HDR increases and, accordingly, profit of the methods grows. For example, if at $Re = 2 \cdot 10^7$ the PEO-use with $M = 4$ mln can give specific effectiveness $5 \cdot 10^8$, at $Re = 4 \cdot 10^8$ we can obtain $\psi/q = 9 \cdot 10^8$. Application of PEO with higher molecular mass promises higher values of specific effectiveness. Using PEO with M=6 mln (BADIMOL) we can obtain $\psi/q = 1.2 \cdot 10^9$ at $Re = 4 \cdot 10^8$.

References

Amirov, A.I., V.M. Kulik & B.N. Semenov 1986 The Toms effect dynamics for poly(ethylene oxide) solutions. In *Thermogasdynamics of turbulent flows*, Novosibirsk, 58 - 75, in Russian.

Bachtyarov, S.I. 1987. On the Toms effect for extremely low additives of polymers. *Izv. Vuzov: neft - gaz*, N 4, 52 - 55, in Russian.

Berman, N.S. 1978 Drag reduction by polymers. In *Ann. Rev. Fluid Mech.* 10, 47 - 64.

Canham, H.J.S., J.P. Catchpole & R.F. Long 1971 Boundary layer additives to reduce ship resistance. *The Naval Architect*, N 2, 187 - 213.

Fabula, A.G. & T.G. Burns 1970 Dilution in a turbulent boundary layer with polymeric friction reduction. In *TP 171*, Naval Undersea Research and Development Center, Pasadena - California.

Hoyt, J.W. 1972 The effect of additives on fluid friction. *Trans. ASME. J.Basic Eng. Ser.D 94*, N 2, 1 - 32.

Kulik, V.M., I.S. Poguda & B.N. Semenov 1984 Experimental study of the effect of one-layer viscoelastic coatings on the turbulent friction and pressure pulsations at the wall. *J. Eng. Physics*, 47, N 2, 189 - 196.

Kulik, V.M. & B.N. Semenov 1987 The initial dependence of the Toms effect in poly(ethylene oxide) solutions on time. *Soviet J. Appl. Physics*, 1, N 3, 63 - 68.

Kutateladze, S.S. & A.I. Leontyev 1972 Heat and mass transfer and friction in turbulent boundary layers. *Energiya, Mockva*, in Russian.

Maksimovic, C. 1985 Turbulence structure of a developing duct flow with near wall injection of drag reducing polymers. In *The influence of polymer additives on velocity and temperature fields.* Springer Verlag, 359 - 368.

Metzner, A.B. 1977 Polymer solution and fiber suspension rheology and their relationship to turbulent drag reduction. *Phys. Fluids*, 20, N 10, pt 2, 145 - 149.

Ramu, K.L.V. & J.P. Tullis 1976 Drag reduction and velocity distribution in developing pipe flow. *J. Hydronaut*, 10, N 2, 55 - 61.

Sedov, L.I., N.G. Vasetskaya & V.A. Ioselevich 1974 Prediction of turbulent boundary layers with small polymer additives. In *Turbulent flows*, Nauka - Moscow, 205 - 220, in Russian.

Sedov, L.I., N.G. Vasetskaya, V.A. Ioselevich & V.N. Pilipenko 1980 On drag reduction by the polymer additives. In *The mechanics of turbulent flows*, Nauka - Moscow, 7 - 28, in Russian.

Semenov, B.N., A.I. Amirov, V.M. Kulik & O.N. Marennikova 1988 Effect of supermolecular structures in poly(ethylene oxide) solutions on drag reduction. In *Near-wall and free turbulent flows*, Novosibirsk, 20 - 52, in Russian.

Semenov, B., P. Zlatev & Y. Yovev 1988 Evaluation of the feasibility of applying additives in hydrofoil craft propulsion. In *Proc. SMSSH'88*, Varna, 3, 82.1 - 82.4.

Voitkunsky, Ya.I., R.Ya. Pershitz & I.A. Titov. Handbook on theory of a ship, Sudpromgiz - Leningrad, in Russian.

Initial section of time-dependence of the Toms effect for solutions of poly (ethylene oxide)

V. M. KULIK and B. N. SEMENOV

Institute of Thermophysics
USSR Academy of Sciences, Siberian Branch
Novosibirsk, USSR

K.-S. Choi (ed.), Recent Developments in Turbulence Management, 309–321.

Abstract

The results of experimental study of the drag reduction change of turbulent flow of PEO poly(ethylene oxide) solutions are presented. As distinct from previous studies, a special attention was paid to initial time interval of the polymer additive presence in a flow, where was found the drag reduction effectiveness growth. The hypothesis explaining this phenomenon as a destruction of stable supermolecular structures was suggested. The effect of a renewal of the section of the drag reduction growth in the time with the flow renewal was found. It was shown the destruction of supermolecular structures in injected PEO - solutions is the real way to increase a profit of the polymer additive application. For example, the method of the action on dynamics of the drag reduction change using low-molecular addition to concentrated solutions of PEO are shown.

1. Introduction

The addition of small quantity of high-molecular polymer in a flow is the effective method to reduce turbulent skin-friction. Many works were devoted to study this phenomenon, but practical use is held back by unsufficient profitableness. In many cases, for example, in short pipes or under conditions of outer flow the presence time of polymer additions near wall is very small. That's why, the interest in problem of hydrodynamic activity of polymeric solutions during initial time interval of its presence in turbulent flow came up.

Fischer and Rodriguez (1971) had found the section of the efficiency growth in the pipe flow of PEO solution (Fig.1,a). Afterwards many researches had observed the analogous section of the efficiency growth presenting the experimental data of reusable runs of polymeric solutions (at concentration more then 1000 ppm) in pipes, see, for example, Sedov *et al.* (1980). When solution was flowing between coaxial cylinders

(Belokon and Kalashnikov 1977) the initial section of growth was also fixed (Fig.1,b). Here it is showed the friction moment changing on the time. But all the authors had not seen a special importance of this fact

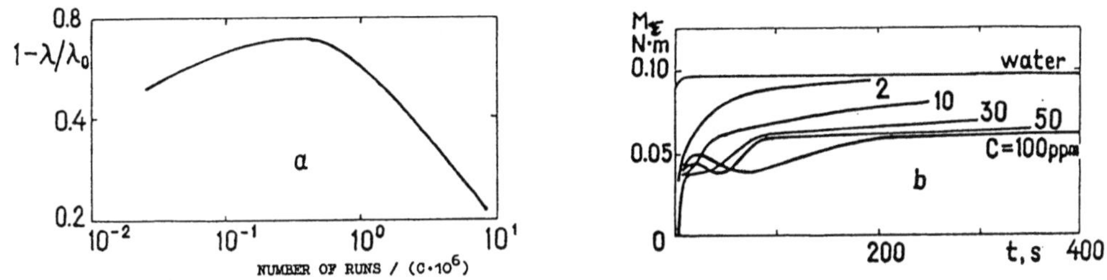

Fig.1. The examples of the observation of initial drag reduction
a - the pipe, D=3.46 mm, PEO, M=4.6 mln, Re=8630 (Fischer and Rodrigues); b - ICC, WSR-301, Re =150000 (Belokon and Kalashnikov).

and had not tried to study it in detail. In our opinion, the study of regularities of the Toms effect during initial time-interval of the polymer additive presence in a flow enables to increase its efficiency and makes its use more profitable (Kulik and Semenov 1986).

2. Experimental technique

The installation with coaxial cylinders (ICC) was modified (Kulik 1981) to study the time-dependence of the Toms effect. Important merits of this simple and handy unit are very wide range of shear stresses ($10 \le \tau_w \le 250$) Pa and the possibility to have constant operating regime for a long time. Besides, the advantage of it is the possibility to use a small volume (300 ml) of moving liquid that allows to use a small quantity of polymer.

The modified installation was used in the given experiment (Fig.2). The essentially new element is a sectional inner cylinder consisting of three parts along axie with small clearances between the parts. The upper (2) and the lower (4) disks are joined stiffly with unmovable base, and the central part (3) is joined with torque meter (8). Such construction reduces significantly the effect of the liquid volume on readings of dynamometer. The 1 ml-difference in the liquid volume leads to the error about 0.5 %, which is less by a factor of 20 than the measurement result with a solid inner cylinder. The outer diameter of the inner (fixed) cylinder is equal to 90.0 mm and the clearance between cylinders is H = 10 mm. Friction coefficient C_f for new construction is 30 % less than earlier. It can be explained by the displacement of the action zone of the edge effects from central part of inner cylinder to the upper and lower disks. Here friction coefficients $C_f = 2\tau_v/\rho U^2$ defined the same Reynolds number $Re = UH/\nu$ were compared, ρ, ν are density and viscosity of moving fluid, U is circular velocity of inner surface of outer cylinder.

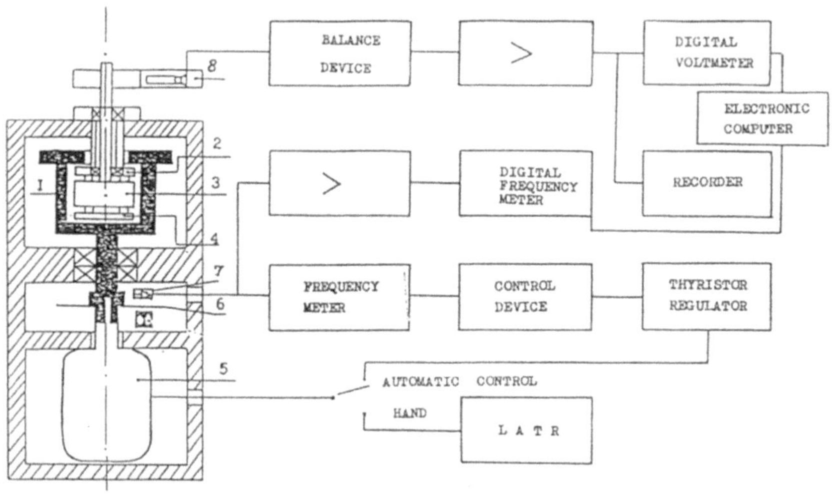

Fig. 2. Block-diagram of installation (ICC). 1-outer rotation cylinder; 2-upper disk; 3-measuring section; 4-lover disk; 5-electro engine; 6-disk vith openings; 7-photodiode; 8-strain element.

The outer (rotating) cylinder was made from stainless steel to improve heat exchange. This results that the temperature growth did not appear at $U < 10$ m/s and at maximal velocity ($U \simeq 35$ m/s) the solution temperature grew only in 5-7 $^{\circ}$C and was constant owing to the heat balance between the heat generation in the clearance and its transfer through the cylinder walls. Automatic regime of the velocity stabilization was realized. The construction of the torque meter had a special device to control "zero" in the measurement process. This gives the possibility to carry out rather time-consuming experiments (about several hours) maintaining constant parameters.

This ICC was certified by a series of measurements using Newtonian liquids of different viscosity. There are three different regions of a flow in dependence on the Reynolds number as shown in Fig.3. The experimental data are in good agreement with theoretical formula $C_f = 2.2/Re$ in laminary flow. The transition region is at $2.5\cdot10^{3} < Re < 3.4\cdot10^{4}$. In this region the turbulence arises in the narrow layer near the wall of the rotating cylinder and when velocity increases it propagates over all width of the clearance (Ustimenko 1977). That is why, the turbulent region of a flow for small clearances begins at smaller Reynolds numbers. So, in Fig.3 it is seen that a flow is turbulent at operating clearance 2 mm already at $Re = 2\cdot10^{4}$, but at the clearance of 5 mm a flow is turbulent at $Re = 3\cdot10^{4}$. The region of turbulent flow is observed at $Re = 3.5\cdot10^{4}$. The Reynolds number based on the clearance width is the modeling parameter. The approximated formula for the friction coefficient is suggested $C_f = 0.063\cdot Re^{-0.42}$ for studied region of turbulent flow of newtonian liquids at $3.5\cdot10^{4} < Re < 4\cdot10^{5}$.

These and other experiments with polymer solutions were carried out on the ICC with smooth surface of cylinders.

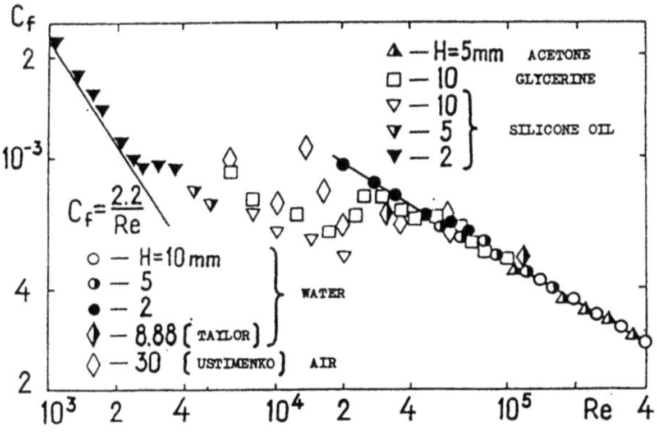

Fig.3. Friction coefficient versus Reynolds number.

3. Experimental results for polimer solutions

The rotation velocity stabilization allows not only to keep constant velocity with the error 0.5 % but to reduce time to 4 s to came the rotating cylinder into the operation mode. The second method reducing significantly the non-controlled initial time interval was the injection of required quantity of concentrated solution into a distillate between the cylinders. This technique was used to study drag reduction of aqueous solution of WSR-301 in dependence on the time of its interaction with the turbulent flow at Reynolds number from $8\cdot10^4$ to $3.5\cdot10^5$. Basic polymer solution of initial concentration 0.5 % was prepared on a magnetic mixer. Lower concentrations of solutions were obtained by dilution of basic solution at minimal stresses to avoid a mechanical destruction of the polymer.

Operating concentrations of PEO solutions from 4 ppm to 1000 ppm were studied. The example of experimental data is shown in Fig.4, where the dependence of drag reduction ψ on time t and concentration C is given at the some constant velocity U. Drag reduction was determined from the formula $\psi = (C_{fo} - C_f)/C_{fo}$, where C_f is friction coefficient of PEO solution flow and C_{fo} is friction coefficient of water taken at the same Reynolds number.

The experimental points obtained for concentrations of PEO from 4 to 100 ppm lay on one curve, if the ratio of the interaction time to concentration is used as an abscissa. This fact is in a good agreement with Belokon and Kalashnikov (1977) and confirms his hypothesis of the concentration-time analogy. But unlike preceding investigations, where the authors had observed in detail only monotonous decrease of drag reduction (corresponding to right part of the curve Fig.4) in given case the initial time interval with the drag reduction growth is found, and the maximum drag reduction is fixed. It must be specially noted the drag reduction decreases is sharper as the solution concentration increases further (C > 200 ppm), and concentration-time analogy is infringed.

Similar experiments were performed at seven velocities. The drag reduction results for the solutions of concentration below 200 ppm are shown in Fig.5,a versus t/C. As it is seen from this figure, all the obtained curves have a domal form. Maximal effectiveness is achieved in some time after injection t_o. This stage is reduced greatly as the flow velocity increases. For instance, at C = 100 ppm this time is equal to

250 s for U = 8.65 m/s, but it is 50 s for U = 17,25 m/s. It is interesting that the maximal drag reduction doesn't change significantly within the range of measured velocities from 8.65 m/s till 35 m/s or within the range of shear stresses from 10 to 100 Pa. This fact can be explained as an analog of a limited drag reduction of Virk in pipes.

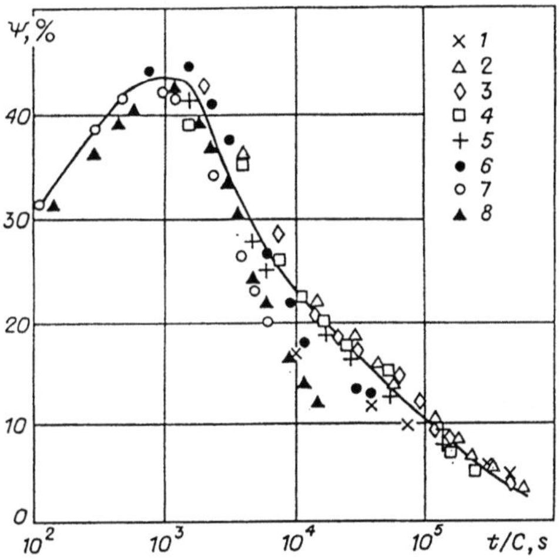

Fig.4. Drag reducton by WSR-301 versus of the interaction time. U=25.9 m/s. 1-C=4; 2-10; 4-40; 5-100; 6-200; 7-500; 8-1000 ppm.

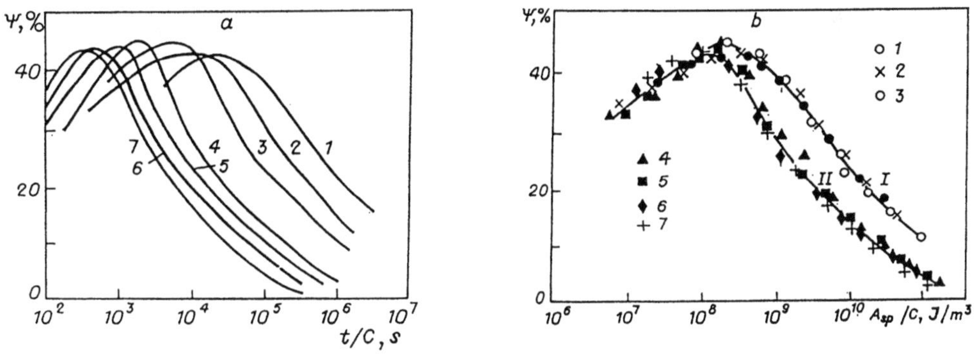

Fig.5. Drag reduction versus the concentration-time parameter (a) and the specific work (b). 1-U=8.65; 2-13; 3-17.25; 4-21.6; 5-25.9; 6-30.25; 7-34.55 m/s.

The dependencies (4-7) were obtained in a month after the experiments (1-3) to reveal the keeping action of concentrated solution. Polymer solution of 0.5 %-concentration was kept in closed glass volume in darkness at temperature of 18 °C. The keeping of concentrated solution didn't change general form of the time dependency of drag reduction.

4. Discussion of results

The productive way to study mechanical destruction of polymers became the use of a specific work of the friction forces as a parameter (Ting and Little 1973). The generalization of experimental curves obtained for various flow velocities is the universal dependence for concrete experimental installation (Fig.5,b). The specific work of friction forces is coordinate modeling dependence on time and parameters of motion for ICC (here M is marked in tne molar form):

$$A_{sp} = A/m_{polymer} = (UM/\rho CH) \int \tau_{w} dt.$$

Unlike Anisimov and Mironov, our study was carried out in the variation range of principal modeling parameter, which was by 10^3 greater and by 10 less in the direction of decrease of former values. Whereas in earlier studies only monotonous decrease of drag reduction was observed as the shear force work increased, in our case the curves have a domal form and maximal effectiveness is achieved only after some specific work applied to polymeric solution $A^{o}_{sp} = (1-2)\cdot 10^{11}$ J/mol. Subsequent decrease of effectiveness over the wide range corresponds to the law: $\psi = \psi_{max} - K\cdot lg(A_{sp}/A^{o}_{sp})$ with gradual decrease of the coefficient K.

Two sets of experiments (1-3) and (4-7) are generalized by two different curves of mechanical destruction. This suggests that mechanic destruction is very effected by changes of supermolecular polymer structure due to the month of the concentrated solution keeping. Resulting change of polymer properties leads to the decrease of the hydrodynamic effectiveness maximum and strong acceleration of mechanic destruction. No visible changes took place on the initial interval ($t<t_o$).

Semenov (1989) had suggested the specific work of the friction forces is no modeling parameter but only normalizing one of work the small-scale pressure pulsations, which cause the degradation of polymer in turbulent flow.

To explain this experimentally observed peculiarity of the Toms phenomenon on the initial interval of the time dependence the most real arguments are the following. As it is well known, high-molecular polymers are able to form supermolecular structures in a static concentrated solution, which are the colloid particles (crystallites, fibrills, lamels and so on). Apparently, the size of macro molecules is quite suitable to show the Toms effect, so the process of association leads to a significant decrease of concentration of effectively working macro molecules. Supermolecular formations are destroyed by turbulent pulsations increasing the number of effectively working macro molecules, if the formations are subjected to the shear (pulsation) stress action.

As energy of the Van-der-Vaals interaction is, at least, by the factor of 10 - 20 less than energy of chemical polymerizational bonds, the process of the supermolecular formation disintegration must predominate over the process of the breaking up of molécules at first, if one considers the number of dissociations. That's why, degradation of macro molecules, as a rule, accompanying by the process of the breaking up of the colloid particles on a certain stage has the least action on the drag reduction change than dissociation of supermolecular structures. Increase of the number of dissociated molecules leads to the drag reduction increase, that is, shows initial section of the time dependence of the Toms effect.

5. New tests for examination of hypothesis

Natural continuation of the described study is the attempt to change supermolecular structures. One of the methods to decrease Van-der-Vaals interaction of macro molecules is the addition of the low-molecular substance in concentrated PEO solution (for example, NaCl). In this experiments were found that injection of PEO solution with NaCl:

- reduces the initial section of the drag reduction growth;
- increases the realization time of maximal effectiveness;
- increases the maximal drag reduction.

And besides, if solution without NaCl was injected into water with equivalent NaCl-concentration, it was no this effect. Therefore, the property of low-molecular additives to reduce the macro molecule interaction acts significantly on dynamics of drag reduction only at the injection of high-concentrated solutions.

New phenomenon of the Toms effect growth renewal confirms a positive role of the breaking of supermolecular structures. So, if specific work of friction forces $A_{sp} < A^{o}_{sp}$ in the PEO solution flow was applied and motion was stopped, the Toms effect growth renewal arises when the ICC is again untwisted up to the same velocity (Fig.6,a). This is explained by conservativeness of the Van-der-Vaals interaction. When the action of the forces breaking supermolecular structures is stopped, the Van-der-Vaals interaction restores the colloid particles.

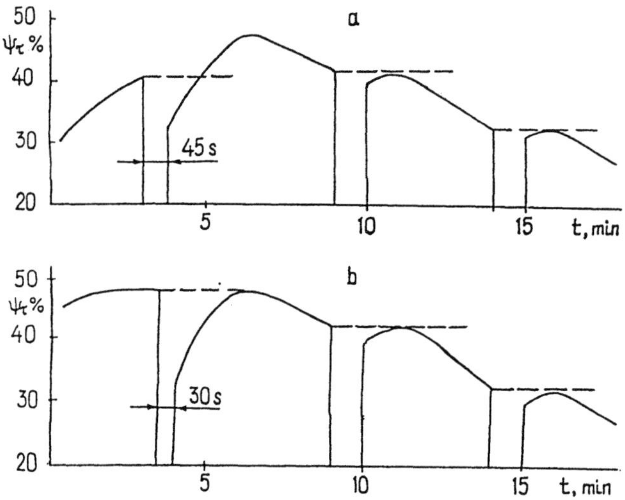

Fig. 6. The drag reduction dynamics for the WSR-301 solution with $C=10^{-3}$, at U = 17.25 m/s. a - freshly-prepared PEO-solution; b - with the [NaCl]/[PEO]=12.

At $A_{sp} > A^{o}_{sp}$ this growth is repeated also, but incompletely even after long stop. Thus, in this case the interaction of polymer with turbulent flow leads to known changes in polymer solution (to its degradation). The polymer molecular mass reduction causes the decrease of intermolecular bonds. And correspondly, the drag reduction growth section is shortened also.

The results of the experiment with injection of highly concentrated PEO solution (WSR-301, C = 0.1 %) with addition of NaCl in the weight ratio [NaCl]/[PEO] = 12 are shown in Fig.6,b. It is seen the drag reduction effectiveness becomes at once practically maximal. With the renewal of a flow after the 30 s - stopping the effectiveness growth section is observed again like with the PEO solution injection without NaCl. The explanation was given earlier. Salt plays the role of dispergator at the 0.1% PEO - concentration in injected solution decreasing intermolecular interaction significantly. As a result of dilution to one-twenty fifth its previous concentration after injection, salt can't have a significant separated effect. We can conclude the reason of observed growth of drag reduction on initial section of interaction of polymer with turbulent flow is reverse destroying of supermolecular structures of polymer.

6. Practical significance

The conditions of water-soluble polymer application in water transport for hydrodynamic drag reduction (HDR) and other purposes suggest the jet injection of prepared polymer solutions into near wall zone, and better, if solutions are the concentrated ones. As their concentration increases the volume for their keeping, the preparation system and so on decreases. It is supposed here, due to intensive turbulent mass transfer in boundary layer downstream from injector the parameters of injection (concentration and volumetric consumption of injected solution) for achievment the required concentration of polymeric additives near a wall are invariant.

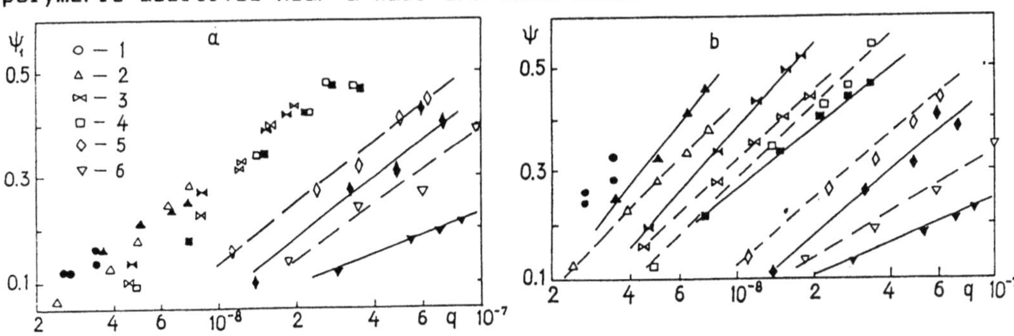

Fig.7. Drag reduction of central part (a) and whole model (b) versus the dimensionless consumption coefficient. Solid symbols - PEO; hollow symbols-[NaCl]/[PEO]=4 1-C=100, 2-200, 3-500, 4-1000, 5-2000, 6-4000 ppm.

Experiments showed that it was valid only below some parameters of injection, above which an invariance is infringed. As an example, in Fig.7 are given the results of the model experiment with the PEO solution injection through the ring slot in width of 1.3 mm placed in 28 mm from leading edge with 20°- slope to a surface. Experiments were made at temperature 6-8 $^{\circ}$C on towed axisymmetrical model of 2 m long. Dimensionless coefficient of consumption $q = C_i Q/SU$ is given in Fig.7, where C_i is relative PEO mass in injected solution, Q is volumetric consumption of solution, S is the surface of model, U = 11 m/s is velocity of towing. Drag reduction $\psi = (1 - C_x/C_{xo})$, where C_x and C_{xo} are drag coefficients with injection and without it. It is seen from this figure:

- the invariance of polymer concentration and volumetric consumption of solutions remains up to 200 ppm, herein more concentrated solutions require significantly greater polymer consumption q for the same HDR;
- at large distances from the injector (see Fig.7,a, where ψ_1 is given only for central part as distinct from Fig.7,b, where ψ - is for total model) these differences are small.

The cause of this is the phenomenon of the drag reduction growth during a flow of polymer solution along a wall. The presence time of polymer in near-wall zone of the model is shares of second. Losses of pressure in tube and the hole of the injector were minimized. So, in this case the action of initial section of the polymer addition effect, as we see, is very considerable.

The data of drag reduction with injection of concentrated solutions PEO + NaCl are also given in this figure. When $C_i < 500$ ppm the NaCl - addition leads to drag reduction decrease. The reason is the globulation process of PEO molecules in low-concentrated solutions. At $C_i > 1000$ ppm the NaCl-addition in injected PEO solution leads to increase of efficiency.

The main problem of further studies is a search for low-molecular substances in order to dispergate supermolecular polymeric structures not leading to globulation of molecules PEO and also the investigation of the effect of other methods of preliminary destruction of supermolecular structures.

References

Anisimov, I. A. & B. P. Mironov 1981 Dependence of the destruction of an aqueous solution of polyethylene oxide on the frictional work. In *Turbulent shear flows of non-newtonian liquids*. Novosibirsk, 14-38, in Russian.

Belokon, V. S. & V. N. Kalashnikov 1977 Hydrodynamics drag and degradation of dilute polymer solutions in turbulent rotating flow between the coaxial cylinders. *Preprint* 91, Institute for Problems in Mechanics, Moscow, in Russian.

Fischer, D. & F. Rodriguez 1971 Degradation of drag-reducing polymers. *J. Apply. Polym. Sci.* **15**, 2975-2985.

Kulik, V. M. 1981 Friction measurements in a flow between coaxial cylinders. In *Turbulent shear flows on non-newtonian liquids*, Novosibirsk, 90-95, in Russian.

Kulik, V. M. & B. N. Semenov 1987 The initial dependence of the Toms effect in polyethylene oxide solutions on time. *Soviet J. Appl. Physics*, 1, N3, 63-68.

Sedov, L. I., N. G. Vasetskaya, V. A. Ioselevich & V. N. Pilipenko 1980 On drag reduction by polymer additives. In *The mechanics of turbulent flows*, Nauka-Moscow, 7-28, in Russian.

Semenov, B. N. 1989 On mechanical destruction of polymer in a flow. *Izvestiya SO AN SSSR, ser. techn. nauk*, vyp. **5**, 73-78, in Russian.

Taylor, G. T. 1936 Fluid friction between rotating cylinders. *Proc. Roy. Soc.*, ser. A, **157**, 546-581.

Ting, P. G. & R. C. Little 1973 Degradation effect of high molecular weight polymers in turbulent pipe flow. *J. Appl. Sci.* **17**, 3345-3356.

Ustimenko, B. P. 1977 Processes of turbulent transfer in rotating flows. *Nauka*, Alma-Ata, in Russian.

Panel Discussions

1. Application of Drag Reducing Devices

Robert (Airbus Industrie, France) began the first part of the discussion by describing the riblet performance tests carried out by Airbus Industrie over the last two years. He pointed out that the friction drag component comprises approximately half the total drag of a civil aircraft. A 2% reduction in friction drag, equivalent to 1% of the total drag represents a substantial fuel-burn saving. With the aid of a short video, Robert described the A320 flight test where 700 m^2 of 3M riblet film was applied on the non-laminar flow regions of the aircraft except for the area near the trailing edge of the wings. He pointed out that the application of this film took two weeks and the riblets were aligned parallel to the fuselage centreline. The flight test data confirmed the model test results conducted by Aerospatiale and ONERA-CERT, showing a 1.5% to 2.0% net drag reduction.

Choi (Nottingham University, U.K.) invited Delachanal of 3M France to comment on any operational difficulties associated with the use of riblets on aircraft. Delachanal (3M, France) referred to the 'expert' advice that for efficient use it is important to maintain the sharpness of the riblet profile. Dirt accretion problem can be kept at or below acceptable limits by the normal cleaning procedures where the aircraft is washed every 40 days using heated water (40°C) under high pressure (70 bars). For cost effectiveness, the riblet film should have a useful lifespan of at least 5 years. Search for more durable materials for the riblet film is in progress.

Choi introduced Girard of GEC Alsthom who described the restrictions imposed on any drag reducing devices for use on railway carriages. Girard (GEC Alsthom, France) disclosed that they are developing the 4th generation of high speed trains where aerodynamic aspects such as drag reduction are being considered. Since the trains are designed to travel in either 'forward' or 'reverse' direction any boundary layer control system must be symmetrical. At a typical speed of 300 km/h the drag associated with the cavity between the carriages is approximately 30% of the friction drag, which itself is approximately 1/3 of the total drag. He speculated that devices similar to LEBUs may be used to reduce this 'cavity' drag. Interference drag associated with pantographs above the trains and aft-body drag are other areas where drag reduction research work is in progress. Bechert (DLR, Germany) added that his work with the German railway produced a leading edge pantograph with reduced interference drag. Since the design has been patented, the idea has not been developed further. Girard explained that the aerodynamic problems of pantographs is of secondary importance compared to the electrical considerations.

K.-S. Choi (ed.), Recent Developments in Turbulence Management, 323–327.

Choi invited delegates from industry to comment on their current and future interests, particularly on their attitude towards the funding of future research. Delachanal stressed that research organisations looking for funding must respond to the current and future development needs of industry. The scientists need to identify and define clearly something credible for the end users. Industry will continue to provide financial support for those research projects which show a good understanding of their requirements and address them specifically.

2. Recent Developments in Drag Reduction Research

Savill (Cambridge University, U.K.) began this part of the discussion by highlighting some of the questions that have been answered since the last drag reduction meeting:

Flight tests	- the A320 fight test has resulted in a lot of publicity which has been good for both Airbus Industrie and the scientific community, bringing the subject into the public domain
Riblets	- riblets are still effective in a 3-D flow with moderate misalignment angles up to 22° - encouraging preliminary results are obtained for heat transfer enhancement - good quality data have been obtained for supersonic flow - there is indirect evidence that transition is delayed by riblets in a relatively high turbulence intensity freestream - turbulence models are beginning to predict drag reductions but levels are still incorrect - some encouraging indications are given on the combined effect with LEBUs
LEBUs	- no net drag reduction - similar in laminar flow - similar in numerical simulation - reduction in pressure fluctuations

The skin friction drag reduction performance of LEBU devices are depressing compared to that of riblets. However, LEBU devices do manipulate the boundary layer in an interesting way. Practical problems are associated with the drag of the device plus support and also vibration. Work on these devices started in turbulent flow and it is curious to see similar results in laminar flow. Some recent Russian work suggests that the transition performance of this type of device is not fully understood. Also computations suggest that optimal parameters for LEBU devices in internal flows are more similar to those in external flows than expected.

Roughness	- the boundary layer over d-type roughness shows some equilibrium behaviour; it may be used in combination with LEBU devices to maintain low skin friction drag levels - drag reduction performance over sandgrain rough surface depends critically on the 3-D shape of the grains; difficult to correlate data; wide range of 3-D shapes to be explored

Savill concluded his presentation by stating that it is time for some lateral thinking, to expand ideas and come up with some devices more clever than riblets to achieve perhaps a 10 fold improvement.

3. Research Requirements for the future

The following is a list of questions, which remained open, prepared by Coustols (ONERA-CERT, France):

Riblets	- alternative types - effect on transition (not only 2-D but also 3-D) and influence on crossflow instability - effect on separation
LEBUs	- alternative types (hydrodynamic and heat transfer applications, effect on separation) - modelling (optimum aerofoil section)
Other surfaces	- compliant surfaces in turbulent flows - curvature (shaping of bodies) - 3-D riblets (roughness, etc.)
Combination of techniques	- transition control with riblets - new techniques - other drag reduction (induced drag in the rear part of fuselage)

Coustols mentioned that at ONERA-CERT they are planning to repeat the riblet performance tests carried out in subsonic three-dimensional flows in the supersonic S2 wind tunnel. Studies on LEBU devices have stopped. The effect of different geometry on crossflow instability will also be examined. In addition, they intend to look at the aft-body (rear fuselage) drag which is the 3rd most important in terms of total drag reduction.

Bechert pointed to their work on riblets over the wings of commuter aircraft which showed no negative effect on separation, rather indicating a 1% increase in lift coefficient. He suggested that this effect should be examined further with thicker aerofoil sections, treating the nose region with care. Choi suggested that application of riblets to vehicles other than aircraft should also be considered to increase lift, in addition to reducing drag.

Bechert described their plans to test 3-D models of shark skin at various angles of attack. The effect of dirt and wear on riblets may also be investigated. It may be possible to study sophisticated riblet geometries to improve on the drag reduction levels further. Bechert pointed out that it may be possible to use riblets to stabilize crossflow instability on swept wings. At present, riblets are being used in turbulent flow only and their use in laminar flow will not be straight forward.

Choi referring to the paper presented by Coustols on riblets in 3-D flows where it was implied that riblets would work if aligned with the external stream direction, stated that the riblet performance is likely to depend on the extent of the three-dimensionality. Results obtained so far cannot be extended to cover highly three-dimensional cases where the flow is not well understood. There is a need for more detailed experiments to determine how best to apply riblets under these conditions. Coustols explained that his results were obtained with a model of the Airbus A320 aircraft where the riblets appeared to work. Choi explained that according to Robert, riblets were not applied to regions near the trailing edge of the wing because of the three-dimensionality problem. He conjectured that by aligning riblets in these regions in some fashion it may be possible to produce better drag reduction results. Robert stated that it is not practical to apply riblets along the local flow direction over an aircraft body. He inferred that the only practical solution is to align the riblets with, say, the centreline of the fuselage. Choi described his work with yachts where by 'cutting and pasting' film the riblets were successfully applied to the surface with compound curvature. For example, with the America's Cup yacht 'White Crusader' the riblets were aligned to the local flow direction which in some regions differed from the external flow direction by as much as 40°. Robert stressed that the time required to apply riblets to an aircraft has to be minimised, since time is money. Airbus Industrie has already decreased this time by a factor of five. Delachanal highlighted the importance of the aesthetic and cosmetic aspects of riblet application. With boats it may be acceptable to 'cut and paste' since the riblets are below the water line. With aircraft any solution to the application problem must take into account the final appearance.

Bonnet (CEAT-LEA, France) warned that we must not abandon the fundamental aspects of these studies since they provide good data bases for testing prediction methods and developing new ideas. For example, testing the performance of turbulence models and the idea of enhancing heat transfer using riblets. First use of second order closure turbulence model on the non-equilibrium flows with external manipulators has provided encouraging results. He urged that low-expense experiments of a fundamental nature should be continued to further study the structure of the boundary layer. For example, the effects of heating the upstream or downstream of the external manipulators may be investigated, alternative uses considered and the results used to test computer codes.

Choi referred to Nguyen's work on LEBU devices which shows that a measure of three-dimensionality develops from such devices. He stated that a similar behaviour has also been observed by other investigators but not all have published their findings. He suggested that if this is the case then measurements at the centreline of such devices would give a false impression. He speculated that this behaviour may be to do with a real instability such as the Karman vortex street. Bonnet suggested that the observed three-dimensionality may be associated with streamwise vortices due perhaps to a secondary instability of mixing layers developed from the LEBU devices. Choi concluded that this three-dimensionality associated with LEBU devices is worth investigating further.

List of Referees

Bandyopadhyay, P. R.	NASA Langley Research Centre Hampton, VA, U.S.A.
Bechert, D. W.	DLR Berlin, Germany
Carpenter, P. W.	University of Warwick Coventry, U.K.
Clark, D.	Queen Mary and Westfield College London, U.K.
Coustols, E.	ONERA/CERT Toulouse, France
Dowling, A. P.	University of Cambridge Cambridge, U.K.
Gajjar, J. S. B.	University of Manchester Manchester, U.K.
Gaudet, L.	Defence Research Agency Bedford, U.K.
Hamid, S.	Defence Research Agency Farnborough, U.K.
Hay, N.	University of Nottingham Nottingham, U.K.
Lewkowicz, A.	University of Liverpool Liverpool, U.K.
Metcalfe, R. W.	University of Houston Houston, TX, U.S.A.

Nieuwstadt, F. T. M.	Delft University of Technology Delft, the Netherlands
Pearcey, H. H.	City University London, U.K
Savill, A. M.	University of Cambridge Cambridge, U.K.
Savory, E.	University of Surrey Guildford, U.K.
Smith, F. T.	University College London London, U.K.
Takagi, S.	National Aerospace Laboratory Tokyo, Japan
Tiederman, W. G.	Purdue University West Lafayette, IN, U.S.A.
Townsin, R. L.	University of Newcastle Newcastle, U.K.
Walker. J. D. A.	Lehigh University Bethlehem, PA, U.S.A.
Wallace, J.	University of Maryland College Park, MD, U.S.A.
Walsh, M. J.	NASA Langley Research Centre Hampton, VA, U.S.A.

List of Participants

U.K.

John Bainbridge	Defence Research Agency Southwell Portland Dorset DT5 2JS
Steve Brown	Research Department BAe Commercial Aircraft Ltd Hatfield Hertfordshire AL10 9TL
Aragon Burlingham	GEC Alsthom Engineering Research Centre Cambridge Road Whetstone Leicester LE8 3LH
Chris Carey	BEG Engineering Research Centre Thermo Fluids Division Cambridge Road Whetstone Leicester LE3 3LH
Peter Carpenter	Department of Engineering University of Warwick Coventry CV4 7A4
Kwing-So Choi	Department of Mechanical Engineering University of Nottingham University Park Nottingham NG7 2RD

Dave Clark	Department of Aeronautics Queen Mary & Westfield College Mile End Road London E1 4NS
Mel Davies	BMT Fluid Mechanics Ltd 1 Waldegrave Road Teddington Middlesex TW11 8LZ
Padelis Fanourakis	Engineering Department Cambridge University Trumpington Street Cambridge CB2 1PZ
Carren Fisher	Aerodynamics Research Department British Aerospace plc Sowerby Research Centre FPC 67, PO Box 5 Filton Bristol BS12 7QW
Jitesh Gajjar	Mathematics Department University of Manchester Oxford Road Manchester M13 9PL
Mike Gaster	Engineering Department Cambridge University Trumpington Street Cambridge CB2 1PZ
Lawrence Gaudet	Defence Research Agency Clapham Bedford MK41 6AE
Shamim Hamid	Defence Research Agency Farnborough Hampshire GU14 6TD

Shaoping Li	UMIST Department of Mechanical Engineering C24, George Begg Building Manchester M60 1QD
Anthony Lucey	Department of Engineering University of Warwick Coventry CV4 7A4
Fouad Mandawali	BAe Commercial Aircraft Ltd Airbus Division Filton House Filton Bristol BS99 7AR
Jimmy Mozaffar	BMT Fluid Mechanics Ltd 1 Waldegrave Road Teddington Middlesex TW11 8LZ
Andy Mullender	Rolls-Royce plc Powerplant Technology (M.L.) PO Box 31 Derby DE2 8BJ
Anthony Pagano	BAe Commercial Aircraft Ltd Airbus Division Filton House Filton Bristol BS99 7AR
Simon Read	Research Department BAe Commercial Aircraft Ltd Hatfield Hertfordshire AL10 9TL
Rico Resing	British Maritime Technology 1 Waldegrave Road Teddington Middlesex TW11 8LZ

Mark Savill — Engineering Department
Cambridge University
Trumpington Street
Cambridge CB2 1PZ

Eric Savory — Department of Civil Engineering
Univeristy of Surrey
Guildford
Surrey GU2 5XH

Yan-Ping Tang — Department of Aeronautics
Queen Mary & Westfield College
Mile End Road
London E1 4NS

Bob Townsin — Department of Marine Technology
University of Newcastle upon Tyne
Newcastle upon Tyne NE1 7RU

France

Jacques-Anche Astolfi — Institut de Mécanique des Fluides
de l'Universite d'Aix-Marseille II
1 rue Honnorat
13003 Marseille Cedex

Eric Augier — Laboratoire de Mécanique des Fluides
Ecole Centrale de Lyon
BP 163
69131 Ecully Cedex

Jean-Paul Bonnet — CERT-LEA
43 route de l'Aerodrome
86036 Poitiers Cedex

Eric Coustols — ONERA/CERT
2 avenue Edouard Belin
31055 Toulouse Cedex

Alain Delachanal — 3M France
Boulevard de l'Oise
95006 Gergy-Pontoise Cedex

Anche Fages
Institut de Mécanique des Fluides
de l'Universite d'Aix-Marseille II
1 rue Honnorat
13003 Maseille Cedex

Herve Girard
Transport Division
GEC Alsthom
Avenue du Commandant Lysiack
BP 359
17001 La Rochelle Cedex

Jean-Pierre Robert
AI/TT-A Airbus Industrie
1 Rond Point Maurice Bellonte
31707 Blagnac Cedex

Jean Luc Teissier
GEC Alsthom
3 Avenue des Trois Chenes
90018 Belfort Cedex

Germany

Dietrich Bechert
DLR
Müller-Breslau-Strasse 8
D-1000 Berlin 12

Franz Münch
Deutsche Airbus GmbH
Hühnefeldstrasse 1-5
D-2800 Bremen 1

Dieter Neumann
Max-Planck-Institut
für Strömunsforschung
Bunsenstrasse 10
D-3400 Göttingen

Michael Schneider
Max-Planck-Institut
für Strömungsforschung
Bunsenstrasse 10
D-3400 Göttingen

Jacoba Van der Hoeven
H.F.I.
Strasse des 17. Juni 135
TU Berlin
D-1000 Berlin 12

The Netherlands

Nigel Hollingworth	Kluwer Academic Publishers Spuiboulevard 50 PO Box 17 3300 AZ Dordrecht
Kalkunte Krishna Prasad	Eindhoven University of Technology Gebouw W&S 0.37 Postbus 513 5600 MB Eindhoven
Frans Nieuwstadt	Lab. Aero- and Hydromechanics Delft University of Technology Rotterdamsweg 145 2628 AL Delft
Adriana Schwarz-van Manen	Eindhoven University of Technology Gebouw W&S 0.47 Postbus 513 5600 MB Eindhoven

U.S.S.R.

Viktor Kulik	USSR Academy of Sciences Institute of Thermophysics Prospekt Lavrentyev 1 Novosibirsk 630090
Boris Semenov	USSR Acadmey of Sciences Institute of Thermophysics Prospekt Lavrentyev 1 Novosibirsk 630090

Italy

G. Iuso	Dipartimento di Ingegneria Aeronautica e Spaziale Politecnico di Torino Corso Ducs degll Abruci 24-10129 Torino

Switzerland

Trong-Vien Truong — DME/IMHEF
EPFL
CH-1015 Lausanne

Sedat Tardu — DME/IMHEF
EPFL
CH-1015 Lausanne

Japan

Shinsuke Mochizuki — Department of Mechanical Engineering
Yamaguchi University
Tokiwadai
Ube-shi 755

Canada

Vinh Duy Nguyen — Institute of Aerospace Research
High-Speed Aerodynamics Laboratory
Montreal Road
Ottawa
Ontario K1A 0RS

Australia

Lyazid Djenidi — Department of Mechanical Engineering
University of Newcastle
Rankin Drive
Newcastle NSW 2308

Author Index

Mechanics

FLUID MECHANICS AND ITS APPLICATIONS

Series Editor: R. Moreau

Aims and Scope of the Series

The purpose of this series is to focus on subjects in which fluid mechanics plays a fundamental role. As well as the more traditional applications of aeronautics, hydraulics, heat and mass transfer etc., books will be published dealing with topics which are currently in a state of rapid development, such as turbulence, suspensions and multiphase fluids, super and hypersonic flows and numerical modelling techniques. It is a widely held view that it is the interdisciplinary subjects that will receive intense scientific attention, bringing them to the forefront of technological advancement. Fluids have the ability to transport matter and its properties as well as transmit force, therefore fluid mechanics is a subject that is particularly open to cross fertilisation with other sciences and disciplines of engineering. The subject of fluid mechanics will be highly relevant in domains such as chemical, metallurgical, biological and ecological engineering. This series is particularly open to such new multidisciplinary domains.

1. M. Lesieur: *Turbulence in Fluids.* 2nd rev. ed., 1990 ISBN 0-7923-0645-7
2. O. Métais and M. Lesieur (eds.): *Turbulence and Coherent Structures.* 1991 ISBN 0-7923-0646-5
3. R. Moreau: *Magnetohydrodynamics.* 1990 ISBN 0-7923-0937-5
4. E. Coustols (ed.): *Turbulence Control by Passive Means.* 1990 ISBN 0-7923-1020-9
5. A. A. Borissov (ed.): *Dynamic Structure of Detonation in Gaseous and Dispersed Media.* 1991 ISBN 0-7923-1340-2
6. K.-S. Choi (ed.): *Recent Developments in Turbulence Management.* 1991 ISBN 0-7923-1477-8

Kluwer Academic Publishers – Dordrecht / Boston / London

Mechanics

SOLID MECHANICS AND ITS APPLICATIONS

Series Editor: G.M.L. Gladwell

Aims and Scope of the Series

The fundamental questions arising in mechanics are: *Why?, How?,* and *How much?* The aim of this series is to provide lucid accounts written by authoritative researchers giving vision and insight in answering these questions on the subject of mechanics as it relates to solids. The scope of the series covers the entire spectrum of solid mechanics. Thus it includes the foundation of mechanics; variational formulations; computational mechanics; statics, kinematics and dynamics of rigid and elastic bodies; vibrations of solids and structures; dynamical systems and chaos; the theories of elasticity, plasticity and viscoelasticity; composite materials; rods, beams, shells and membranes; structural control and stability; soils, rocks and geomechanics; fracture; tribology; experimental mechanics; biomechanics and machine design.

1. R.T. Haftka, Z. Gürdal and M.P. Kamat: *Elements of Structural Optimization.* 2nd rev.ed., 1990 ISBN 0-7923-0608-2
2. J.J. Kalker: *Three-Dimensional Elastic Bodies in Rolling Contact.* 1990 ISBN 0-7923-0712-7
3. P. Karasudhi: *Foundations of Solid Mechanics.* 1991 ISBN 0-7923-0772-0
4. N. Kikuchi: *Computational Methods in Contact Mechanics.* (forthcoming) ISBN 0-7923-0773-9
5. Y.K. Cheung and A.Y.T. Leung: *Finite Element Methods in Dynamics.* (forthcoming) ISBN 0-7923-1313-5
6. J.F. Doyle: *Static and Dynamic Analysis of Structures.* With an Emphasis on Mechanics and Computer Matrix Methods. 1991 ISBN 0-7923-1124-8; Pb 0-7923-1208-2
7. O.O. Ochoa and J.N. Reddy: *Finite Element Modelling of Composite Structures.* (forthcoming) ISBN 0-7923-1125-6
8. M.H. Aliabadi and D.P. Rooke: *Numerical Fracture Mechanics.* ISBN 0-7923-1175-2
9. J. Angeles and C.S. López-Cajún: *Optimization of Cam Mechanisms.* 1991 ISBN 0-7923-1355-0
10. D.E. Grierson, A. Franchi and P. Riva: *Progress in Structural Engineering.* 1991 ISBN 0-7923-1396-8

Kluwer Academic Publishers – Dordrecht / Boston / London

Mechanics

From 1990, books on the subject of *mechanics* will be published under two series:
FLUID **MECHANICS AND ITS APPLICATIONS**
Series Editor: R.J. Moreau
SOLID **MECHANICS AND ITS APPLICATIONS**
Series Editor: G.M.L. Gladwell

Prior to 1990, the books listed below were published in the respective series indicated below.

MECHANICS: DYNAMICAL SYSTEMS
Editors: L. Meirovitch and G.Æ. Oravas

1. E.H. Dowell: *Aeroelasticity of Plates and Shells.* 1975 ISBN 90-286-0404-9
2. D.G.B. Edelen: *Lagrangian Mechanics of Nonconservative Nonholonomic Systems.* 1977 ISBN 90-286-0077-9
3. J.L. Junkins: *An Introduction to Optimal Estimation of Dynamical Systems.* 1978 ISBN 90-286-0067-1
4. E.H. Dowell (ed.), H.C. Curtiss Jr., R.H. Scanlan and F. Sisto: *A Modern Course in Aeroelasticity.* *Revised and enlarged edition see under Volume 11*
5. L. Meirovitch: *Computational Methods in Structural Dynamics.* 1980 ISBN 90-286-0580-0
6. B. Skalmierski and A. Tylikowski: *Stochastic Processes in Dynamics.* Revised and enlarged translation. 1982 ISBN 90-247-2686-7
7. P.C. Müller and W.O. Schiehlen: *Linear Vibrations.* A Theoretical Treatment of Multi-degree-of-freedom Vibrating Systems. 1985 ISBN 90-247-2983-1
8. Gh. Buzdugan, E. Mihăilescu and M. Radeş: *Vibration Measurement.* 1986 ISBN 90-247-3111-9
9. G.M.L. Gladwell: *Inverse Problems in Vibration.* 1987 ISBN 90-247-3408-8
10. G.I. Schuëller and M. Shinozuka: *Stochastic Methods in Structural Dynamics.* 1987 ISBN 90-247-3611-0
11. E.H. Dowell (ed.), H.C. Curtiss Jr., R.H. Scanlan and F. Sisto: *A Modern Course in Aeroelasticity.* Second revised and enlarged edition (of Volume 4). 1989 ISBN Hb 0-7923-0062-9; Pb 0-7923-0185-4
12. W. Szemplińska-Stupnicka: *The Behavior of Nonlinear Vibrating Systems.* Volume I: Fundamental Concepts and Methods: Applications to Single-Degree-of-Freedom Systems. 1990 ISBN 0-7923-0368-7
13. W. Szemplińska-Stupnicka: *The Behavior of Nonlinear Vibrating Systems.* Volume II: Advanced Concepts and Applications to Multi-Degree-of-Freedom Systems. 1990 ISBN 0-7923-0369-5
    Set ISBN (Vols. 12–13) 0-7923-0370-9

MECHANICS OF STRUCTURAL SYSTEMS
Editors: J.S. Przemieniecki and G.Æ. Oravas

1. L. Frýba: *Vibration of Solids and Structures under Moving Loads.* 1970 ISBN 90-01-32420-2
2. K. Marguerre and K. Wölfel: *Mechanics of Vibration.* 1979 ISBN 90-286-0086-8

Mechanics

3. E.B. Magrab: *Vibrations of Elastic Structural Members.* 1979 ISBN 90-286-0207-0
4. R.T. Haftka and M.P. Kamat: *Elements of Structural Optimization.* 1985
   Revised and enlarged edition see under Solid Mechanics and Its Applications, Volume 1
5. J.R. Vinson and R.L. Sierakowski: *The Behavior of Structures Composed of Composite Materials.* 1986 ISBN Hb 90-247-3125-9; Pb 90-247-3578-5
6. B.E. Gatewood: *Virtual Principles in Aircraft Structures.* Volume 1: Analysis. 1989
   ISBN 90-247-3754-0
7. B.E. Gatewood: *Virtual Principles in Aircraft Structures.* Volume 2: Design, Plates, Finite Elements. 1989 ISBN 90-247-3755-9
   Set (Gatewood 1 + 2) ISBN 90-247-3753-2

MECHANICS OF ELASTIC AND INELASTIC SOLIDS

Editors: S. Nemat-Nasser and G.Æ. Oravas

---

1. G.M.L. Gladwell: *Contact Problems in the Classical Theory of Elasticity.* 1980
   ISBN Hb 90-286-0440-5; Pb 90-286-0760-9
2. G. Wempner: *Mechanics of Solids with Applications to Thin Bodies.* 1981
   ISBN 90-286-0880-X
3. T. Mura: *Micromechanics of Defects in Solids.* 2nd revised edition, 1987
   ISBN 90-247-3343-X
4. R.G. Payton: *Elastic Wave Propagation in Transversely Isotropic Media.* 1983
   ISBN 90-247-2843-6
5. S. Nemat-Nasser, H. Abé and S. Hirakawa (eds.): *Hydraulic Fracturing and Geothermal Energy.* 1983 ISBN 90-247-2855-X
6. S. Nemat-Nasser, R.J. Asaro and G.A. Hegemier (eds.): *Theoretical Foundation for Large-scale Computations of Nonlinear Material Behavior.* 1984 ISBN 90-247-3092-9
7. N. Cristescu: *Rock Rheology.* 1988 ISBN 90-247-3660-9
8. G.I.N. Rozvany: *Structural Design via Optimality Criteria.* The Prager Approach to Structural Optimization. 1989 ISBN 90-247-3613-7

MECHANICS OF SURFACE STRUCTURES

Editors: W.A. Nash and G.Æ. Oravas

---

1. P. Seide: *Small Elastic Deformations of Thin Shells.* 1975 ISBN 90-286-0064-7
2. V. Panc: *Theories of Elastic Plates.* 1975 ISBN 90-286-0104-X
3. J.L. Nowinski: *Theory of Thermoelasticity with Applications.* 1978
   ISBN 90-286-0457-X
4. S. Łukasiewicz: *Local Loads in Plates and Shells.* 1979 ISBN 90-286-0047-7
5. C. Fiřt: *Statics, Formfinding and Dynamics of Air-supported Membrane Structures.* 1983 ISBN 90-247-2672-7
6. Y. Kai-yuan (ed.): *Progress in Applied Mechanics.* The Chien Wei-zang Anniversary Volume. 1987 ISBN 90-247-3249-2
7. R. Negruţiu: *Elastic Analysis of Slab Structures.* 1987 ISBN 90-247-3367-7
8. J.R. Vinson: *The Behavior of Thin Walled Structures.* Beams, Plates, and Shells. 1988
   ISBN Hb 90-247-3663-3; Pb 90-247-3664-1

Mechanics

MECHANICS OF FLUIDS AND TRANSPORT PROCESSES
Editors: R.J. Moreau and G.Æ. Oravas

1. J. Happel and H. Brenner: *Low Reynolds Number Hydrodynamics.* With Special Applications to Particular Media. 1983 ISBN Hb 90-01-37115-9; Pb 90-247-2877-0
2. S. Zahorski: *Mechanics of Viscoelastic Fluids.* 1982 ISBN 90-247-2687-5
3. J.A. Sparenberg: *Elements of Hydrodynamics Propulsion.* 1984 ISBN 90-247-2871-1
4. B.K. Shivamoggi: *Theoretical Fluid Dynamics.* 1984 ISBN 90-247-2999-8
5. R. Timman, A.J. Hermans and G.C. Hsiao: *Water Waves and Ship Hydrodynamics.* An Introduction. 1985 ISBN 90-247-3218-2
6. M. Lesieur: *Turbulence in Fluids.* Stochastic and Numerical Modelling. 1987 ISBN 90-247-3470-3
7. L.A. Lliboutry: *Very Slow Flows of Solids.* Basics of Modeling in Geodynamics and Glaciology. 1987 ISBN 90-247-3482-7
8. B.K. Shivamoggi: *Introduction to Nonlinear Fluid-Plasma Waves.* 1988 ISBN 90-247-3662-5
9. V. Bojarevičs, Ya. Freibergs, E.I. Shilova and E.V. Shcherbinin: *Electrically Induced Vortical Flows.* 1989 ISBN 90-247-3712-5
10. J. Lielpeteris and R. Moreau (eds.): *Liquid Metal Magnetohydrodynamics.* 1989 ISBN 0-7923-0344-X

MECHANICS OF ELASTIC STABILITY
Editors: H. Leipholz and G.Æ. Oravas

1. H. Leipholz: *Theory of Elasticity.* 1974 ISBN 90-286-0193-7
2. L. Librescu: *Elastostatics and Kinetics of Aniosotropic and Heterogeneous Shell-type Structures.* 1975 ISBN 90-286-0035-3
3. C.L. Dym: *Stability Theory and Its Applications to Structural Mechanics.* 1974 ISBN 90-286-0094-9
4. K. Huseyin: *Nonlinear Theory of Elastic Stability.* 1975 ISBN 90-286-0344-1
5. H. Leipholz: *Direct Variational Methods and Eigenvalue Problems in Engineering.* 1977 ISBN 90-286-0106-6
6. K. Huseyin: *Vibrations and Stability of Multiple Parameter Systems.* 1978 ISBN 90-286-0136-8
7. H. Leipholz: *Stability of Elastic Systems.* 1980 ISBN 90-286-0050-7
8. V.V. Bolotin: *Random Vibrations of Elastic Systems.* 1984 ISBN 90-247-2981-5
9. D. Bushnell: *Computerized Buckling Analysis of Shells.* 1985 ISBN 90-247-3099-6
10. L.M. Kachanov: *Introduction to Continuum Damage Mechanics.* 1986 ISBN 90-247-3319-7
11. H.H.E. Leipholz and M. Abdel-Rohman: *Control of Structures.* 1986 ISBN 90-247-3321-9
12. H.E. Lindberg and A.L. Florence: *Dynamic Pulse Buckling.* Theory and Experiment. 1987 ISBN 90-247-3566-1
13. A. Gajewski and M. Zyczkowski: *Optimal Structural Design under Stability Constraints.* 1988 ISBN 90-247-3612-9

Mechanics

MECHANICS: ANALYSIS

Editors: V.J. Mizel and G.Æ. Oravas

1. M.A. Krasnoselskii, P.P. Zabreiko, E.I. Pustylnik and P.E. Sbolevskii: *Integral Operators in Spaces of Summable Functions*. 1976 ISBN 90-286-0294-1
2. V.V. Ivanov: *The Theory of Approximate Methods and Their Application to the Numerical Solution of Singular Integral Equations*. 1976 ISBN 90-286-0036-1
3. A. Kufner, O. John and S. Pučík: *Function Spaces*. 1977 ISBN 90-286-0015-9
4. S.G. Mikhlin: *Approximation on a Rectangular Grid*. With Application to Finite Element Methods and Other Problems. 1979 ISBN 90-286-0008-6
5. D.G.B. Edelen: *Isovector Methods for Equations of Balance*. With Programs for Computer Assistance in Operator Calculations and an Exposition of Practical Topics of the Exterior Calculus. 1980 ISBN 90-286-0420-0
6. R.S. Anderssen, F.R. de Hoog and M.A. Lukas (eds.): *The Application and Numerical Solution of Integral Equations*. 1980 ISBN 90-286-0450-2
7. R.Z. Has'minskiĭ: *Stochastic Stability of Differential Equations*. 1980 ISBN 90-286-0100-7
8. A.I. Vol'pert and S.I. Hudjaev: *Analysis in Classes of Discontinuous Functions and Equations of Mathematical Physics*. 1985 ISBN 90-247-3109-7
9. A. Georgescu: *Hydrodynamic Stability Theory*. 1985 ISBN 90-247-3120-8
10. W. Noll: *Finite-dimensional Spaces*. Algebra, Geometry and Analysis. Volume I. 1987 ISBN Hb 90-247-3581-5; Pb 90-247-3582-3

MECHANICS: COMPUTATIONAL MECHANICS

Editors: M. Stern and G.Æ. Oravas

1. T.A. Cruse: *Boundary Element Analysis in Computational Fracture Mechanics*. 1988 ISBN 90-247-3614-5

MECHANICS: GENESIS AND METHOD

Editor: G.Æ. Oravas

1. P.-M.-M. Duhem: *The Evolution of Mechanics*. 1980 ISBN 90-286-0688-2

MECHANICS OF CONTINUA

Editors: W.O. Williams and G.Æ. Oravas

1. C.-C. Wang and C. Truesdell: *Introduction to Rational Elasticity*. 1973 ISBN 90-01-93710-1
2. P.J. Chen: *Selected Topics in Wave Propagation*. 1976 ISBN 90-286-0515-0
3. P. Villaggio: *Qualitative Methods in Elasticity*. 1977 ISBN 90-286-0007-8

Mechanics

MECHANICS OF FRACTURE

Editors: G.C. Sih

1. G.C. Sih (ed.): *Methods of Analysis and Solutions of Crack Problems.* 1973 ISBN 90-01-79860-8
2. M.K. Kassir and G.C. Sih (eds.): *Three-dimensional Crack Problems.* A New Solution of Crack Solutions in Three-dimensional Elasticity. 1975 ISBN 90-286-0414-6
3. G.C. Sih (ed.): *Plates and Shells with Cracks.* 1977 ISBN 90-286-0146-5
4. G.C. Sih (ed.): *Elastodynamic Crack Problems.* 1977 ISBN 90-286-0156-2
5. G.C. Sih (ed.): *Stress Analysis of Notch Problems.* Stress Solutions to a Variety of Notch Geometries used in Engineering Design. 1978 ISBN 90-286-0166-X
6. G.C. Sih and E.P. Chen (eds.): *Cracks in Composite Materials.* A Compilation of Stress Solutions for Composite System with Cracks. 1981 ISBN 90-247-2559-3
7. G.C. Sih (ed.): *Experimental Evaluation of Stress Concentration and Intensity Factors.* Useful Methods and Solutions to Experimentalists in Fracture Mechanics. 1981 ISBN 90-247-2558-5

MECHANICS OF PLASTIC SOLIDS

Editors: J. Schroeder and G.Æ. Oravas

1. A. Sawczuk (ed.): *Foundations of Plasticity.* 1973 ISBN 90-01-77570-5
2. A. Sawczuk (ed.): *Problems of Plasticity.* 1974 ISBN 90-286-0233-X
3. W. Szczepiński: *Introduction to the Mechanics of Plastic Forming of Metals.* 1979 ISBN 90-286-0126-0
4. D.A. Gokhfeld and O.F. Cherniavsky: *Limit Analysis of Structures at Thermal Cycling.* 1980 ISBN 90-286-0455-3
5. N. Cristescu and I. Suliciu: *Viscoplasticity.* 1982 ISBN 90-247-2777-4

Kluwer Academic Publishers - Dordrecht / Boston / London

MIX
Papier aus verantwortungsvollen Quellen
Paper from responsible sources
FSC® C105338

If you have any concerns about our products,
you can contact us on
ProductSafety@springernature.com

In case Publisher is established outside the EU,
the EU authorized representative is:
Springer Nature Customer Service Center GmbH
Europaplatz 3, 69115 Heidelberg, Germany

Printed by Libri Plureos GmbH
in Hamburg, Germany